Probability and its Applications

Series Editors

Charles Newman
Sidney Resnick

For further volumes:
www.springer.com/series/4893

B.L.S. Prakasa Rao

Associated Sequences, Demimartingales and Nonparametric Inference

 Birkhäuser

B.L.S. Prakasa Rao
Department of Mathematics and Statistics
University of Hyderabad
Hyderabad, India

ISBN 978-3-0348-0239-0   e-ISBN 978-3-0348-0240-6
DOI 10.1007/978-3-0348-0240-6
Springer Basel Dordrecht Heidelberg London New York

Library of Congress Control Number: 2011941605

Mathematics Subject Classification (2010): 60G48

Printed on acid-free paper

Springer Basel AG is part of Springer Science+Business Media (www.birkhauser-science.com)

In memory of my mother
Bhagavatula Saradamba
(1926–2002)
for her abiding trust in me
and for her love and affection

Contents

Preface

One of the basic aims of theory of probability and statistics is to build stochastic models which explain the phenomenon under investigation and explore the dependence among various covariates which influence this phenomenon. Classic examples are the concepts of Markov dependence or of mixing for random processes. Esary, Proschan and Walkup introduced the concept of association for random variables. In several situations, for example, in reliability and survival analysis, the random variables of lifetimes for components are not independent but associated. Newman and Wright introduced the notion of a demimartingale whose properties extend those of a martingale. It can be shown that the partial sums of mean zero associated random variables form a demimartingale. The study of demimartingales and related concepts is the subject matter of this book. Along with a discussion on demimartingales and related concepts, we review some recent results on probabilistic inequalities for sequences of associated random variables and methods of nonparametric inference for such sequences.

The idea to write this book occurred following the invitation of Professor Tasos Christofides to me to visit the University of Cyprus, Nicosia in October 2010. Professor Tasos Christofides has contributed extensively to the subject matter of this book. It is a pleasure to thank him for his invitation. Professor Isha Dewan of the Indian Statistical Institute and I have been involved in the study of probabilistic properties of associated sequences and in developing nonparametric inference for such processes during the years 1990-2004 while I was at the Indian Statistical Institute. This book is a culmination of those efforts and I would like to thank Professor Isha Dewan for her collaboration over the years.

Thanks are due to my wife Vasanta Bhagavatula for her continued support to pursue my academic interests even after my retirement.

B.L.S. Prakasa Rao
Hyderabad, July 1, 2011

Chapter 1

Associated Random Variables and Related Concepts

1.1 Introduction

In classical statistical inference, the observed random variables of interest are generally assumed to be independent and identically distributed. However in some real life situations, the random variables need not be independent. In reliability studies, there are structures in which the components share the load, so that failure of one component results in increased load on each of the remaining components. Minimal path structures of a coherent system having components in common behave in a 'similar' manner. Failure of a component will adversely effect the performance of all the minimal path structures containing it. In both the examples given above, the random variables of interest are not independent but are "associated" , a concept we will define soon. This book is concerned with the study of properties of stochastic processes termed as demimartingales and N-demimartingales and related concepts. As we will see in the next chapter, an important example of a demimartingale is the sequence of partial sums of mean zero associated random variables. We will now briefly review some properties of associated random variables. We will come back to the study of these sequences again in Chapter 6.

We assume that all the expectations involved in the following discussions exist.

Hoeffding (1940) (cf. Lehmann (1966)) proved the following result.

Theorem 1.1.1. *Let (X, Y) be a bivariate random vector such that $E(X^2) < \infty$ and $E(Y^2) < \infty$. Then*

$$\text{Cov}(X, Y) = \int\limits_{-\infty}^{\infty} \int\limits_{-\infty}^{\infty} H(x, y) dx dy \qquad (1.1.1)$$

B.L.S. Prakasa Rao, *Associated Sequences, Demimartingales and Nonparametric Inference*, Probability and its Applications, DOI 10.1007/978-3-0348-0240-6_1, © Springer Basel AG 2012

where,

$$H(x,y) = P[X > x, Y > y] - P[X > x]P[Y > y]$$
$$= P[X \le x, Y \le y] - P[X \le x]P[Y \le y]. \qquad (1.1.2)$$

Proof. Let (X_1, Y_1) and (X_2, Y_2) be independent and identically distributed random vectors. Then,

$$2[E(X_1 Y_1) - E(X_1)E(Y_1)]$$
$$= E[(X_1 - X_2)(Y_1 - Y_2)]$$
$$= E[\int_{-\infty}^{\infty} \int_{-\infty}^{\infty} [I(u, X_1) - I(u, X_2)][I(v, Y_1) - I(v, Y_2)] du dv]$$
$$= \int_{-\infty}^{\infty} \int_{-\infty}^{\infty} E([I(u, X_1) - I(u, X_2)][I(v, Y_1) - I(v, Y_2)]) du dv$$

where $I(u, a) = 1$ if $u \le a$ and 0 otherwise. The last equality follows as an application of Fubini's theorem.   □

Relation (1.1.1) is known as the *Hoeffding identity*. A generalized Hoeffding identity has been proved by Block and Fang (1983) for multidimensional random vectors. Newman (1980) showed that for any two functions $h(\cdot)$ and $g(\cdot)$ with $E[h(X)]^2 < \infty$ and $E[g(Y)]^2 < \infty$ and finite derivatives $h'(\cdot)$ and $g'(\cdot)$,

$$\text{Cov}(h(X), g(Y)) = \int\limits_{-\infty}^{\infty} \int\limits_{-\infty}^{\infty} h'(x)g'(y)H(x,y)dx \, dy \qquad (1.1.3)$$

hereinafter called Newman's identity. Yu (1993) extended the relation (1.1.3) to even-dimensional random vectors. Prakasa Rao (1998) further extended this identity following Queseda-Molina (1992).

As a departure from independence, a bivariate notion of positive quadrant dependence was introduced by Lehmann (1966).

Definition. A pair of random variables (X, Y) is said to be *positively quadrant dependent* (PQD) if

$$P[X \le x, Y \le y] \ge P[X \le x]P[Y \le y] \quad \forall \, x, y \qquad (1.1.4)$$

or equivalently

$$H(x, y) \ge 0, \quad x, y \in R. \qquad (1.1.5)$$

It can be shown that the condition (1.1.5) is equivalent to the following: for any pair of nondecreasing functions h and g on R,

$$\text{Cov}(h(X), g(Y)) \ge 0. \qquad (1.1.6)$$

A stronger condition is that, for a pair of random variables (X, Y) and for any two real componentwise nondecreasing functions h and g on R^2,

$$\operatorname{Cov}(h(X,Y), g(X,Y)) \geq 0. \tag{1.1.7}$$

As a natural multivariate extension of (1.1.7), the following concept of association was introduced by Esary, Proschan and Walkup (1967).

Definition. A finite collection of random variables $\{X_j, 1 \leq j \leq n\}$ is said to be *associated* if for every every choice of componentwise nondecreasing functions h and g from R^n to R,

$$\operatorname{Cov}(h(X_1, \ldots, X_n), g(X_1, \ldots, X_n)) \geq 0 \tag{1.1.8}$$

whenever it exists; an infinite collection of random variables $\{X_n, n \geq 1\}$ is said to be *associated* if every finite sub-collection is associated.

It is easy to see that any set of independent random variables is associated (cf. Esary, Proschan and Walkup (1967)). Associated random variables arise in widely different areas such as reliability, statistical mechanics, percolation theory etc. The basic concept of association has appeared in the context of percolation models (cf. Harris (1960)) and later applied to the Ising models in statistical mechanics in Fortuin et al. (1971) and Lebowitz (1972).

Examples.   (i) Let $\{X_i', i \geq 1\}$ be independent random variables and Y be another random variable independent of $\{X_i', i \geq 1\}$. Then the random variables $\{X_i = X_i' + Y, i \geq 1\}$ are associated. Thus, independent random variables subject to the same stress are associated (cf. Barlow and Proschan (1975)). For an application of this observation to modelling dependent competing risks, see Bagai and Prakasa Rao (1992).

(ii) Order statistics corresponding to a finite set of independent random variables are associated.

(iii) Let $\{X_i, 1 \leq i \leq n\}$ be associated random variables. Let $S_i = X_1 + \cdots + X_i$, $1 \leq i \leq n$. Then the set $\{S_i, 1 \leq i \leq n\}$ is associated.

(iv) Positively correlated normal random variables are associated (cf. Pitt (1982)).

(v) Suppose a random vector $(X_1, \ldots, X_m)$ has a multivariate exponential distribution $F(x_1, \ldots, x_m)$ (cf. Marshall and Olkin (1967)) with

$$
\begin{aligned}
&1 - F(x_1, \ldots, x_m) \\
&= \exp[-\sum_{i=1}^{m} \lambda_i x_i - \sum_{i<j} \lambda_{ij} \max(x_i, x_j) \\
&\quad - \sum_{i<j<k} \lambda_{ijk} \max(x_i, x_j, x_k) - \ldots - \lambda_{12\ldots m} \max(x_1, \ldots, x_m)],
\end{aligned}
$$

$x_i > 0, \ 1 \le i \le m.$

Then the components $X_1, \ldots, X_m$ are associated.

(vi) Let $\{X_1, \ldots, X_n\}$ be jointly α-stable random variables, $0 < \alpha < 2$. Then Lee, Rachev and Samorodnitsky (1990) discussed necessary and sufficient conditions under which $\{X_1, \ldots, X_n\}$ are associated.

(vii) Let $\{e_k : k = \ldots, -1, 0, 1, \ldots\}$ be a sequence of independent random variables with zero mean and unit variance. Let $\{w_j : j = 0, 1, \ldots\}$ be a sequence of nonnegative real numbers such that $\sum_{j=0}^{\infty} w_j < \infty$. Define $X_k = \sum_{j=0}^{\infty} w_j e_{k-j}$. Then the sequence $\{X_k\}$ is associated. (Nagaraj and Reddy (1993)).

(viii) Let the process $\{X_k\}$ be a stationary autoregressive process of order p given by $X_n = \phi_1 X_{n-1} + \ldots + \phi_p X_{n-p} + e_n$ where $\{e_n\}$ is a sequence of independent random variables with zero mean and unit variance. Then $\{X_k\}$ is associated if $\phi_i \ge 0, \ 1 \le i \le p$. Suppose $p = 1$ and $\phi_1 < 0$. Then $\{X_{2k}\}$ and $\{X_{2k+1}\}$ are associated sequences (Nagaraj and Reddy (1993)).

(ix) Consider the following network. Suppose customers arrive according to a Poisson process with rate λ and all the customers enter the node 1 initially. Further suppose that the service times at the nodes are mutually independent and exponentially distributed and customers choose either the route $r_1 : 1 \to 2 \to 3$ or $r_2 : 1 \to 3$ according to a Bernoulli process with probability p of choosing r_1. The arrival and the service processes are mutually independent. Let S_1 and S_3 be the sojourn times at the nodes 1 and 3 of a customer that follows route r_1. Foley and Kiessler (1989) showed that S_1 and S_3 are associated.

(x) Let the process $\{X_n\}$ be a discrete time homogeneous Markov chain. It is said to be a monotone Markov chain if $\Pr[X_{n+1} \ge y | X_n = x]$ is nondecreasing in x for each fixed y. Daley (1968) showed that a monotone Markov chain is associated.

(xi) Consider a system of k components $1, \ldots, k$, all new at time 0 with life lengths $T_1, \ldots, T_k$. Arjas and Norros (1984) discussed a set of conditions under which the life lengths are associated.

(xii) Consider a system of N non-renewable components in parallel. Let T_i denote the life length of a component i, $i = 1, \ldots, N$. Suppose the environment is represented by a real-valued stochastic process $Y = \{Y_t, \ t \ge 0\}$ which is external to the failure mechanism. Assume that given Y, the life lengths T_i are independent and let

$$\lim_{\tau \to 0} \frac{1}{\tau} \Pr[t \le T_i \le t + \tau | T_i > t, Y] = \eta_i(t, Y_t), \ i = 1, \ldots, N$$

where each $\eta_i(t, y)$ is a positive continuous function of $t > 0$ and real y. Assume that $\eta_i(t, y)$ are all increasing (or decreasing) in y. Further, let

$$Q_i(t_i) = \int_0^{t_i} \eta_i(u, Y_u) du, \ t_i \ge 0, \ i = 1, \ldots, N.$$

Thus $Q_i(t_i)$ is the total risk incurred by the component i from the starting time to time t_i. Then Lefevre and Milhaud (1990) showed that, if Y is associated , then the life lengths $T_1, \ldots, T_n$, are associated as well as the random variables $Q_1(t_1), \ldots, Q_N(t_N)$ are associated.

Remarks. The concept of FKG inequalities, which is connected with statistical mechanics and percolation theory, is related to the concept of association. It started from the works of Harris (1960), Fortuin, Kastelyn and Ginibre (1971), Holley (1974), Preston (1974), Batty (1976), Kemperman (1977) and Newman (1983). For the relationship between the two concepts, see Karlin and Rinott (1980) and Newman (1984). They observed the following - a version of the FKG inequality is equivalent to

$$\frac{\partial^2}{\partial x_i \partial x_j} \log f \ge 0 \text{ for } i \ne j \ 1 \le i, j \le n \tag{1.1.9}$$

when $f(x_1, \ldots, x_n)$, the joint density of $X_1, \ldots, X_n$, is strictly positive on R^n. This is a sufficient but not a necessary condition for association of $(X_1, X_2, \ldots, X_n)$. For example, if (X_1, X_2) is a bivariate normal vector whose covariance matrix Σ is not the inverse of a matrix with non-positive off diagonal entries, then the pair (X_1, X_2) is associated (Pitt (1982)) but the density of (X_1, X_2) does not satisfy the condition (1.1.9).

1.2   Some Probabilistic Properties of Associated Sequences

Esary, Proschan and Walkup (1967) studied the fundamental properties of association. They showed that the association of a set of random variables is preserved under some operations, for instance,

(i) any subset of associated random variables is associated;

(ii) union of two independent sets of associated random variables is a set of associated random variables;

(iii) a set consisting of a single random variable is associated;

(iv) nondecreasing functions of associated random variables are associated; and

(v) if $X_1^{(k)}, \ldots, X_n^{(k)}$ are associated for each k, and if $\underline{X}^{(k)} = (X_1^{(k)}, \ldots, X_n^{(k)}) \to \underline{X} = (X_1, \ldots, X_n)$ in distribution as $k \to \infty$, then the set of random variables $\{X_1, \ldots, X_n\}$ is associated.

(vi) A sequence of random variables $\{X_n,\ n \geq 1\}$ is said to be *stochastically increasing* if, for each $n > 1$ and every x, the sequence

$$P(X_n \leq x | X_1 = x_1, \ldots, X_{n-1} = x_{n-1})$$

is nonincreasing in $x_1, \ldots, x_{n-1}$. It can be shown that, if a sequence $\{X_n, n \geq 1\}$ is stochastically increasing, then it is associated.

(vii) Let the set $\{X_i,\ i = 1, \ldots, n\}$ be a set of associated random variables and $\{X_i^*,\ i = 1, \ldots, n\}$ be independent random variables such that X_i^* and X_i have the same distribution for each $i = 1, \ldots, n$. Let $X_{(1)} \leq X_{(2)} \leq \ldots, \leq X_{(n)}$ and $X_{(1)}^* \leq X_{(2)}^* \leq \ldots, \leq X_{(n)}^*$. Let $F_{(i)}(x)$ and $F_{(i)}^*(x)$ denote the distribution functions of $X_{(i)}$ and $X_{(i)}^*$, respectively. Then Hu and Hu (1998) showed that

$$(F_{(1)}(t), F_{(2)}(t), \ldots, F_{(n)}(t)) \overset{m}{>} (F_{(1)}^*(t), F_{(2)}^*(t), \ldots, F_{(n)}^*(t)), \text{ for all } t \in R,$$

and for a monotone function h,

$$(Eh(X_{(1)}), Eh(X_{(2)}), \ldots, Eh(X_{(n)})) \overset{m}{>} (Eh(X_{(1)}^*), Eh(X_{(2)}^*), \ldots, Eh(X_{(n)}^*)),$$

where, for $\underline{a} = (a_1, \ldots, a_n)$ and $\underline{b} = (b_1, \ldots, b_n)$, we say $\underline{b} \overset{m}{>} \underline{a}$ if $\sum_{i=1}^{n} a_i = \sum_{i=1}^{n} b_i$ and $\sum_{i=1}^{k} a_{(i)} \geq \sum_{i=1}^{k} b_{(i)}$ for $k = 1, \ldots, n$ and $a_{(1)} \leq a_{(2)} \leq \ldots \leq a_{(n)}$ and $b_{(1)} \leq b_{(2)} \leq \ldots \leq b_{(n)}$ denote the ordered values of $a_i's$ and $b_i's$.

For example, in an animal genetic selection problem where $X_1, \ldots, X_n$ are the phenotypes of animals, the best k of n animals (with scores $(X_{(n-k+1)}, \ldots, X_{(n)})$) are kept for breeding (Shaked and Tong (1985)). The partial sum of $\sum_{i=n-k+1}^{n} X_{(i)}$ is the selection differential used by geneticists. If the $X_i's$ have the same mean μ, then the total expected gain of the genetic selection project, $\sum_{i=n-k+1}^{n} X_{(i)}$ is less significant for associated samples when compared with independent samples. For other applications, see Szekli (1995).

Esary et al. (1967) have also developed a simple criterion for establishing association. Instead of checking the condition (1.1.8) for arbitrary nondecreasing functions h and g, one can restrict to nondecreasing test functions h and g which are binary or functions h and g which are nondecreasing, bounded and continuous. In addition, they obtained bounds for the joint distribution function of associated random variables in terms of the joint distribution function of the components under independence.

Some Probability Inequalities

Theorem 1.2.1. *If the random variables $X_1, \ldots, X_n$ are associated, then*

$$P[X_i > x_i,\ i = 1, \ldots, n] \geq \prod_{i=1}^{n} P[X_i > x_i],$$

and

$$P[X_i \le x_i, \ i = 1, \ldots, n] \ge \prod_{i=1}^{n} P[X_i \le x_i]. \tag{1.2.1}$$

The concept of association is useful in the study of approximate independence. This follows from a basic distribution function inequality due to Lebowitz (1972).

Define, for A and B, subsets of $\{1, 2, \ldots, n\}$ and real x_j's,

$$H_{A,B}(x_j, j \in A \cup B) = P[X_j > x_j; j \in A \cup B]$$
$$- P[X_k > x_k, k \in A] P[X_\ell > x_\ell, \ell \in B]. \tag{1.2.2}$$

Observe that the function $H(x, y)$ in (1.1.2) is a special case of this definition.

Theorem 1.2.2 (Lebowitz (1972)). *If the random variables $X_j, 1 \le j \le n$, are associated, then*

$$0 \le H_{A,B} \le \sum_{i \in A} \sum_{j \in B} H_{\{i\},\{j\}}. \tag{1.2.3}$$

Proof. Let $Z_i = I(X_i \ge x_i)$ where $I(G)$ denote the indicator function of a set G. Define

$$U(A) = \prod_{i \in A} Z_i, \text{ and } V(A) = \sum_{i \in A} Z_i.$$

Then

$$H_{A,B} = \text{Cov}\ (U(A), U(B)),$$

and

$$\text{Cov}(V(A), V(B)) = \sum_{i \in A} \sum_{j \in B} H_{\{i\},\{j\}}.$$

Observe that $V(A) - U(A)$ and $V(B)$ are nondecreasing functions of $Z_i, 1 \le i \le n$. Since Z_i's are associated, it follows that

$$\text{Cov}(V(A) - U(A), V(B)) \ge 0.$$

Similarly, $V(B) - U(B)$ and $U(A)$ are nondecreasing functions of $Z_i, 1 \le i \le n$ and

$$\text{Cov}(V(B) - U(B), U(A)) \ge 0.$$

Hence

$$\text{Cov}(U(A), U(B)) \le \text{Cov}(U(A), V(B)) \le \text{Cov}(V(A), V(B)). \qquad \square$$

As an immediate consequence of the above theorem we have the following result.

Theorem 1.2.3 (Joag-Dev (1983), Newman (1984)). *Suppose the random variables* $X_1, \ldots, X_n$ *are associated. Then, the set* $\{X_k, \ k \in A\}$ *is independent of the set* $\{X_j, \ j \in B\}$ *if and only if* $\mathrm{Cov}(X_k, X_j) = 0$ *for all* $k \in A$, $j \in B$ *and the set* X_j, $1 \leq j \leq n$*'s are jointly independent if and only if* $\mathrm{Cov}(X_k, X_j) = 0$ *for all* $k \neq j$, $1 \leq k, j \leq n$. *Thus, a set of uncorrelated associated random variables are independent.*

A fundamental inequality, known as Newman's inequality, which is useful in proving several probabilistic results involving associated random variables, is given below.

Theorem 1.2.4 (Newman (1980)). *Let the pair* (X, Y) *be associated random variables with* $E(X^2) < \infty$ *and* $E(Y^2) < \infty$. *Then, for any two differentiable functions* h *and* g,

$$|\mathrm{Cov}(h(X), g(Y))| \leq \sup_x |h'(x)| \ \sup_y |g'(y)| \mathrm{Cov}(X, Y) \qquad (1.2.4)$$

where h' *and* g' *denote the derivatives of* h *and* g, *respectively.*

This result is an immediate consequence of the Hoeffding identity given by (1.1.3).

Using the above inequality, we get

$$|\mathrm{Cov}(\exp(irX), \exp(isY))| \leq |r||s|\mathrm{Cov}(X, Y) \qquad (1.2.5)$$

for $-\infty < r, s < \infty$. This leads to the following inequality for characteristic functions.

Theorem 1.2.5 (Newman and Wright (1981)). *Suppose* $X_1, \ldots, X_n$ *are associated random variables with the joint and the marginal characteristic functions* $\phi(r_1, \ldots, r_n)$ *and* $\phi_j(r_j)$, $1 \leq j \leq n$, *respectively. Then*

$$\left| \phi(r_1, \ldots, r_n) - \prod_{j=1}^{n} \phi_j(r_j) \right| \leq \frac{1}{2} \sum_{j \neq k} \sum |r_j| \, |r_k| \, \mathrm{Cov}(X_j, X_k). \qquad (1.2.6)$$

The theorem gives an alternate method to show that a set of associated random variables which are uncorrelated are jointly independent.

The following inequality is due to Bagai and Prakasa Rao (1991) and it is a consequence of a result due to Sadikova (1966). Sadikova's result is a two-dimensional analogue of an inequality due to Esseen. We will discuss Sadikova's result in more detail in Chapter 6. The inequality in Theorem 1.2.6 given below has been used by Matula (1996) and Dewan and Prakasa Rao (1997b) for proving their results dealing with probabilistic and statistical inferential aspects for associated random variables.

Theorem 1.2.6 (Bagai and Prakasa Rao (1991)). *Suppose the pair X and Y are associated random variables with bounded continuous densities f_X and f_Y, respectively. Then there exists an absolute constant $c > 0$ such that*

$$\sup_{x,y} |P[X \leq x, Y \leq y] - P[X \leq x]P[Y \leq y]|$$

$$\leq c\{\max(\sup_x f_X(x), \sup_x f_Y(x))\}^{2/3}(\text{Cov}(X,Y))^{1/3}. \tag{1.2.7}$$

The covariance structure of an associated sequence $\{X_n, n \geq 1\}$ plays a significant role in studying the probabilistic properties of the associated sequence $\{X_n, n \geq 1\}$. Let

$$u(n) = \sup_{k \geq 1} \sum_{j:|j-k| \geq n} \text{Cov}(X_j, X_k), \quad n \geq 0. \tag{1.2.8}$$

Then, for any stationary associated sequence $\{X_j\}$, the sequence $u(n)$ is given by

$$u(n) = 2 \sum_{j=n+1}^{\infty} \text{Cov}(X_1, X_j).$$

Moment Bounds

Birkel (1988a) observed that moment bounds for partial sums of associated sequences also depend on the rate of decrease of $u(n)$.

Theorem 1.2.7 (Birkel (1988a)). *Let the sequence $\{X_n, n \geq 1\}$ be a sequence of associated random variables with $EX_j = 0$, $j \geq 1$ and suppose that*

$$\sup_{j \geq 1} E|X_j|^{r+\delta} < \infty \text{ for some } r > 2, \; \delta > 0.$$

Assume that

$$u(n) = O(n^{-(r-2)(r+\delta)/2\delta}).$$

Then, there is a constant $B > 0$ not depending on n such that for all $n \geq 1$,

$$\sup_{m \geq 0} E|S_{n+m} - S_m|^r \leq B n^{r/2} \tag{1.2.9}$$

where $S_n = \sum_{j=1}^{n} X_j$.

If the X_j's are uniformly bounded, then the following result holds.

Theorem 1.2.8 (Birkel(1988a)). *Let the sequence $\{X_n, n \geq 1\}$ be a sequence of associated random variables satisfying $EX_j = 0$ and $|X_j| \leq c < \infty$ for $j \geq 1$. Assume that*

$$u(n) = O(n^{-(r-2)/2}).$$

Then (1.2.9) holds.

The result stated above can easily be generalized to obtain the following result by methods in Birkel (1988a).

Theorem 1.2.9 (Bagai and Prakasa Rao (1991)). *For every $\alpha \in I$, an index set, let $\{X_n(\alpha), \ n \geq 1\}$ be an associated sequence with $EX_n(\alpha) = 0$ and*

$$\sup_{\alpha \in I} \sup_{n \geq 1} |X_n(\alpha)| \leq A < \infty.$$

Let

$$S_n(\alpha) = \sum_{j=1}^{n} X_j(\alpha),$$

and

$$u(n, \alpha) = \sup_{k \geq 1} \sum_{j:|j-k| \geq n} \mathrm{Cov}(X_j(\alpha), X_k(\alpha)).$$

Suppose there exists $b > 0$, independent of $\alpha \in I$ and $n \geq 1$, such that for some $r > 2$, and all $\alpha \in I$ and $n \geq 1$,

$$u(n, \alpha) \leq bn^{-(r-2)/2}.$$

Then, there exists a constant C, not depending on n and α, such that for all $n \geq 1$,

$$\sup_{\alpha \in I} \sup_{m \geq 0} E|S_{n+m}(\alpha) - S_n(\alpha)|^r \leq Cn^{r/2}.$$

The result stated above is useful in the nonparametric estimation of the survival function for associated random variables (cf. Bagai and Prakasa Rao (1991)).

Bagai and Prakasa Rao (1995) generalized Theorem 1.2.7 and Theorem 1.2.8 to functions of associated random variables when the functions are of bounded variation and then used the results in the study of nonparametric density estimation for stationary associated sequences.

Theorem 1.2.10 (Bagai and Prakasa Rao (1995)). *For every $\alpha \in J$, an index set, let $\{X_j(\alpha), j \geq 1\}$ be an associated sequence. Let $\{f_n, \ n \geq 1\}$ be functions of bounded variation which are differentiable and suppose that $\sup_{n \geq 1} \sup_x |f_n'(x)| \leq c < \infty$. Let $E(f_n(X_j(\alpha))) = 0$ for every $n \geq 1$, $j \geq 1$ and $\alpha \in J$. Suppose there exist $r > 2$ and $\delta > 0$ (independent of α, j and n) such that*

$$\sup_{n \geq 1} \sup_{\alpha \in J} \sup_{j \geq 1} E|f_n(X_j(\alpha))|^{r+\delta} < \infty. \tag{1.2.10}$$

Let

$$u(n, \alpha) = \sup_{k \geq 1} \sum_{j:|j-k| \geq n} \mathrm{Cov}(X_j(\alpha), X_k(\alpha)). \tag{1.2.11}$$

Suppose that there exists $c > 0$ independent of $\alpha \in J$ such that

$$u(n, \alpha) \leq cn^{-(r-2)(r+\delta)/2\delta}.$$

Then there exists a constant B not depending on n, m and α, such that

$$\sup_{m \geq 1} \sup_{\alpha \in J} \sup_{k \geq 0} E|S_{n+k,m}(\alpha) - S_{k,m}(\alpha)|^r \leq B n^{r/2} \qquad (1.2.12)$$

where

$$S_{m_n,n}(\alpha) = \sum_{j=1}^{m_n} f_n(X_j(\alpha)).$$

Theorem 1.2.11 (Bagai and Prakasa Rao (1995)). *For any $\alpha \in J$, an index set, let the sequence $\{X_j(\alpha), \; j \geq 1\}$ be an associated sequence. Let $\{f_n, \; n \geq 1\}$ be functions of bounded variation which are differentiable and suppose that $\sup_{n \geq 1} \sup_{x} |f_n(x)| < \infty$, and $\sup_{n \geq 1} \sup_{x} |f'_n(x)| \leq c < \infty$. Let $E(f_n(X_j(\alpha)) = 0$, $n \geq 1, \alpha \in J$ and $j \geq 1$. Assume that there exists $r > 2$ such that*

$$u(n, \alpha) = O(n^{-(r-2)/2}).$$

Then there exists a constant B not depending on n, m and α, such that

$$\sup_{m \geq 1} \sup_{\alpha \in J} \sup_{k \geq 0} E|S_{n+k,m}(\alpha) - S_{k,m}(\alpha)|^r \leq B n^{r/2} \qquad (1.2.13)$$

where $u(n, \alpha)$ is defined by (1.2.11).

Bulinski (1993) generalized Birkel's results to random fields of associated variables. Bulinski and Shaskin (2007) give a comprehensive survey of limit theorems for associated random fields and related systems. Shao and Yu (1996) obtained some Rosenthal-type moment inequalities for associated sequences useful in their study of empirical processes for associated sequences.

Theorem 1.2.12 (Shao and Yu (1996)). *Let $2 < p < r \leq \infty$. Let f be an absolutely continuous function satisfying $\sup_{x \in R} |f'| \leq B$ and let $\{X_n, \; n \geq 1\}$ be a sequence of associated random variables with $E f(X_n) = 0$ and $\|f(X_n)\|_r = (E|f(X_n)|^r)^{1/r} < \infty$. Let*

$$u(n) = \sup_{i \geq 1} \sum_{j:|j-i| \geq n} \mathrm{Cov}(X_i, X_j) < \infty, \; n \geq 0.$$

Suppose that

$$u(n) \leq C n^{-\theta} \qquad (1.2.14)$$

for some $C > 0$ and $\theta > 0$. Then, for any $\varepsilon > 0$, there exists $K = K(\varepsilon, r, p, \theta) < \infty$ such that

$$E|\sum_{i=1}^{n} f(X_i)|^p \leq K \left(n^{1+\varepsilon} \max_{i \leq n} E|f(X_i)|^p \right.$$

$$\left. + (n \max_{i \leq n}(\sum_{j=1}^{n} |\mathrm{Cov}(f(X_i), f(X_j))|))^{p/2} \right.$$

$$+ n^{(r(p-1)-p+\theta(p-r)/(r-2)\vee(1+\varepsilon)}$$

$$\times \max_{i\leq n} \|f(X_i)\|_r^{r(p-2)/(r-2)} (B^2 C)^{(r-p)/(r-2)}\bigg). \qquad (1.2.15)$$

Bernstein-type Inequality

Prakasa Rao (1993) obtained a Bernstein-type inequality applicable for sums of finite sequences of associated random variables.

Theorem 1.2.13 (Prakasa Rao (1993)). *For every $n \geq 1$, let $\{Z_i^{(n)}, 1 \leq i \leq l_n\}$ be associated random variables such that $E(Z_i^{(n)}) = 0$, $|Z_i^{(n)}| \leq d_n \leq \infty$. Define*

$$\psi^{(n)}(m) = \sup_{1\leq i_1, i_2 \leq l_n} \sum_{\substack{i_1 \leq j, k \leq i_2 \\ 0 < |j-k| \leq m}} \mathrm{Cov}(Z_j^{(n)}, Z_k^{(n)}). \qquad (1.2.16)$$

Let

$$S_n = \sum_{i=1}^{l_n} Z_i^{(n)}. \qquad (1.2.17)$$

Define $1 \leq m_n \leq l_n$ and α by the relation

$$\alpha\, m_n\, d_n \leq \frac{1}{4}. \qquad (1.2.18)$$

Then, for any $\epsilon > 0$,

$$P(|S_n| \geq \epsilon) \leq 2\ exp\{6\alpha^2 l_n C_n + \frac{3}{4}\sqrt{e}\alpha^{-2} l_n m_n^{-2} C_n^{-1} - \alpha\epsilon\} \qquad (1.2.19)$$

where

$$C_n = (d_n^2 + \psi^{(n)}(m_n)). \qquad (1.2.20)$$

Strong Laws of Large Numbers

Strong laws of large numbers for associated sequences have been obtained by Newman (1984) and Birkel (1989), the former for the stationary case and the latter for the non-stationary case.

Theorem 1.2.14 (Newman (1984)). *Let the sequence $\{X_n, n \geq 1\}$ be a stationary sequence of associated random variables with $E(X_1^2) < \infty$. If*

$$\frac{1}{n} \sum_{j=1}^{n} \mathrm{Cov}(X_1, X_j) \to 0\ as\ n \to \infty,$$

then

$$\frac{1}{n}(S_n - E(S_n)) \to 0\ a.s.\ as\ n \to \infty. \qquad (1.2.21)$$

Theorem 1.2.15 (Birkel (1989)). *Let the sequence $\{X_n, n \geq 1\}$ be a sequence of associated random variables with $E(X_n^2) < \infty$, $n \geq 1$. Assume that*

$$\sum_{j=1}^{\infty} \frac{1}{j^2} \operatorname{Cov}(X_j, S_j) < \infty.$$

Then (1.2.21) holds.

This theorem has been generalized in Bagai and Prakasa Rao (1995) to functions of associated random variables.

Theorem 1.2.16 (Bagai and Prakasa Rao (1995)). *Let the sequence $\{X_n, n \geq 1\}$ be a stationary sequence of associated random variables. Let $S_{k,n} = \sum_{j=1}^{k} f_n(X_j)$ where f_n is differentiable with $\sup_n \sup_x |f_n'(x)| < \infty$. Suppose $E[f_n(X_1)] = 0$, $\operatorname{Var}[f_n(X_1)] < \infty$ and*

$$\sum_{j=1}^{\infty} \operatorname{Cov}(X_1, X_j) < \infty.$$

Then,

$$\frac{S_{n,n}}{n} \to 0 \text{ a.s. as } n \to \infty.$$

Remarks. Theorem 1.2.16 can be used to prove the pointwise strong consistency of kernel type nonparametric density estimators for density estimation for stationary associated sequences (cf. Bagai and Prakasa Rao (1995)).

The following result is a strong law of large numbers for a triangular array of associated random variables which is useful in nonparametric density estimation.

Theorem 1.2.17 (Dewan and Prakasa Rao (1997a)). *Let the set $\{X_{nj}, 1 \leq j \leq k_n, n \geq 1\}$ be a triangular array of random variables such that the sequence $\{X_{nj}, 1 \leq j \leq k_n\}$ is strictly stationary and associated for every $n \geq 1$ with $E[X_{n1}] = 0$ and $\operatorname{Var}(X_{n1}) < \infty$ for all $n \geq 1$. Suppose that $k_n = O(n^\gamma)$ for some $0 \leq \gamma < \frac{3}{2}$ and the following condition holds:*

$$\sum_{j=1}^{k_n} \operatorname{Cov}(X_{n1}, X_{nj}) < \infty. \tag{1.2.22}$$

Let $S_{n,l} = \sum_{j=1}^{l} X_{nj}$. Further suppose that

$$E[\max_{n^2 < j \leq (n+1)^2} |S_{j,k_j} - S_{n^2,k_{n^2}}|]^2 = O(n^{4-\delta}) \tag{1.2.23}$$

for some $\delta > 1$. Then,

$$\frac{S_{n,k_n}}{n} \to 0 \text{ a.s. as } n \to \infty. \tag{1.2.24}$$

The following result, due to Lebowitz (1972), deals with a necessary and sufficient condition for erogodicity of a stationary sequence of associated random variables.

Theorem 1.2.18 (Lebowitz (1972), Newman (1980)). *Let the sequence $\{X_n, n \geq 1\}$ be a stationary sequence of associated random variables. Then the process $\{X_n, n \geq 1\}$ is ergodic if and only if*

$$\lim_{n \to \infty} n^{-1} \sum_{j=1}^{n} \text{Cov}(X_1, X_j) = 0. \tag{1.2.25}$$

In particular, if (1.2.25) holds, then, for any real-valued function $f(.)$ such that $E|f(X_1)| < \infty$,

$$\lim_{n \to \infty} n^{-1} \sum_{j=1}^{n} f(X_j) = E[f(X_1)] \quad a.s. \tag{1.2.26}$$

Central Limit Theorems

The next theorem was the original application (and motivation) of the characteristic function inequality given in Theorem 1.2.5. It gives the central limit theorem for partial sums of stationary associated random variables.

Theorem 1.2.19 (Newman (1980, 1984)). *Let the sequence $\{X_n, n \geq 1\}$ be a stationary associated sequence of random variables with $E[X_1^2] < \infty$ and $0 < \sigma^2 = \text{Var}(X_1) + 2 \sum_{j=2}^{\infty} \text{Cov}(X_1, X_j) < \infty$. Then, $n^{-1/2}(S_n - E(S_n)) \xrightarrow{\mathcal{L}} N(0, \sigma^2)$ as $n \to \infty$.*

Since σ is not known in practice, it needs to be estimated. Peligrad and Suresh (1995) obtained a consistent estimator of σ. Let the sequence $\{l_n, n \geq 1\}$ be a sequence of positive integers with $1 \leq l_n \leq n$. Set $S_j(k) = \sum_{i=j+1}^{j+k} X_i$, $\bar{X}_n = \frac{1}{n} \sum_{i=1}^{n} X_i$, and $B_n = \frac{1}{n-l}[\sum_{j=0}^{n-l} \frac{|S_j(l) - l\bar{X}_n|}{\sqrt{l}}]$ where we write $l = l_n$ for convenience.

Theorem 1.2.20. *Let the sequence $\{X_n, n \geq 1\}$ be a stationary associated sequence of random variables satisfying $E(X_1) = \mu$, $E(X_1^2) < \infty$. Let $l_n = o(n)$ as $n \to \infty$. Assume that $\sum_{i=2}^{\infty} \text{Cov}(X_1, X_i) < \infty$. Then*

$$B_n \to \sigma \sqrt{2/\pi} \text{ in } L_2 \text{ as } n \to \infty.$$

In addition, if $l_n = O(n/\log n)^2)$ as $n \to \infty$, then the convergence above is almost sure.

A local limit theorem of the type due to Shepp (1964) was proved for stationary associated sequences by Wood (1985). We do not discuss these results here.

Cox and Grimmett (1984) proved a central limit theorem for double sequences and used it in percolation theory and the voter model. It generalizes the central limit theorem due to Newman (1984) discussed earlier.

Theorem 1.2.21 (Cox and Grimmett (1984)). *Let the set $\{X_{nj}, 1 \leq j \leq n, n \geq 1\}$ be a triangular array of random variables associated row-wise satisfying the following conditions:*

(i) *there are strictly positive, finite constants c_1, c_2 such that*

$$\text{Var}(X_{nj}) \geq c_1, \ E[|X_{nj}|^3] \leq c_2 \quad \forall j \ and \ n.$$

(ii) *there is a function $u : \{0, 1, 2, \ldots, \} \to \mathcal{R}$ such that $u(r) \to 0$ as $r \to \infty$ and*

$$\sum_{j:|k-j|\geq r} \text{Cov}(X_{nj}, X_{nk}) \leq u(r) \ for \ all \ k, n \ and \ r \geq 0.$$

Let $S_{n,n} = \sum_{j=1}^{n} X_{nj}$. Then the sequence $\{S_{n,n}, n \geq 1\}$ satisfies the central limit theorem.

Roussas (1994) established asymptotic normality of random fields of partial sums of positively (as well as negatively) associated processes. We will discuss concepts of positive and negative association later in this chapter.

Berry-Esseen Type Bound

The next, natural, question is the rate of convergence in the central limit theorem. We have the following versions of the Berry-Esseen theorem for associated sequences. Hereafter we denote the standard normal distribution function by $\Phi(x)$.

Theorem 1.2.22 (Wood (1983)). *Suppose the sequence $\{X_n, n \geq 1\}$ is a stationary sequence of associated random variables satisfying $E[X_n] = 0$, $E(X_n^2) < \infty$, $E[|X_n|^3] < \infty$ for all n and $0 < \sigma^2 = \text{Var}(X_1) + 2\sum_{j=2}^{\infty} \text{Cov}(X_1, X_j) < \infty$. Then, for $n = mk$,*

$$|F_n(x) - \Phi(\frac{x}{\sigma})| \leq \frac{16\bar{\sigma}_k^4 m(\sigma^2 - \bar{\sigma}_k^2)}{9\pi\rho_k^2} + \frac{3\rho_k}{\sigma k^3 \sqrt{m}},$$

where $\bar{S}_n = \sum_{i=1}^{n} \frac{X_i}{\sqrt{n}}$, $\bar{\sigma}_n^2 = E(\bar{S}_n^2)$, $\rho_n = E[|\bar{S}_n|^3]$ and $F_n(x) = P[\bar{S}_n \leq x]$.

The rate obtained above by Wood (1983) at its best is of $O(n^{-1/5})$ which is far from the optimal Berry-Esseen rate $O(n^{-1/2})$ in the classical Berry-Esseen bound for sums of independent and identically distributed random variables. An improvement of the same is given below.

Theorem 1.2.23 (Birkel (1988b)). *Let the sequence $\{X_n, n \geq 1\}$ be an associated sequence with $E[X_n] = 0$, satisfying*

(i) $u(n) = O(e^{-\lambda n})$, $\lambda \geq 0$,

(ii) $\displaystyle\inf_{n \geq 1} \frac{\sigma_n^2}{n} > 0$, $\sigma_n^2 = E[S_n^2]$, *and*

(iii) $\displaystyle\sup_{n \geq 1} E[|X_n|^3] < \infty$. *where $u(n)$ is as defined by (1.2.8).*

Then there exists a constant B not depending on n such that, for all $n \geq 1$,

$$\Delta_n \equiv \sup_{x \in R} |P\{\sigma_n^{-1} S_n \leq x\} - \Phi(x)| \leq B n^{-1/2} \log^2 n.$$

If, instead of (iii) in Theorem 1.2.23, we assume that

(iii)' $\displaystyle\sup_{j \geq 1} E|X_j|^{3+\delta} < \infty$ for some $\delta > 0$,

then there exists a constant B not depending on n such that, for all $n \geq 1$,

$$\Delta_n \leq B n^{-1/2} \log n.$$

Even though Birkel (1988b) obtained an improved rate of $O(n^{-1/2} log^2 n)$, yet it is not clear how the constant B involved in the bound depends on the moments of the random variables $\{X_n\}$. The following result is an attempt in this direction.

Theorem 1.2.24 (Dewan and Prakasa Rao (1997b)). *Let the set $\{X_i, 1 \leq i \leq n\}$ be a set of stationary associated random variables with $E[X_1] = 0$, $\mathrm{Var}[X_1] = \sigma_0^2 > 0$ and $E[|X_1|^3] < \infty$. Suppose the distribution of X_1 is absolutely continuous. Let*
$$S_n = \sum_{i=1}^{n} X_i \text{ and } \sigma_n^2 = \mathrm{Var}(S_n).$$
Suppose that $\frac{\sigma_n^2}{n} \to \sigma_0^2$ as $n \to \infty$. Let $F_n(x)$ be the distribution function of $\frac{S_n}{\sigma_n}$ and $F_n^*(.)$ be the distribution function of $\frac{\sum_{i=1}^{n} Z_i}{\sigma_n}$ where Z_i, $1 \leq i \leq n$ are i.i.d. with distribution function the same as that of X_1. Let m_n be a bound on the derivative of F_n^*. Then there exist absolute constants $B_i > 0$, $1 \leq i \leq 3$, such that

$$\sup_{x} |F_n(x) - \Phi(x)| \leq B_1 \frac{d_n^{1/3} m_n^{2/3}}{\sigma_n^{2/3}} + B_2 \frac{E|X_1|^3}{\sqrt{n}\, \sigma_0^3}$$

$$+ B_3 \left(\frac{\sigma_n}{\sigma_0 \sqrt{n}} - 1 \right) \tag{1.2.27}$$

where

$$d_n = \sum_{j=2}^{n} (n - j + 1) \, \mathrm{Cov}(X_1, X_j).$$

Remarks. The bound given above can be made more explicit by bounding m_n in (1.2.27) if we assume that the characteristic function of X_1 is absolutely integrable. Then, for large n,

$$
m_n \leq \left(\frac{\sigma_n}{\sigma_0 \sqrt{n}} - 1\right) \sup_x g_n\left(\frac{\sigma_n x}{\sigma_0 \sqrt{n}}\right) + \sup_x g_n\left(\frac{\sigma_n x}{\sigma_0 \sqrt{n}}\right)
$$

$$
\leq \frac{2}{\sqrt{2\pi}} \left(\frac{\sigma_n}{\sigma_0 \sqrt{n}} - 1\right) + \frac{2}{\sqrt{2\pi}} \tag{1.2.28}
$$

Bulinski (1995) established the rate of convergence of standardized sums of associated random variables to the normal law for a random field of associated random variables.

Invariance Principle

Let the sequence $\{X_n, n \geq 1\}$ be a sequence of random variables with $EX_n = 0$, $EX_n^2 < \infty$, $n \geq 1$. Let

$$
S_0 = 0, \quad S_n = \sum_{k=1}^{n} X_k, \quad \sigma_n^2 = ES_n^2, \quad n \geq 1.
$$

Assume that $\sigma_n^2 > 0$, $n \geq 1$. Let $\{k_n,\ n \geq 0\}$ be an increasing sequence of real numbers such that

$$
0 = k_0 < k_1 < k_2 < \dots \tag{1.2.29}
$$

and

$$
\lim_{n \to \infty} \max_{1 \leq i \leq n} (k_i - k_{i-1})/k_n = 0. \tag{1.2.30}
$$

Define $m(t) = \max\{i : k_i \leq t\}$, $t \geq 0$, and

$$
W_n(t) = S_{m_n(t)}/\sigma_n, \quad t \in [0,1], \quad n \geq 1, \tag{1.2.31}
$$

where $m_n(t) = m(tk_n)$. Consider the process

$$
W_n^*(t) = \frac{S_{[nt]}}{\sigma_n}, \quad t \in [0,1]. \tag{1.2.32}
$$

When $k_n = n$, $n \geq 1$ the processes defined by (1.2.31) and (1.2.32) are equivalent.

Theorem 1.2.25 (Newman and Wright (1981)). *Let the sequence $\{X_n, n \geq 1\}$ be a strictly stationary sequence of associated random variables with $EX_1 = 0$ and $EX_1^2 < \infty$. If $0 < \sigma^2 = \mathrm{Var}(X_1) + 2\sum_{n=2}^{\infty} \mathrm{Cov}(X_1, X_n) < \infty$, then, $W_n^* \overset{\mathcal{L}}{\to} W$ as $n \to \infty$, where W is a standard Wiener process.*

An invariance principle for a non-stationary associated process has been studied by Birkel (1988c).

Theorem 1.2.26 (Birkel (1988c)). *Let the sequence $\{X_n, n \geq 1\}$ be a sequence of associated random variables with $E(X_n) = 0$, $E(X_n^2) < \infty$ for $n \geq 1$. Assume that*

(i) $\lim\limits_{n \to \infty} \sigma_n^{-2} E(U_{nk} U_{n\ell}) = \min(k, \ell)$ *for $k, \ell \geq 1$ and $U_{m,n} = S_{m+n} - S_m$; and*

(ii) $\sigma_n^{-2}(S_{n+m} - S_n)^2$, $m \geq 0$, $n \geq 1$ *is uniformly integrable.*

Then
$$W_n^* \overset{\mathcal{L}}{\to} W \quad as \quad n \to \infty.$$

Birkel's result was generalized by Matula and Rychlik (1990). They observed that if $W_n^* \overset{\mathcal{L}}{\to} W$ as $n \to \infty$, then

$$\sigma_n^2 = nh(n), \tag{1.2.33}$$

where $h : R^+ \to R^+$ is slowly varying. They proved an invariance principle for sequences $\{X_n, n \geq 1\}$ which do not satisfy the condition (1.2.33).

Theorem 1.2.27 (Matula and Rychlik (1990)). *Let the sequence $\{X_n, n \geq 1\}$ be a sequence of associated random variables with $E(X_n) = 0$, $E(X_n^2) < \infty$ for $n \geq 1$. Let $\{k_n, n \geq 1\}$ be a sequence of real numbers satisfying (1.2.29) and (1.2.30). Assume that*

(i) $\lim\limits_{n \to \infty} \sigma_n^{-2} E(S_{m_n(p)} S_{m_n(q)}) = \min(p, q)$, *for $p, q \geq 1$,*

(ii) $(\sigma_{n+m}^2 - \sigma_m^2)^{-1}(S_{n+m} - S_n)^2$, $m \geq 0$, $n \geq 1$ *is uniformly integrable.*

Then
$$W_n \overset{\mathcal{L}}{\to} W \quad as \quad n \to \infty.$$

Strong Invariance Principle

Yu (1996) proved a strong invariance principle for associated sequences. We now briefly discuss this result.

Let the sequence $\{X_n, n \geq 1\}$ be an associated sequence with $E(X_n) = 0$ and define

$$u(n) = \sup_{k \geq 1} \sum_{j:|j-k|>n} \mathrm{Cov}(X_j, X_k).$$

Define blocks H_k and I_k of consecutive positive integers leaving no gaps between the blocks. The order is $H_1, I_1, H_2, I_2, \ldots$. The lengths of the blocks are defined by

$$\mathrm{card}\{H_k\} = [k^\alpha], \; \mathrm{card}\{I_k\} = [k^\beta]$$

for some suitably chosen real numbers $\alpha > \beta > 0$ with $\mathrm{card}\{K\}$ denoting the number of integers in K. Let

$$U_k = \sum_{i \in H_k} X_i, \lambda_k = E(U_k^2)$$

and

$$V_k = \sum_{i \in I_k} X_i, \tau_k = E(V_k^2)$$

for $k \geq 1$. Let the sequence $\{W_k, k \geq 1\}$ be a sequence of independent $N(0, \frac{\tau_k^2}{2})$ distributed random variables independent of the sequence $\{U_k, k \geq 1\}$. Define

$$\xi_k = (U_k + W_k)/(\lambda_k^2 + \frac{\tau_k^2}{2})^{1/2}, \ k \geq 1.$$

Let F_k be the distribution function of ξ_k. Note that the function F_k is continuous. Let

$$\eta_k = \Phi^{-1}(F_k(\xi_k)), \ k \geq 1,$$

where Φ^{-1} denotes the inverse of the standard normal distribution function Φ. Note that each η_k has a standard normal distribution and the sequence $\{\eta_k, k \geq 1\}$ is an associated sequence. Furthermore the covariances of the sequences $\{\eta_k, k \geq 1\}$ are controlled by the sequence $\{X_n, n \geq 1\}$. The following theorem is due to Yu (1996).

Theorem 1.2.28. *Let the sequence $\{X_n, n \geq 1\}$ be an associated sequence satisfying $E(X_n) = 0$, $\inf_{n \geq 1} E(X_n^2) > 0$, and*

$$\sup_{n \geq 1} |X_n|^{2+r+\delta} < \infty \tag{1.2.34}$$

for some $r, \delta > 0$. Further suppose that

$$u(n) = O(n^{-\gamma}), \ \gamma = \frac{r(2+r+\delta)}{2\delta} > 1. \tag{1.2.35}$$

If moreover $5\beta/3 > \alpha > \beta > 0$, then for any $0 < \theta < (1/2)$ and for all $i \neq j$,

$$0 \leq E\eta_i\eta_j \leq C((ij)^{-\alpha/2} E(U_i U_j))^{\theta/(1+\theta)} \tag{1.2.36}$$

for some constant C not depending on i, j. Furthermore there exists real numbers $\alpha > \beta > 1$ and some $\varepsilon > 0$ such that, for k satisfying $N_k < N \leq N_{k+1}$,

$$|\sum_{j=1}^{N} X_j - \sum_{i=1}^{k}(\lambda_i^2 + \frac{\tau_i^2}{2})^{1/2}\eta_i| \leq C_1 N^{(1/2)-\varepsilon} \ a.s., \tag{1.2.37}$$

for some constant C_1 not depending on N.

Based on this theorem, Yu (1996) established the following strong invariance principle for associated sequences.

Theorem 1.2.29. *Let the sequence $\{X_n, n \geq 1\}$ be an associated sequence satisfying $E(X_n) = 0$, $\inf_{n \geq 1} E(X_n^2) > 0$, and*

$$\sup_{n \geq 1} |X_n|^{2+r+\delta} < \infty \tag{1.2.38}$$

for some $r, \delta > 0$. Further suppose that

$$u(n) = O(e^{-\lambda n}) \qquad (1.2.39)$$

for some $\lambda > 0$. Then without changing its distribution we can redefine the sequence $\{X_n, \ n \geq 1\}$ on another probability space with a standard Wiener process $\{W(t), \ t \geq 0\}$ such that , for some $\varepsilon > 0$,

$$\sum_{j=1}^{N} X_j - W(\sigma_N^2) = O(N^{(1/2)-\varepsilon}) \ a.s., \qquad (1.2.40)$$

where $\sigma_N^2 = \mathrm{Var}(\sum_{j=1}^{N} X_j)$.

As a special case of the above theorem, it follows that

$$\liminf_{n \to \infty} \left[\frac{8 \log \log n}{\pi^2 \sigma_n^2} \right]^{1/2} \sup_{1 \leq i \leq n} |S_n| = 1 \ \ a.s., \qquad (1.2.41)$$

where $S_n = \sum_{i=1}^{n} X_i$, under the conditions stated above.

Law of the Iterated Logarithm

Dabrowski (1985) proved the law of the iterated logarithm for stationary associated sequences.

Theorem 1.2.30 (Dabrowski (1985)). *Let the sequence $\{X_n, \ n \geq 1\}$ be a sequence of stationary associated random variables with $E(X_1) = 0$, $\sigma^2 = 1$, $\sup(E|S_k/k^{1/2}|^3 : k \geq 1) < \infty$, and*

$$1 - E(S_n^2/n) = O(n^{-\delta}) \ \ for \ some \ \delta > 0. \qquad (1.2.42)$$

Then the sequence $\{X_n, \ n \geq 1\}$ satisfies the functional law of the iterated logarithm, that is, let

$$\chi_n(t) = \begin{cases} S_k (2n \log \log n)^{-1/2} & if \ 0 \leq k \leq n \ and \ t = k/n, \\ linear & otherwise \ for \ 0 \leq t \leq 1. \end{cases}$$

Then, with probability 1, χ_n is equicontinuous and the set of its limit points (in the supremum norm on $C[0,1]$) coincides with the set

$$\{k \in C[0,1] : k \ is \ absolutely \ continuous \ in \ [0,1], \ k(0) = 0 \ and \ \int_0^1 (k(t))^2 dt \leq 1\}.$$

In particular,

$$P\left[\lim_{n \to \infty} \sup(S_n/\sqrt{2n \log \log n}) = 1 \right] = 1.$$

Bounds on Expectations of Partial Sums

We will now discuss some inequalities which are useful in finding bounds for expectations of partial sums of functions of associated random variables.

Theorem 1.2.31 (Bakhtin's lemma)). *Let (X, Y) be an associated bivariate random vector with $E(X) = 0$. Suppose that $f(.)$ is a convex function such that $E[f(X+Y)]$ exists. Then*

$$E[f(Y)] \le E[f(X+Y)].$$

Proof. We will sketch the proof assuming that the function f is bounded below, nondecreasing and has a continuous derivative f' and Y is a bounded random variable. For the general case, see Bulinski and Shaskin (2007). Without loss of generality, we assume that $f \ge 0$. Let

$$g(x, y) = \begin{cases} \dfrac{f(x+y) - f(y)}{x} & \text{if } x \ne 0 \\ f'(y) & \text{if } x = 0 \end{cases}.$$

Then g is a continuous and componentwise nondecreasing function on R^2. Furthermore, for $x \ne 0$,

$$\frac{\partial g(x, y)}{\partial y} = \frac{f'(x+y) - f'(y)}{x} \ge 0$$

and

$$\frac{\partial g(x, y)}{\partial x} = \frac{f(x+y) - f(y) - xf'(y)}{x^2} \ge 0.$$

Note that f' is a nondecreasing function since f is convex. Furthermore

$$E|g(X, Y)| = E(|g(X, Y)I_{[|X| \le 1]}) + E(|g(X, Y)I_{[|X| > 1]})$$
$$\le E(\sup_{|x| \le 1} f'(Y + x)) + E|f(X+Y)| + E|f(Y)| < \infty$$

since the random variable Y is bounded and hence the random variable $\sup_{|x| \le 1} f'(Y + x)$ is also bounded. By the association property of the random vector (X, Y), it follows that

$$\text{Cov}(g(X, Y), X) \ge 0.$$

Since $E(X) = 0$, it follows that

$$\text{Cov}(g(X, Y), X) = E[g(X, Y)X] - E[g(X, Y)] \, E(X) = E[g(X, Y)X]$$
$$= E[f(X+Y) - f(Y)].$$

Hence

$$E[f(X+Y)] - E[f(Y)] \ge 0. \qquad \square$$

Definition. A function $f : R^n \to R$ is called *supermodular* if is bounded on any bounded subset of R^n and

$$f(\mathbf{x} \vee \mathbf{y}) + f(\mathbf{x} \wedge \mathbf{y}) \geq f(\mathbf{x}) + f(\mathbf{y}), \quad \mathbf{x}, \mathbf{y} \in R^n$$

where

$$\mathbf{x} \vee \mathbf{y} = (\max(x_1, y_i), \ldots, \max(x_n, y_n))$$

and

$$\mathbf{x} \wedge \mathbf{y} = (\min(x_1, y_i), \ldots, \min(x_n, y_n)).$$

Here $\mathbf{x} = (x_1, \ldots, x_n)$ and $\mathbf{y} = (y_1, \ldots, y_n)$. A function f is called *submodular* if the reverse inequality holds.

A function f is supermodular if and only if the function $-f$ is submodular. If f is twice differentiable, then f is supermodular if and only if

$$\frac{\partial^2 f(\mathbf{x})}{\partial x_i \partial x_j} \geq 0$$

for all $1 \leq i \neq j \leq n$ and $\mathbf{x} \in R^n$ (cf. Marshall and Olkin (1979)).

It is easy to check that the following functions $f : R^n \to R$ are supermodular.

(i) $\sum_{i=1}^{n} x_i$;

(ii) $\max_{k=1,\ldots,n} \sum_{i=1}^{k} x_i$;

(iii) $-\max_{k=1,\ldots,n} x_k$;

(iv) $-\sum_{k=1}^{n} (x_k - \bar{x})^2$ where $\bar{x} = n^{-1} \sum_{i=1}^{n} x_i$.

Furthermore, if $g : R^n \to R$ is supermodular and $h_i : R \to R$ are nondecreasing for $i = 1, \ldots, n$, then the function $f(x) = g(h_1(x), \ldots, h_n(x))$ is supermodular and if $h : R^n \to R$ is supermodular and componentwise nondecreasing and $g : R \to R$ is a nondecreasing convex function, then $f(\mathbf{x}) = g(h(\mathbf{x}))$ is supermodular.

Definition. A random vector $\mathbf{X} = (X_1, \ldots, X_n)$ is said to be smaller than another random vector $\mathbf{Y} = (Y_1, \ldots, Y_n)$ in the *supermodular* order, denoted by $\mathbf{X} \ll_{sm} \mathbf{Y}$, if $E[f(\mathbf{X})] \leq E[f(\mathbf{Y})]$ for all supermodular functions f for which the expectations exist.

The following results are due to Christofides and Vaggelatou (2004).

Theorem 1.2.32. *Let* $\mathbf{X}$ *and* $\mathbf{Y}$ *be two n-dimensional random vectors with the same marginal distributions. If* $E[f(\mathbf{x})] \leq E[f(\mathbf{Y})]$ *for all f twice differentiable increasing supermodular functions, then* $\mathbf{X} \ll_{sm} \mathbf{Y}$.

Theorem 1.2.33. *Let the set $\{X_i, 1 \leq i \leq n\}$ be a set of associated random variables and $\{X_i^*, 1 \leq i \leq n\}$ be another set of independent random variables independent of the sequence $\{X_i, 1 \leq i \leq n\}$ such that the distributions of X_i and X_i^* are the same for $i = 1, \ldots, n$. Then*

$$(X_i, \ldots X_n) \gg_{sm} (X_1^*, \ldots, X_m^*).$$

Remarks. The result stated above also holds if the set of random variables $\{X_i, 1 \leq i \leq n\}$ is positively associated and the reverse inequality holds if the set $\{X_i, 1 \leq i \leq n\}$ is negatively associated. For the definition of negative association, see the next section. A set $\{X_i, 1 \leq i \leq n\}$ is said to be *positively associated* if, for every pair of disjoint subsets A_1, A_2 of $\{1, \ldots, n\}$,

$$\mathrm{Cov}(f(X_i, \ i \in A_1), g(X_i, \ i \in A_2)) \geq 0,$$

for every pair of componentwise nondecreasing functions f, g of $\{x_i, i \in A_1\}$ and $\{x_i, i \in A_2\}$ respectively.

We have given a brief review of probability inequalities and limit theorems connected with associated random variables. We will come back to the more recent results in this area later in Chapter 6 of this book. We now discuss some other concepts related to the concept of associated random variables.

1.3 Related Concepts of Association

Negative Association

The concept of negative association as introduced by Joag-Dev and Proschan (1983) is not a dual of the theory and applications of association, but differs in several aspects.

Definition. A finite set of random variables $\{X_1, \ldots, X_n\}$ is said to be *negatively associated* (NA) if, for every pair of disjoint subsets A_1, A_2 of $\{1, 2, \ldots, n\}$,

$$\mathrm{Cov}(h(X_i, \ i \in A_1), \ g(X_j, \ j \in A_2)) \leq 0, \tag{1.3.1}$$

whenever h and g are nondecreasing componentwise; an infinite collection is said to be *negatively associated* if every finite sub-collection is negatively associated.

Examples. A set of independent random variables is negatively associated. Other examples of negatively associated random variables are sets of components of random vectors whose distributions are (a)multinomial (b) multivariate hypergeometric (c) Dirichlet and (d) Dirichlet compound multinomial distributions. However, the most interesting case is that of models of categorical data analysis where negative association (NA) and association exist side by side. Consider a model where the individuals are classified according to two characteristics. Suppose the marginal totals are fixed. Then, the marginal distributions of row (column) vectors possess NA property, and the marginal distribution of a set of cell frequencies such that no pair of cells is in the same row or in the same column (for example, the diagonal cells) are associated. It will be shown that the partial sums of mean zero negatively associated random variables form an N-demimartingale . Properties of N-demimartingales will be studied later in this book but we will not discuss other probabilistic properties of negatively associated random variables in any detail.

Remarks. There are other related concepts of weak association for a finite collection of random variables such as weak association and associated measures. We will discuss these concepts briefly.

Weak Association

Burton, Dabrowski and Dehling (1986) defined weak association of random vectors.

Definition. Let the set $X_1, X_2, \ldots, X_m$ be R^d-valued random vectors. They are said to be *weakly associated* if whenever π is a permutation of $\{1, 2, \ldots, m\}$, $1 \leq k < m$, and $h : R^{kd} \to R$, $g : R^{(m-k)d} \to R$ are componentwise nondecreasing, then

$$\mathrm{Cov}(h(X_{\pi(1)}, \ldots, X_{\pi(k)}), g(X_{\pi(k+1)}, \ldots, X_{\pi(m)}) \geq 0$$

whenever it is defined. An infinite family of R^d-valued random vectors is *weakly associated* if every finite subfamily is weakly associated.

Weak association defines a strictly larger class of random variables than does association. Burton et al. (1986) proved a functional central limit theorem for such sequences. Dabrowski and Dehling (1988) proved a Berry-Esseen theorem and a functional Law of iterated logarithm for weakly associated sequences.

Processes with Associated Increments

Glasserman (1992) defined a class of processes with associated increments. Recall that a stochastic process $X = \{X_t, t \geq 0\}$ is said to have independent increments if, for all $n > 0$ and all $0 \leq t_0 < t_1 < \ldots < t_n$, the random variable $\Delta_0 = X_0$ and the increments

$$\Delta_1 = X_{t_1} - X_{t_0}, \ldots, \Delta_n = X_{t_n} - X_{t_{n-1}}$$

are independent.

Definition. A process $\{X_t, t \geq 0\}$ is said to have *associated increments* if the random variables $\{\Delta_i, i = 0, \ldots, n\}$ are associated, that is, for all bounded functions f and g nondecreasing componentwise,

$$\mathrm{Cov}[f(\Delta_0, \ldots, \Delta_n), g(\Delta_0, \ldots, \Delta_n)] \geq 0,$$

for all $n > 0$ and all $0 \leq t_0 < t_1 < \ldots < t_n$.

Glasserman (1992) derives sufficient conditions under which a process has associated increments and describes transformations under which this property is preserved. He derives sufficient conditions, for a Markov process with a generator Q and initial distribution p_0, to have associated increments. It is clear that all processes with independent increments have associated increments trivially from the definition of association. Suppose we consider a pure birth process where the birth rate is λ_n when the population is of size n and further suppose that λ_n is increasing in n and bounded with an arbitrary initial distribution on $\{0, 1, 2, \ldots\}$. Then this process has associated increments.

If $\{N_t, t \geq 0\}$ is a Poisson process, then the process $\{e^{N_t}, t \geq 0\}$ has associated increments. In general if $\{X_t, t \geq 0\}$ is nondecreasing and has associated increments and that g is nondecreasing and directionally convex, then $\{g(X_t), t \geq 0\}$ has associated increments (cf. Glasserman (1992)). For the definition of directional convexity of a function $g : R^d \to R$, $d \geq 1$ see Shaked and Shantikumar (1990). If $d = 1$, then this property coincides with the usual convexity.

Associated Measures

Burton and Waymire (1985) defined associated measures. A random measure X is associated if and only if the family of random variables $\mathcal{F} = \{X(B) : B$ a Borel set $\}$ is associated. They discussed some basic properties of associated measures. Lindqvist (1988) defined a notion of association of probability measures on partially ordered spaces and discussed its applications to stochastic processes with both discrete and continuous time parameters on partially ordered state spaces, and to mixtures of statistical experiments. Evans (1990) showed that each infinitely divisible random measure is associated. However, there are random measures which are not infinitely divisible but are associated. For instance, if μ is a fixed Radon measure and Y is a nonnegative random variable, then it can be shown that the random measure $Y\mu$ is associated.

Association in Time

Hjort, Natvig and Funnemark (1985) considered a multi-state system with states $S = \{0, 1, \ldots, m\}$. Here m indicates perfect functioning and 0 indicates complete failure . Let $C = \{1, 2, \ldots, n\}$ denote the set of components of the system.

Definition. The performance process of the i-th component is a stochastic process $\{X_i(t), t \in \tau\}$ where for each fixed $t \in \tau$, $X_i(t)$ denotes the state of component i at time t. The joint performance process of the components is given by $\{\underline{X}(t), t \in \tau\} = \{(X_1(t), \ldots, X_n(t)), t \in \tau\}$.

Let $I = [t_A, t_B] \subset [0, \infty), \tau(I) = \tau \cap I$.

Definition. The joint performance process $\{\underline{X}(t), t \in \tau\}$ of the components is said to be associated in time interval I if, for any integer m and $\{t_1, \ldots, t_m\} \subset \tau(I)$, the random variables in the array

$$
\begin{array}{ccc}
X_1(t_1) & \dots & X_1(t_m) \\
\vdots & \vdots & \vdots \\
X_n(t_1) & \dots & X_n(t_m)
\end{array}
$$

are associated.

Let $X = \{X(t),\, t \in \tau\}$ be a Markov process with state space $\{0, 1, \dots, m\}$. Define the transition probabilities as

$$P_{ij}(s,t) = P(X(t) = j | X(s) = i), \quad s \le t, \tag{1.3.2}$$

and the transition probability matrix as

$$P(s,t) = \{P_{ij}(s,t)\}_{\substack{i=0,1,\dots,m. \\ j=0,1,\dots,m.}}. \tag{1.3.3}$$

Let

$$\tau(I) = (0, \infty). \tag{1.3.4}$$

The transition intensity is defined as

$$\mu_{ij}(s) = \lim_{h \to 0^+} \frac{P_{ij}(s, s+h)}{h}, \quad i \ne j. \tag{1.3.5}$$

Let

$$
\begin{aligned}
P_{i,\ge j}(s,t) &= \sum_{v=j}^{k} P_{i,v}(s,t) \\
&= P[X(t) \ge j | X(s) = i],
\end{aligned}
\tag{1.3.6}
$$

$$\mu_{i,\ge j}(s) = \sum_{v=j}^{k} \mu_{iv}(s), \quad i < j, \tag{1.3.7}$$

and

$$\mu_{i,<j}(s) = \sum_{v=0}^{j-1} \mu_{iv}(s), \quad i \ge j. \tag{1.3.8}$$

Theorem 1.3.1. *Let X be a continuous time Markov process with state space $\{0, 1, \dots, m\}$ and transition probability matrix $P(s,t)$. Assume the transition intensities to be continuous. Consider the following statements about X :*

(i) *X is associated in time,*

(ii) *X is conditionally, stochastically, nondecreasing in time, that is,*

$$P[X(t) \ge j | X(s_1) = i_1, \dots, X(s_n) = i_n]$$

is nondecreasing in $i_1, \dots, i_n$ for each j and for each choice of $s_1 < s_2 < \dots < s_n < t,\ n \ge 1$,

(iii) $P_{i,\geq j}(s,t)$ *is nondecreasing in* i *for each* j *and for each* $s < t$,

(iv) *for each* j *and* s,

$$\mu_{i,\geq j}(s) \text{ is nondecreasing in } i \in \{0,1,\ldots,j-1\}$$

and

$$\mu_{i,<j}(s) \text{ is nondecreasing in } i \in \{j,j+1,\ldots,m\}.$$

Then (ii), (iii) *and* (iv) *are equivalent and each of them implies* (i).

For the binary case $(m = 1)$ it is easily seen that the statement (iii) of the above theorem is equivalent to

$$P_{1,1}(s,t) + P_{0,0}(s,t) \geq 1, \quad \text{for each } s < t. \tag{1.3.9}$$

This was the sufficient condition given by Esary and Proschan (1970) for X to be associated in time. Furthermore when $\mu_{1,0}(s)$ and $\mu_{0,1}(s)$ are continuous, then the statement (iv) of the above theorem is always satisfied and the corresponding Markov Process is always associated in time. It is also true for a general birth and death process (cf. Keilson and Kester (1977) and Kirstein (1976)).

Kuber and Dharmadikari (1996) discussed association in time for semi-Markov processes. Let $(\Omega, \mathcal{F}, \mathcal{P})$ be a probability space and $E = \{0,1,\ldots,k\}$. Define measurable functions $X_n : \Omega \to E$, $T_n : \Omega \to R^+$, $n \in N$, so that $0 = T_0 \leq T_1 \leq T_2 \leq \ldots$. Then the sequence $\{(X_n, T_n),\, n \geq 1\}$ is said to form a Markov renewal process with state space E if for all $n \in N$, $j \in E$, $t \in R^+$, we have

$$P[X_{n+1} = j, T_{n+1} - T_n \leq t | X_0, \ldots, X_n; T_0, \ldots, T_n]$$
$$= P[X_{n+1} = j, T_{n+1} - T_n \leq t | X_n].$$

Assume that $P[X_{n+1} = j, T_{n+1} - T_n \leq t | X_n = i] = Q_{ij}(t)$ is independent of n.

Definition. The semi-Markov Process (SMP) $\{Y(t),\, t \geq 0\}$ corresponding to a Markov renewal process (X, T) is defined as $Y(t) = X_n$ for $t \in [T_n, T_{n+1}]$, $n \geq 0$. Define, for $i, j \in E$ and $0 \leq u < s < t$,

$$P_{ij}(u,(s,t)) = P[Y(t) = j | Y(u) = i, Z_i > s - u],$$

where Z_i is the waiting time in state i. Suppose the transition intensity

$$\mu_{ij}(u,s) = \lim_{h \to 0^+} \frac{P_{ij}(u,(s,s+h))}{h}, \quad i \neq j \tag{1.3.10}$$

is finite. Let

$$P_{i,\geq j}(u,(s,t)) = \sum_{v=j}^{k} P_{i,v}(u,(s,t)), \tag{1.3.11}$$

$$\mu_{i,\geq j}(u,s) = \sum_{v=j}^{k} \mu_{iv}(u,s), \quad i < j, \tag{1.3.12}$$

and

$$\mu_{i,<j}(s) = \sum_{v=0}^{j-1} \mu_{iv}(u,s), \quad i \geq j. \tag{1.3.13}$$

Theorem 1.3.2. *Let* $\{Y(t), t \geq 0\}$ *be a semi-Markov process with state space* E, *waiting times* Z_i *having bounded transition intensities* $\mu_{ij}(s)$, $i, j \in E$ *which are continuous in* s *uniformly in* u *for each* $u > 0$. *Consider the following statements.*

(i) $Y(t)$ *is associated in time.*

(ii) $P_{i,\geq j}(u,(s,t))$ *is increasing in* i *and decreasing in* u *on* $(0,s]$.

(iii) *For each* j *and* s,

$\mu_{i,\geq j}(u,s)$ *is increasing in* $i \in \{0,1,\ldots,j-1\}$ *and decreasing in* u *on* $(0,s]$,

and

$\mu_{i,<j}(u,s)$ *is decreasing in* $i \in \{j,j+1,\ldots,m\}$ *and increasing in* u *on* $(0,s]$.

(iv) *For fixed* j *and each choice of* $t_1 < \ldots < t_n < t$, $n \geq 1$,

$$P[Y(t) > j | Z_{i_n} > t_n - u,\ Y(u) = i_n,\ Y(t_1) = i_1,\ldots,Y(t_\ell) = i_\ell]$$

is increasing in $i_1,\ldots,i_n$ *and decreasing in* $u \in (t_\ell, t_n]$, $\ell \in \{0,\ldots,n-1\}$.

Then (ii), (iii) *and* (iv) *are equivalent and each of them implies* (i).

If $Y(t)$ is a Markov Process with state space $E = \{0,\ldots,k\}$, the statements (ii) and (iii) in Theorem 1.3.2 simplify to statements (iii) and (iv) in Theorem 1.3.1.

Association for Jointly Stable Laws

Pitt (1982) proved the following result.

Theorem 1.3.3 (Pitt (1982)). *Let* $\underline{X} = (X_1,\ldots,X_k)$ *be* $N_k(0,\Sigma)$, $\Sigma = ((\sigma_{ij}))$ *where* $\mathrm{Cov}(X_i, X_j) = \sigma_{ij}$. *Then a necessary and sufficient condition for* $(X_1,\ldots,X_k)$ *to be associated is that* $\sigma_{ij} \geq 0$, $1 \leq i, j \leq k$.

Definition (Weron (1984)). Let $\underline{X} = (X_1,\ldots,X_k)$. Then $\underline{X}$ is said to be jointly stable with index α, $0 < \alpha < 2$, if the characteristic function of $\underline{X}$ is of the form

$$\phi_{\underline{X}}(\underline{t}) = E[e^{i(\underline{X},\underline{t})}]$$

$$= \exp\{-\int_{S_k} |(\underline{s},\underline{t})|^{\alpha}(1 - i\,\mathrm{sgn}((\underline{s},\underline{t})Q(\alpha;\underline{s},\underline{t}))\Gamma(ds) + i(\underline{\mu},\underline{t})\}, \tag{1.3.14}$$

where $\underline{t} = (t_1, \ldots, t_k) \in R^k$, S_k is the unit sphere in R^k, $\underline{s} = (s_1, \ldots, s_k) \in S_k$, Γ is a finite measure on S_k, $\underline{\mu} = (\mu_1, \ldots, \mu_k) \in R^k$ and

$$Q(\alpha; \underline{s}, \underline{t}) = \begin{cases} \tan \frac{\pi\alpha}{2} & \text{if } \alpha \neq 1, \\ -\frac{2}{\pi} \log |(\underline{s}, \underline{t})| & \text{if } \alpha = 1. \end{cases}$$

Remarks. There is a one to one correspondence between the distributions of jointly α-stable random vectors $\underline{X} = (X_1, \ldots, X_k)$ and the finite Borel measures Γ in (1.3.14). Γ is called the spectral measure of the α-stable vector $\underline{X}$.

Theorem 1.3.4 (Lee, Rachev and Samorodnitsky (1990)). *Let $\underline{X} = (X_1, \ldots, X_k)$ be jointly α-stable with $0 < \alpha < 2$ with the characteristic function given by (1.3.14). Then $(X_1, \ldots, X_k)$ is associated if and only if the spectral measure Γ satisfies the condition*

$$\Gamma(S_{k-}) = 0, \tag{1.3.15}$$

where $S_{k-} = \{\underline{s} = (s_1, \ldots, s_k) \in S_k : \text{for some } i \text{ and } j \{1, \ldots, k\}, s_i > 0 \text{ and } s_j < 0\}$.

Let $\underline{X} = (X_1, \ldots, X_k)$ be an infinitely divisible random vector with the characteristic function

$$\phi_{\underline{X}}(\underline{t}) = \exp\{\int_{R^k - \{0\}} (e^{i(\underline{t}, \underline{x})} - 1 - i \, I(\|\underline{x}\| \leq 1)(\underline{t}, \underline{x}))\nu(d\underline{x}) + i(\underline{t}, \underline{\mu})\}. \tag{1.3.16}$$

Here ν is called the Levy measure of $\underline{X}$ and $\underline{\mu} = (\mu_1, \ldots, \mu_k) \in R^k$. Resnick (1988) has proved that a sufficient condition for $(X_1, \ldots, X_k)$ to be associated is that

$$\nu\{\underline{x} = (x_1, \ldots, x_k) : x_i x_j < 0 \text{ for some } i \neq j, \, 1 \leq i, j \leq k\} = 0. \tag{1.3.17}$$

In other words, the Levy measure is concentrated on the positive (R_+^k) and the negative (R_-^k) quadrants of R^k. The result given above in Theorem 1.3.4 due to Lee et al. (1990) proves that for α-stable random vectors, the condition (1.3.15) is necessary and sufficient for association. Samorodnitsky (1995) showed that there is an infinitely divisible random vector $\underline{X}$ taking values in R^2 which is associated with a Levy measure ν such that

$$\nu\{\underline{x} = (x_1, x_2) : x_1 x_2 < 0\} > 0$$

leading to the fact that condition (1.3.17) is not necessary for the association of an infinitely divisible random vector.

Note that if $\underline{X}$ is infinitely divisible with the characteristic function $\phi_{\underline{X}}(\underline{t})$, then, for every $\gamma > 0$, $\phi_{\underline{X}}(\underline{t})^\gamma$ is also a characteristic function in R^k. Let $\underline{X}^{*\gamma}$ be an infinitely divisible random vector with this characteristic function. It is clear that $\underline{X}$ and $\underline{X}^{*1}$ have the same distribution. Samorodnitsky (1995) proved that the condition (1.3.15) is equivalent to the fact $\underline{X}^{*\gamma}$ is associated for every $\gamma > 0$.

An infinitely divisible random vector $\underline{X}$ is said to be *r-semistable index* α, $0 < r < 1$, $0 < \alpha < 2$, if for every $n \geq 1$, there is a non-random vector $\underline{d_n} \in R^k$ such that

$$\underline{X}^{*r^n} \overset{\mathcal{L}}{=} r^{n/\alpha}\underline{X} + d_n$$

where $\underline{X} \overset{\mathcal{L}}{=} \underline{Y}$ indicates that $\underline{X}$ and $\underline{Y}$ have the same distribution. If $\underline{X}$ is r-semistable index α for all $0 < r < 1$, then it is jointly stable with index α (cf. Chung et al (1982)). The following result extends Theorem 1.3.4 from stable to semistable random vectors.

Theorem 1.3.5. *A random vector $\underline{X} = (X_1, \ldots, X_k)$ which is r-semistable index α, $0 < r < 1$, $0 < \alpha < 2$ is associated if and only if its Levy measure is concentrated on $R_+^k \cup R_-^k$.*

Chapter 2

Demimartingales

2.1 Introduction

Let $(\Omega, \mathcal{F}, P)$ be a probability space and suppose that a set of random variables $X_1, \ldots, X_n$ defined on the probability space $(\Omega, \mathcal{F}, P)$ have mean zero and are associated. Let $S_0 = 0$ and $S_j = X_1 + \ldots + X_j$, $j = 1, \ldots, n$. Then it follows that, for any componentwise nondecreasing function g,

$$E((S_{j+1} - S_j)g(S_1, \ldots, S_j)) \geq 0, \quad j = 1, \ldots, n \qquad (2.1.1)$$

provided the expectation exists.

Throughout this chapter, we assume that all the relevant expectations exist unless otherwise specified. Recall that a sequence of random variables $\{S_n, n \geq 1\}$ defined on a probability space $(\Omega, \mathcal{F}, P)$ is a martingale with respect to the natural sequence $\mathcal{F}_n = \sigma\{S_1, \ldots, S_n\}$ of σ-algebras if $E(S_{n+1}|S_1, \ldots, S_n) = S_n$ a.s. for $n \geq 1$. Here $\sigma\{S_1, \ldots, S_n\}$ denotes the σ-algebra generated by the random sequence $S_1, \ldots, S_n$. An alternate way of defining the martingale property of the sequence $\{S_n, n \geq 1\}$ is that

$$E((S_{n+1} - S_n)g(S_1, \ldots, S_n)) = 0, \quad n \geq 1$$

for all measurable functions $g(x_1, \ldots, x_n)$ assuming that the expectations exist.

Newman and Wright (1982) introduced the notion of demimartingales.

Definition. A sequence of random variables $\{S_n, n \geq 1\}$ in $L^1(\Omega, \mathcal{F}, P)$ is called a *demimartingale* if, for every componentwise nondecreasing function g,

$$E((S_{j+1} - S_j)g(S_1, \ldots, S_j)) \geq 0, \quad j \geq 1 \qquad (2.1.2)$$

Remarks. If the sequence $\{X_n, n \geq 1\}$ is an L^1, mean zero sequence of associated random variables and $S_j = X_1 + \ldots + X_j$ with $S_0 = 0$, then the sequence $\{S_n, n \geq 1\}$ is a demimartingale.

B.L.S. Prakasa Rao, *Associated Sequences, Demimartingales and Nonparametric Inference*, Probability and its Applications, DOI 10.1007/978-3-0348-0240-6_2, © Springer Basel AG 2012

If the function g is required to be nonnegative (resp., non-positive) and nondecreasing in (2.1.2), then the sequence will be called a *demisubmartingale* (resp., *demisupermartingale*).

A martingale $\{S_n, \mathcal{F}_n, n \geq 1\}$ with the natural choice of σ-algebras $\{\mathcal{F}_n, n \geq 1\}$, $\mathcal{F}_n = \sigma\{S_1, \ldots, S_n\}$ is a demimartingale. This can be seen by noting that

$$E((S_{j+1} - S_j)g(S_1, \ldots, S_j)) = E[E((S_{j+1} - S_j)g(S_1, \ldots, S_j)|\mathcal{F}_j)] \qquad (2.1.3)$$
$$= E[g(S_1, \ldots, S_j)E((S_{j+1} - S_j)|\mathcal{F}_j)]$$
$$= 0$$

by the martingale property of the process $\{S_n, \mathcal{F}_n, n \geq 1\}$. Similarly it can be seen that every submartingale $\{S_n, \mathcal{F}_n, n \geq 1\}$ with the natural choice of σ-algebras $\{\mathcal{F}_n, n \geq 1\}$, $\mathcal{F}_n = \sigma(S_1, \ldots, S_n)$ is a demisubmartingale. However a demisubmartingale need not be a submartingale. This can be seen by the following example (cf. Hadjikyriakou (2010)).

Example 2.1.1. Let the random variables $\{X_1, X_2\}$ be such that

$$P(X_1 = -1, X_2 = -2) = p \text{ and } P(X_1 = 1, X_2 = 2) = 1 - p$$

where $0 \leq p \leq \frac{1}{2}$. Then the finite sequence $\{X_1, X_2\}$ is a demisubmartingale since for every nonnegative nondecreasing function $g(.)$,

$$E[(X_2 - X_1)g(X_1)] = -pg(-1) + (1 - p)g(1) \qquad (2.1.4)$$
$$\geq -pg(-1) + pg(1) \text{ (since } p \leq \frac{1}{2})$$
$$= p(g(1) - g(-1)) \geq 0.$$

However the sequence $\{X_1, X_2\}$ is not a submartingale since

$$E(X_2|X_1 = -1) = \sum_{x_2 = -2, 2} x_2 P(X_2 = x_2|X_1 = -1) = -2 < -1.$$

As remarked earlier, the sequence of partial sums of mean zero associated random variables is a demimartingale. However the converse need not hold. In other words, there exist demimartingales such that the demimartingale differences are not associated. This can be seen again by the following example (cf. Hadjikyriakou (2010)).

Example 2.1.2. Let X_1 and X_2 be random variables such that

$$P(X_1 = 5, X_2 = 7) = \frac{3}{8}, \quad P(X_1 = -3, X_2 = 7) = \frac{1}{8}$$

and

$$P(X_1 = -3, X_2 = 7) = \frac{4}{8}.$$

Let g be a nondecreasing function. Then the finite sequence $\{X_1, X_2\}$ is a demi-martingale since

$$E[(X_2 - X_1)g(X_1)] = \frac{6}{8}[g(5) - g(-3)] \geq 0.$$

Let f be a nondecreasing function such that

$$f(x) = 0 \text{ for } x < 0, f(2) = 2, f(5) = 5 \text{ and } f(10) = 20.$$

It can be checked that

$$\text{Cov}(f(X_1), f(X_2 - X_1)) = -\frac{75}{32} < 0$$

and hence X_1 and $X_2 - X_1$ are not associated.

Christofides (2004) constructed another example of a demimartingale. We now discuss his example.

Example 2.1.3. Let $X_1, \ldots, X_n$ be associated random variables, let $h(x_1, \ldots, x_m)$ be a "kernel" mapping R^m to R where $1 \leq m \leq n$. Without loss of generality, we assume that h is a symmetric function. Define the U-statistic

$$U_n = \binom{n}{m}^{-1} \sum_{1 \leq i_1 < \ldots < i_m \leq n} h(X_{i_1}, \ldots, X_{i_m}) \tag{2.1.5}$$

where $\sum_{1 \leq i_1 < \ldots < i_m \leq n} h(X_{i_1}, \ldots, X_{i_m})$ denotes the summation over the $\binom{n}{m}$ combinations of m distinct elements $\{i_1, \ldots, i_m\}$ from $\{1, \ldots, n\}$. Suppose that the function h is componentwise nondecreasing and $E(h) = 0$. Then the sequence $\{S_n = \binom{n}{m}U_n, n \geq m\}$ is a demimartingale. This can be checked in the following way.

Note that

$$S_{n+1} - S_n = \sum_{1 \leq i_1 < \ldots < i_m \leq n+1} h(X_{i_1}, \ldots, X_{i_m}) - \sum_{1 \leq i_1 < \ldots < i_m \leq n} h(X_{i_1}, \ldots, X_{i_m})$$

$$= \sum_{1 \leq i_1 < \ldots < i_{m-1} \leq n} h(X_{i_1}, \ldots, X_{i_{m-1}}, X_{n+1}) \tag{2.1.6}$$

Then, for any componentwise nondecreasing function g, and for $n \geq m$,

$$E[(S_{n+1} - S_n)g(S_m, \ldots, S_n)] \tag{2.1.7}$$

$$= E[\sum_{1 \leq i_1 < \ldots < i_{m-1} \leq n} h(X_{i_1}, \ldots, X_{i_{m-1}}, X_{n+1})g(S_m, \ldots, S_n)]$$

$$= \sum_{1 \leq i_1 < \ldots < i_{m-1} \leq n} E[h(X_{i_1}, \ldots, X_{i_{m-1}}, X_{n+1})g(S_m, \ldots, S_n)]$$

$$= \sum_{1 \le i_1 < \ldots < i_{m-1} \le n} E[h(X_{i_1}, \ldots, X_{i_{m-1}}, X_{n+1}) v(X_1, \ldots, X_n)]$$

$$\ge 0$$

where, for $n \ge m$,

$$v(x_1, \ldots, x_n) = g\Big(h(x_1, \ldots, x_m), \sum_{1 \le i_1 < \ldots < i_m \le m+1} h(x_{i_1}, \ldots, x_{i_m}), \ldots,$$

$$\sum_{1 \le i_1 < \ldots < i_m \le n} h(x_{i_1}, \ldots, x_{i_m})\Big).$$

Note that the function $v(x_1, \ldots, x_n)$ is componentwise nondecreasing. The last inequality in equation (2.1.7) follows from the fact that the sequence $\{X_i, i \ge 1\}$ is an associated random sequence and the function $v(x_1, \ldots, x_n)$ is componentwise nondecreasing. Hence the sequence $\{S_n, n \ge m\}$ is a demimartingale.

Example 2.1.4 (Hadjikyriakou (2010)). Suppose the random sequence $\{X_n, n \ge 1\}$ is a sequence of associated identically distributed random variables such that the probability density function of X_1 is $f(x, \theta)$ with respect to a σ-finite measure μ and the support of the density function $f(x, \theta)$ does not depend on θ. Suppose the function

$$h(x) = \frac{f(x, \theta_1)}{f(x, \theta_0)}$$

is nondecreasing in x for any fixed θ_0 and θ_1. Define

$$L_n = \prod_{k=1}^{n} \frac{f(X_k, \theta_1)}{f(X_k, \theta_0)}.$$

Then, under the hypothesis $H_0 : \theta = \theta_0$, the sequence $\{L_n, n \ge 1\}$ is a demisubmartingale. This can be checked in the following manner. Observe that

$$L_{n+1} - L_n = [\frac{f(X_{n+1}, \theta_1)}{f(X_{n+1}, \theta_0)} - 1] L_n$$

and, for any componentwise nonnegative nondecreasing function g,

$$E[(L_{n+1} - L_n) g(L_1, \ldots, L_n)] = E([\frac{f(X_{n+1}, \theta_1)}{f(X_{n+1}, \theta_0)} - 1] L_n g(L_1, \ldots, L_n)) \quad (2.1.8)$$

$$\ge E([\frac{f(X_{n+1}, \theta_1)}{f(X_{n+1}, \theta_0)} - 1]) E(L_n g(L_1, \ldots, L_n))$$

$$= 0.$$

Note that the inequality in the above relations is a consequence of the fact that the random sequence $\{X_n, n \ge 1\}$ is associated and the last equality follows from the observation that

$$E([\frac{f(X_{n+1}, \theta_1)}{f(X_{n+1}, \theta_0)}] = 1$$

under $H_0; \theta = \theta_0$. Observe that L_n is the likelihood ratio when the random sequence $\{X_n, n \geq 1\}$ is an independent sequence of random variables.

The following result, due to Christofides (2000), shows that if the random sequence $\{S_n, n \geq 1\}$ is a demisubmartingale or a demimartingale, then the random sequence $\{g(S_n), n \geq 1\}$ is a demisubmartingale if g is a nondecreasing convex function.

Theorem 2.1.1. *Let the random sequence $\{S_n, n \geq 1\}$ be a demisubmartingale (or a demimartingale) and g be a nondecreasing convex function. Then the random sequence $\{g(S_n), n \geq 1\}$ is a demisubmartingale.*

Proof. Define

$$h(x) = \lim_{x \to y-0} \frac{g(x) - g(y)}{x - y}.$$

From the convexity of the function g, it follows that the function h is a nonnegative nondecreasing function. Furthermore

$$g(y) \geq g(x) + (y - x)h(x).$$

Suppose that f is a nonnegative componentwise nondecreasing function. Then

$$
\begin{aligned}
&E[(g(S_{n+1}) - g(S_n))f(g(S_1), \ldots, g(S_n))] \\
&\geq E[(S_{n+1} - S_n)h(S_n)f(g(S_1), \ldots, g(S_n))] \\
&= E[(S_{n+1} - S_n)f^*(S_1, \ldots, S_n)]
\end{aligned}
\tag{2.1.9}
$$

where $f^*(x_1, \ldots, x_n) = h(x_n)f(g(x_1), \ldots, g(x_n))$. Note that f^* is a componentwise nondecreasing and nonnegative function. Since the sequence $\{S_n, n \geq 1\}$ is a demimartingale, it follows that the last term is nonnegative and hence the sequence $\{g(S_n), n \geq 1\}$ is a demimartingale. $\qquad\square$

As an application of the above theorem, we have the following result.

Theorem 2.1.2. *If the sequence $\{S_n, n \geq 1\}$ is a demimartingale, then the sequence $\{S_n^+, n \geq 1\}$ is a demisubmartingale and the sequence $\{S_n^-, n \geq 1\}$ is also a demisubmartingale.*

Proof. Since the function $g(x) \equiv x^+ = \max(0, x)$ is nondecreasing and convex, it follows that the sequence $\{S_n^+, n \geq 1\}$ is a demisubmartingale from the previous theorem. Let $Y_n = -S_n, n \geq 1$. It is easy to see that the sequence $\{Y_n, n \geq 1\}$ is also a demimartingale and $Y_n^+ = S_n^-$ where $x^- = \max(0, -x)$. Hence, as an application of the first part of the theorem, it follows that the sequence $\{S_n^-, n \geq 1\}$ is a demisubmartingale. $\qquad\square$

Suppose the sequence $\{S_n, n \geq 1\}$ is a demimartingale. The following result due to Hu et al. (2010) gives sufficient conditions for a stopped demisubmartingale to be a demisubmartingale.

Theorem 2.1.3. *Let the sequence $\{S_n,\ n \geq 1\}$ be a demisubmartingale, $S_0 = 0$, and τ be a positive integer-valued random variable. Furthermore suppose that the indicator function $I_{[\tau \leq j]} = h_j(S_1, \ldots, S_j)$ is a componentwise nonincreasing function of $S_1, \ldots, S_j$ for $j \geq 1$. Then the random sequence $\{S_j^* = S_{\min(\tau,j)}, j \geq 1\}$ is a demisubmartingale.*

Proof. Note that

$$S_j^* = S_{\min(\tau,j)} = \sum_{i=1}^{j} (S_i - S_{i-1}) I_{[\tau \geq i]}.$$

We have to show that

$$E[(S_{j+1}^* - S_j^*) f(S_1^*, \ldots, S_j^*)] \geq 0, \quad j \geq 1$$

for any f which is componentwise nondecreasing and nonnegative. Since

$$g_j(S_1, \ldots, S_j) \equiv 1 - I_{[\tau \leq j]} = 1 - h_j(S_1, \ldots, S_j)$$

is a componentwise nondecreasing and and nonnegative function, we get that

$$u_j(S_1, \ldots, S_j) \equiv g_j(S_1, \ldots, S_j) f(S_1, \ldots, S_j)$$

is a componentwise nondecreasing nonnegative function. By the demisubmartingale property, we get that

$$\begin{aligned}
E[(S_{j+1}^* - S_j^*) f(S_1^*, \ldots, S_j^*)] &= E[(S_{j+1} - S_j) I_{[\tau \geq j+1]} f(S_1^*, \ldots, S_j^*)] \\
&= E[(S_{j+1} - S_j) I_{[\tau \geq j+1]} f(S_1, \ldots, S_j)] \\
&= E[(S_{j+1} - S_j) u_j(S_1, \ldots, S_j)] \geq 0 \quad (2.1.10)
\end{aligned}$$

for $j \geq 1$. Hence the sequence $\{S_j^*, j \geq 1\}$ is a demisubmartingale. $\square$

We now obtain some consequences of this theorem.

Theorem 2.1.4. *Let the sequence $\{S_n,\ n \geq 1\}$ be a demisubmartingale. and τ be a positive integer-valued random variable. Furthermore suppose that the indicator function $I_{[\tau \leq j]} = h_j(S_1, \ldots, S_j)$ is a componentwise nonincreasing function of $S_1, \ldots, S_j$ for $j \geq 1$. Then, for any $1 \leq n \leq m$,*

$$E(S_{\min(\tau,m)}) \geq E(S_{\min(\tau,n)}) \geq E(S_1), \quad\quad (2.1.11)$$

Suppose the sequence $\{S_n,\ n \geq 1\}$ is a demimartingale and the indicator function $I_{[\tau \leq j]} = h_j(S_1, \ldots, S_j)$ is a componentwise nondecreasing function of $S_1, \ldots, S_j$ for $j \geq 1$. Then, for any $1 \leq n \leq m$,

$$E(S_{\min(\tau,m)}) \leq E(S_{\min(\tau,n)}) \leq E(S_1). \quad\quad (2.1.12)$$

Proof. Suppose that the sequence $\{S_n, n \geq 1\}$ is a demisubmartingale and the indicator function $I_{[\tau \leq j]} = h_j(S_1, \ldots, S_j)$ is a componentwise nonincreasing function of $S_1, \ldots, S_j$ for $j \geq 1$. Then the sequence $\{S_n^*, n \geq 1\}$ is a demisubmartingale by Theorem 2.1.3. The inequalities stated in equation (2.1.11) follow from the demisubmartingale property by choosing $f \equiv 1$ in equation (2.1.10).

Suppose the sequence $\{S_n, n \geq 1\}$ is a demimartingale and that the indicator function $I_{[\tau \leq j]} = h_j(S_1, \ldots, S_j)$ is a componentwise nondecreasing function of $S_1, \ldots, S_j$ for $j \geq 1$. Since the sequence $\{S_n, n \geq 1\}$ is a demimartingale, we note that

$$-E(S_{j+1}^* - S_j^*) = -E[(S_{j+1} - S_j)I_{[\tau \geq j+1]}]$$
$$= E[(S_{j+1} - S_j)(h_j(S_1, \ldots, S_j) - 1)]$$
$$\geq 0 \tag{2.1.13}$$

for $j \geq 1$ from the demimartingale property. This in turn proves (2.1.12). □

2.2   Characteristic Function Inequalities

For any two complex-valued functions $f(x_1, \ldots, x_m)$ and $f_1(x_1, \ldots, x_m)$ defined on R^m, define $f \ll f_1$ if $f_1 - Re(e^{i\alpha}f)$ is componentwise nondecreasing for all real α. Note that

$$f_1 = \frac{(f_1 - Re(f)) + (f_1 - Re(-f))}{2}$$

and hence f_1 is real-valued and nondecreasing. Furthermore $f \ll f_1$ for real f if and only if $f_1 + f$ and $f_1 - f$ are both nondecreasing. We write $f \ll_A f_1$ if $f \ll f_1$ and both the functions f and f_1 depend only on x_j, $j \in A$. The following results are due to Newman (1984).

Lemma 2.2.1. *Suppose h is real and $h \ll h_1$ and ϕ is a complex-valued function on R such that*

$$|\phi(t) - \phi(s)| \leq |t - s|, \quad -\infty < t, s < \infty.$$

Then $\phi(h) \ll h_1$. In particular, this property holds for the function $\phi(h) = e^{ih}$.

Proof. For any function $g(x_1, \ldots, x_m)$, defined on R^m, let Δg denote the increment in the function whenever one or more of the components x_j, $1 \leq j \leq m$ is increased. We have to prove that

$$\Delta[h_1 - Re(e^{i\alpha}\phi(h))] \geq 0 \tag{2.2.1}$$

for any real α. Note that

$$|\Delta Re(e^{i\alpha}\phi(h))| \leq |\Delta(e^{i\alpha}\phi(h))| = |\Delta\phi(h)| \leq |\Delta h|$$

from the property of the function ϕ and

$$|\Delta h| \leq \Delta h_1$$

since $h \ll h_1$. □

Lemma 2.2.2. *Suppose $f \ll_A f_1$ and $g \ll_B g_1$. Define*

$$\langle f, g \rangle = \text{Cov}(f(X_1, X_2, \ldots), g(X_1, X_2, \ldots))$$

where the X_j's are either associated or negatively associated. In the negatively associated case, assume in addition that the sets A and B are disjoint. Then

$$|\langle f, g \rangle| \leq \langle f_1, g_1 \rangle \text{ if } f \text{ and/or } g \text{ real}$$

and

$$|\langle f, g \rangle| \leq 2 \langle f_1, g_1 \rangle \text{ otherwise.}$$

Proof. Suppose that f is real. Since $|\langle f, g \rangle| = \sup_{\alpha \in R} Re(e^{i\alpha} \langle f, g \rangle)$, it is sufficient to show that $Re(e^{i\alpha} \langle f, g \rangle) \leq \langle f_1, g_1 \rangle$ for every $\alpha \in R$. This can be seen from the observation that $h \equiv Re(e^{i\alpha} g) \ll g_1$ and $f \ll f_1$ and the identities

$$\langle f_1, g_1 \rangle - \langle f, h \rangle = \frac{1}{2}[\langle f_1 + f, g_1 - h \rangle + \langle f_1 - f, g_1 + h \rangle] \geq 0 \qquad (2.2.2)$$

in the associated case and

$$-\langle f_1, g_1 \rangle - \langle f, h \rangle = \frac{1}{2}[\langle f_1 + f, g_1 + h \rangle + \langle f_1 - f, g_1 - h \rangle] \geq 0 \qquad (2.2.3)$$

in the negatively associated case. If g is real, the arguments are the same. If neither f nor g is real, then

$$|\langle f, g \rangle| = |\langle Re\ f, h \rangle + i\langle Im\ f, g \rangle| \leq |\langle Re\ f, g \rangle| + |\langle Im\ f, g \rangle| \qquad (2.2.4)$$

and the required inequality follows from the inequality for the real f case discussed above.   □

As a consequence of Lemma 2.2.1 and arguments similar to those given in the proof of Lemma 2.2.2, Newman (1984) proved the following theorem.

Theorem 2.2.3. *Let $\{S_n, n \geq 1\}$ be a demimartingale, $S_0 = 0$, and f and f_1 be complex-valued functions on R^j such that $f \ll f_1$. Then*

$$|E((S_{j+1} - S_j)f(S_1, \ldots, S_j))| \leq E((S_{j+1} - S_j)f_1(S_1, \ldots, S_j)). \qquad (2.2.5)$$

In particular, this is the case for the function $f(x_1, \ldots, x_j) = \exp\{i \sum_{k=1}^{j} r_k x_k\}$ and $f_1(x_1, \ldots, x_j) = \sum_{k=1}^{j} |r_k| x_k$.

The next result due to Newman (1984) gives sufficient conditions for a demimartingale to be a martingale with respect to the natural sequence of sub-σ-algebras.

Theorem 2.2.4. *Let $S_0 = 0$, and the sequence $\{S_n, n \geq 1\}$ be an L^2-demimartingale. Let $\mathcal{F}_n$ be the σ-algebra generated by the sequence $\{S_1, \ldots, S_n\}$. If the random sequence $\{S_n, n \geq 1\}$ has uncorrelated increments, that is, if*

$$\text{Cov}((S_{j+1} - S_j), (S_{k+1} - S_k)) = 0, \quad 0 \leq k \leq j,$$

then the sequence $\{S_n, \mathcal{F}_n, n \geq 1\}$ is a martingale.

Proof. For any $0 \leq k \leq j$,

$$\text{Cov}(S_{j+1} - S_j, S_k) = E[(S_{j+1} - S_j)S_k] - E[(S_{j+1} - S_j)]E[S_k]$$
$$= E[(S_{j+1} - S_j)S_k] \tag{2.2.6}$$

since $E(S_j)] = E[S_{j+1}]$ by the demimartingale property of the sequence $\{S_n, n \geq 1\}$. From the uncorrelated increment hypothesis, we get that

$$E[(S_{j+1} - S_j)S_k] = 0, \quad 1 \leq k \leq j.$$

Applying Theorem 2.2.3, we get that

$$E[(S_{j+1} - S_j) \exp(i \sum_{k=1}^{j} r_k S_k)] = 0$$

for $r_1, \ldots, r_k \in R$. This in turn proves that

$$E[S_{j+1} - S_j | \mathcal{F}_j] = 0 \ \text{ a.s } j \geq 1.$$

Hence the random sequence $\{S_n, \mathcal{F}_n, n \geq 1\}$ is a martingale. $\square$

Remarks. Suppose $S_0 \equiv 0$ and $S_n = X_1 + \ldots + X_n$ where $\{X_n, n \geq 1\}$ is a Gaussian process. Then the sequence $\{S_n, n \geq 1\}$ is a *martingale* if and only if $\text{Cov}(X_k, X_j) = 0$ for all $k > j$ and $E[X_j] = 0$ for all $j \geq 1$. It is the partial sum sequence of *associated random variables* if $\text{Cov}(X_k, X_j) \geq 0$ for all $k > j$ and the sequence $\{S_n, n \geq 1\}$ is a *demimartingale* if and only if $\sum_{j=1}^{n} \text{Cov}(X_k, X_j) \geq 0$ for all $k > n$ and $E[X_j] = 0$ for all $j \geq 1$.

2.3 Doob Type Maximal Inequality

We will now discuss a Doob type maximal inequality for demisubmartingales due to Newman and Wright (1982). Let

$$S_n^* = \max(S_1, \ldots, S_n). \tag{2.3.1}$$

Define the rank orders $R_{n,j}$ by

$$R_{n,j} = \begin{cases} j\text{-th largest of } (S_1, \ldots, S_n), & \text{if } j \leq n, \\ \min(S_1, \ldots, S_n) = R_{n,n}, & \text{if } j > n. \end{cases}$$

The following theorem is due to Newman and Wright (1982).

Theorem 2.3.1. *Suppose $S_1, S_2, \ldots$ is a demimartingale (resp., demisubmartingale) and m is a nondecreasing (resp., nonnegative and nondecreasing) function on $(-\infty, \infty)$ with $m(0) = 0$; then, for any n and j,*

$$E(\int_0^{R_{n,j}} u \, dm(u)) \leq E(S_n m(R_{n,j})); \tag{2.3.2}$$

and, for any $\lambda > 0$,

$$\lambda\, P(R_{n,j} \geq \lambda) \leq \int_{[R_{n,j} \geq \lambda]} S_n dP. \tag{2.3.3}$$

Proof. For fixed n and j, let $Y_k = R_{k,j}$ and $Y_0 = 0$. Then

$$S_n m(Y_n) = \sum_{k=0}^{n-1} S_{k+1}(m(Y_{k+1}) - m(Y_k)) + \sum_{k=1}^{n-1}(S_{k+1} - S_k)m(Y_k). \tag{2.3.4}$$

From the definition of $R_{n,j}$, it follows that

$$\text{for } k < j, \text{ either } Y_k = Y_{k+1} \text{ or } S_{k+1} = Y_{k+1};$$

and

$$\text{for } k \geq j, \text{ either } Y_k = Y_{k+1} \text{ or } S_{k+1} \geq Y_{k+1}.$$

Hence, for any k,

$$S_{k+1}(m(Y_{k+1} - m(Y_k)) \geq Y_{k+1}(m(Y_{k+1} - m(Y_k)) \geq \int_{Y_k}^{Y_{k+1}} u\, dm(u). \tag{2.3.5}$$

Combining the inequalities in equations (2.3.4) and (2.3.5), we get that

$$S_n m(Y_n) \geq \int_0^{Y_n} u\, dm(u) + \sum_{k=1}^{n-1}(S_{k+1} - S_k)m(Y_k). \tag{2.3.6}$$

Note that

$$E[(S_{k+1} - S_k)m(Y_k)] \geq 0, \quad 1 \leq k \leq n-1 \tag{2.3.7}$$

by the definition of demimartingale (resp., demisubmartingale) since the function $m(Y_k)$ is a nondecreasing (resp., nonnegative and nondecreasing) function of $S_1, \ldots, S_k$. Taking the expectations on both sides of the inequality (2.3.6) and applying the inequality (2.3.7), we obtain the inequality (2.3.2) stated in the theorem by observing that $Y_n = R_{n,j}$. . The inequality (2.3.3) is an easy consequence of (2.3.2) by choosing the function $m(u)$ to be the indicator function $I_{[u \geq \lambda]}$. $\quad\square$

We obtain some corollaries to the theorem stated above.

Theorem 2.3.2. *If the sequence $\{S_n, n \geq 1\}$ is an L^2-demimartingale, then*

$$E((R_{n,j} - S_n)^2) \leq E(S_n^2) \tag{2.3.8}$$

and if the sequence $\{S_n, n \geq 1\}$ is an L^2-demisubmartingale, then

$$E((R_{n,j}^+ - S_n)^2) \leq E(S_n^2). \tag{2.3.9}$$

Proof. In the demimartingale case, the corollary is proved by choosing $m(u) = u$ to obtain the inequality

$$E(R_{n,j}^2/2) \le E(S_n R_{n,j})$$

which is equivalent to the inequality stated in (2.3.8). In the demisubmartingale case, we prove (2.3.9) by choosing $m(u) = uI_{[u \ge 0]}$. $\qquad\square$

The following result gives a maximal inequality for the sequence of partial sums of mean zero associated random variables. This can be derived as a corollary to the theorem proved earlier. We omit the proof of this result. For details, see Newman and Wright (1982).

Theorem 2.3.3. *Suppose the sequence $\{X_n, n \ge 1\}$ is a mean zero associated sequence of random variables and $S_n = X_1 + \ldots + X_n$ with $S_0 = 0$. Then the sequence $\{S_n, n \ge 1\}$ is a demimartingale and*

$$E(R_{n,j}^2) \le E(S_n^2) = \sigma_n^2, \qquad (2.3.10)$$

and, for $\lambda_1 < \lambda_2$,

$$(1 - \sigma_n^2/(\lambda_2 - \lambda_1)^2)P(S_n^* \ge \lambda_2) \le P(S_n \ge \lambda_1), \qquad (2.3.11)$$

so that for $\alpha_1 < \alpha_2$ with $\alpha_2 - \alpha_1 > 1$,

$$P(max(|S_1|, \ldots, |S_n|) \ge \alpha_2 \sigma_n) \le \frac{(\alpha_2 - \alpha_1)^2}{(\alpha_2 - \alpha_1)^2 - 1} P(|S_n| \ge \alpha_1 \sigma_n). \qquad (2.3.12)$$

Note that Theorem 2.3.3 holds for partial sums of mean zero associated random variables which form a demimartingale. However the following maximal inequality is valid for any demimartingale.

Theorem 2.3.4. *Suppose the sequence $\{S_n, n \ge 1\}$ is an L^2-demisubmartingale. Let $E(S_n^2) = \sigma_n^2$. Then, for $0 \le \lambda_1 < \lambda_2$,*

$$P(S_n^* \ge \lambda_2) \le (\sigma_n/(\lambda_2 - \lambda_1))(P(S_n \ge \lambda_1))^{1/2}. \qquad (2.3.13)$$

If $S_1, S_2, \ldots$ is an L^2 demimartingale, then for $0 \le \alpha_1 < \alpha_2$,

$$P(max(|S_1|, \ldots, |S_n|) \ge \alpha_2 \sigma_n) \le \sqrt{2}(\alpha_2 - \alpha_1)^{-1}(P(|S_n| \ge \alpha_1 \sigma_n))^{1/2}. \quad (2.3.14)$$

Proof. Let $\lambda = \lambda_2$ and $j = 1$ in the inequality proved in (2.3.3). Then, it follows that

$$\lambda_2 \, P(S_n^* \ge \lambda_2) \le \int_{[S_n^* \ge \lambda_2]} S_n \, dP$$

$$\le \int_{[S_n \ge \lambda_1]} S_n \, dP + \int_{[S_n^* \ge \lambda_2, S_n < \lambda_1]} S_n \, dP$$

$$\leq \int_{[S_n \geq \lambda_1]} S_n \, dP + \lambda_1 P(S_n^* \geq \lambda_2) \qquad (2.3.15)$$

which implies that

$$P(S_n^* \geq \lambda_2) \leq (\lambda_2 - \lambda_1)^{-1} E(S_n I_{[S_n \geq \lambda_1]}). \qquad (2.3.16)$$

Applying the Cauchy-Schwartz inequality to the term on the right-hand side of the above inequality, we get the inequality (2.3.13). In order to derive inequality (2.3.14), we take $\lambda_i = \alpha_i \sigma_n$ and add to (2.3.13) the corresponding inequality with all the S_i's replaced by $-S_i$'s which also forms a demimartingale. $\qquad \square$

Chow (1960) proved a maximal inequality for submartingales. Christofides (2000) obtained an analogue of this inequality for demisubmartingales. Prakasa Rao (2002, 2007) and Wang (2004) derived other maximal inequalities for demisubmartingales. We will discuss these results later in this chapter.

2.4   An Upcrossing Inequality

The following theorem, due to Newman and Wright (1982), extends Doob's upcrossing inequality for submartingales to demisubmartingales. Given a set of random variables $S_1, S_2, \ldots, S_n$ and $a < b$, we define a sequence of stopping times $J_0 = 0, J_1, J_2, \ldots$ as follows (for $k = 1, 2, \ldots$):

$$J_{2k-1} = \begin{cases} n+1 & \text{if } \{j : J_{2k-2} < j \leq n \text{ and } S_j \leq a\} \text{ is empty,} \\ \min\{j : J_{2k-2} < j \leq n \text{ and } S_j \leq a\}, & \text{otherwise,} \end{cases}$$

and

$$J_{2k-1} = \begin{cases} n+1 & \text{if } \{j : J_{2k-2} < j \leq n \text{ and } S_j \geq b\} \text{ is empty,} \\ \min\{j : J_{2k-2} < j \leq n \text{ and } S_j \geq b\}, & \text{otherwise.} \end{cases}$$

The number of complete upcrossings of the interval $[a, b]$ by $S_1, \ldots, S_n$ is denoted by $U_{a,b}$ where

$$U_{a,b} = \max\{k : J_{2k} < n+1\}. \qquad (2.4.1)$$

Theorem 2.4.1. *If the finite sequence $S_1, S_2, \ldots, S_n$ is a demisubmartingale, then for $a < b$,*

$$E(U_{a,b}) \leq \frac{E((S_n - a)^+) - E((S_1 - a)^+)}{b - a}. \qquad (2.4.2)$$

The following theorem gives sufficient conditions for the almost sure convergence of a demimartingale. It is a consequence of the upcrossing inequality stated in Theorem 2.4.1 as in the case of martingales.

Theorem 2.4.2. *If the sequence $\{S_n, n \geq 1\}$ is a demimartingale and*

$$\limsup_{n \to \infty} E|S_n| < \infty,$$

then the S_n converge a.s. to a random variable X such that $E|X| < \infty$.

We now prove the upcrossing inequality.

Proof of Theorem 2.4.1. For $1 \leq j \leq n-1$, define

$$\epsilon_j = \begin{cases} 1 & \text{if for some } k = 1, 2, \ldots, \ J_{2k-2} \leq j < J_{2k-1} \\ 0 & \text{if for some } k = 1, 2, \ldots, \ J_{2k-1} \leq j < J_{2k} \end{cases} \quad (2.4.3)$$

so that $1 - \epsilon_j$ is the indicator function of the event that the time interval $[j, j+1)$ is a part of an upcrossing possibly incomplete; equivalently

$$\epsilon_j = \begin{cases} 1 & \text{if either } S_i > a \text{ for } i = 1, \ldots, j \text{ or} \\ & \text{for some } i = 1, \ldots, j, \ S_i \geq b \text{ and } S_k > a \text{ for } k = i+1, \ldots, j \quad (2.4.4) \\ = 0 & \text{otherwise.} \end{cases}$$

Let Λ be the event that the sequence $S_1, \ldots, S_n$ ends with an incomplete upcrossing, that is, $\tilde{J} \equiv J_{2U_{a,b}+1} < n$. Note that

$$(S_n - a)^+ - (S_1 - a)^+ = \sum_{j=1}^{n-1} [(S_{j+1} - a)^+ - (S_j - a)^+] = M_u + M_d \quad (2.4.5)$$

where

$$M_d = \sum_{j=1}^{n-1} \epsilon_j [(S_{j+1} - a)^+ - (S_j - a)^+] \geq \sum_{j=1}^{n-1} \epsilon_j (S_{j+1} - S_j). \quad (2.4.6)$$

This inequality follows from the observation that

$$(S_{j+1} - a)^+ \geq S_{j+1} - a$$

and

$$\epsilon_j (S_j - a)^+ = \epsilon_j (S_j - a)$$

since $\epsilon_j = 1$ implies that $S_j > a$ from the definition of ϵ_j. Observe that

$$M_u = \sum_{j=1}^{n-1} (1 - \epsilon_j)[(S_{j+1} - a)^+ - (S_j - a)^+] \quad (2.4.7)$$

$$= \sum_{k=1}^{U_{a,b}} \sum_{j=J_{2k-1}}^{J_{2k}-1} [(S_{j+1} - a)^+ - (S_j - a)^+] + \sum_{j=\tilde{J}}^{n-1} [(S_{j+1} - a)^+ - (S_j - a)^+]$$

$$= \sum_{k=1}^{U_{a,b}} [(S_{J_{2k}} - a)^+ - (S_{J_{2k-1}} - a)^+] + [(S_n - a)^+ - (S_{\tilde{J}} - a)^+] I_\Lambda$$

$$= \sum_{k=1}^{U_{a,b}} (S_{J_{2k}} - a)^+ + (S_n - a)^+] I_\Lambda$$

$$\geq (b - a) U_{a,b}.$$

Combining equations (2.4.5), (2.4.6) and (2.4.7) and taking expectations, we get that

$$E[(S_n - a)^+ - (S_1 - a)^+] \geq (b - a)E[U_{a,b}] + \sum_{j=1}^{n-1} E[\epsilon_j(S_{j+1} - S_j)]. \quad (2.4.8)$$

Note that ϵ_j is a nonnegative nondecreasing function of $S_1, \ldots, S_n$ from the definition of ϵ_j. Since $\{S_j, j = 1, \ldots, n\}$ is a demisubmartingale, it follows that

$$E[\epsilon_j(S_{j+1} - S_j)] \geq 0, \quad j = 1, \ldots, n-1.$$

Hence

$$E[(S_n - a)^+ - (S_1 - a)^+] \geq (b - a)E[U_{a,b}] \quad (2.4.9)$$

which implies the upcrossing inequality stated in the theorem. $\square$

2.5   Chow Type Maximal Inequality

We now derive some more maximal inequalities for demimartingales which can be used to derive strong laws of large numbers for demimartingales. The following result, due to Christofides (2000), is an analogue of the maximal inequality for submartingales proved by Chow (1960).

Theorem 2.5.1. *Let the sequence $\{S_n, n \geq 1\}$ be a demisubmartingale with $S_0 = 0$. Let $\{c_n, n \geq 0\}$ be a nonincreasing sequence of positive numbers, Then, for any $\epsilon > 0$,*

$$\epsilon \ P[\max_{1 \leq k \leq n} c_k S_k \geq \epsilon] \leq \sum_{j=1}^{n} c_j \ E[S_j^+ - S_{j-1}^+] \quad (2.5.1)$$

where $x^+ = \max\{0, x\}$.

Proof. Let $\epsilon > 0$. Let $A = [\max_{1 \leq k \leq n} c_k S_k \geq \epsilon]$ and

$$A_j = [\max_{1 \leq i < j} c_i S_i < \epsilon, \ c_j \ S_j \geq \epsilon], \quad j = 1, \ldots, n.$$

Let E^c denote the complement of a set E. Let $I_{[E]}$ denote the indicator function of a set E. Note that the events A_j, $j = 1, \ldots, n$ are disjoint and $A = \cup_{j=1}^{n} A_j$. Hence

$$\epsilon \ P(A) = \epsilon \sum_{j=1}^{n} P(A_j) \quad (2.5.2)$$

$$= \sum_{j=1}^{n} E[\epsilon I_{[A_j]}]$$

$$\leq \sum_{j=1}^{n} E[c_j \, S_j I_{[A_j]}]$$

$$\leq \sum_{j=1}^{n} E[c_j \, S_j^+ I_{[A_j]}]$$

$$= c_1 E[S_1^+] - c_1 E[S_1^+ I_{[A_1^c]}] + \sum_{j=2}^{n} E[c_j \, S_j^+ I_{[A_j]}]$$

$$\leq c_1 E[S_1^+] - c_2 E[S_1^+ I_{[A_1^c]}] + c_2 E[S_2^+ I_{[A_2]}] + \sum_{j=3}^{n} E[c_j \, S_j^+ I_{[A_j]}]$$

$$= c_1 E[S_1^+] + c_2 E[(S_2^+ - S_1^+)I_{[A_1^c]}] - c_2 E[S_2^+ I_{[A_1^c \cap A_2^c]}] + \sum_{j=3}^{n} E[c_j \, S_j^+ I_{[A_j]}].$$

The last equality follows from the fact that $I_{[A_2]} = I_{[A_1^c]} - I_{[A_1^c \cap A_2^c]}$ which in turn holds since $A_2 \subset A_1^c$. The expression on the right-hand side of the last equality can be written in the form

$$c_1 E[S_1^+] + c_2 E[S_2^+ - S_1^+] - c_2 E[(S_2^+ - S_1^+)I_{[A_1]}] - c_2 E[S_2^+ I_{[A_1^c \cap A_2^c]}] + \sum_{j=3}^{n} E[c_j \, S_j^+ I_{[A_j]}].$$

Let $h(y) = \lim_{x \to y-o} \frac{x^+ - y^+}{x - y}$. Then $h(.)$ is a nonnegative nondecreasing function by the convexity of the function $x^+ = \max\{0, x\}$ and we have

$$S_2^+ - S_1^+ \geq (S_2 - S_1)h(S_1).$$

Therefore

$$E[(S_2^+ - S_1^+)I_{[A_1]}] \geq E[(S_2 - S_1)h(S_1)I_{[A_1]}].$$

Since $h(S_1)I_{[A_1]}$ is a nonnegative nondecreasing function of S_1, it follows that

$$E[(S_2 - S_1)h(S_1)I_{[A_1]}] \geq 0$$

by the demisubmartingale property of the sequence $\{S_n, n \geq 1\}$ which in turn shows that

$$E[(S_2^+ - S_1^+)I_{[A_1]}] \geq 0.$$

Hence the expression on the right-hand side of the last inequality in (2.5.2) is bounded above by J where

$$J = c_1 E[S_1^+] + c_2 E[S_2^+ - S_1^+] - c_2 E[S_2^+ I_{[A_1^c \cap A_2^c]}] + \sum_{j=3}^{n} E[c_j \, S_j^+ I_{[A_j]}].$$

Since the sequence c_k is a nondecreasing sequence,

$$J \leq c_1 E[S_1^+] + c_2 E[S_2^+ - S_1^+] - c_3 E[S_2^+ I_{[A_1^c \cap A_2^c]}] + \sum_{j=3}^{n} E[c_j \, S_j^+ I_{[A_j]}] \qquad (2.5.3)$$

$$= c_1 E[S_1^+] + c_2 E[S_2^+ - S_1^+] + c_3 E[(S_3^+ - S_2^+)I_{[A_1^c \cap A_2^c]}]$$

$$- c_3 E[S_3^+ I_{[A_1^c \cap A_2^c \cap A_3^c]}] + \sum_{j=4}^{n} E[c_j \, S_j^+ I_{[A_j]}]$$

and the last equality follows from the fact that $I_{[A_3]} = I_{[A_1^c \cap A_2^c]} - I_{[A_1^c \cap A_2^c \cap A_3^c]}$ which in turn holds since $A_3 \subset A_1^c \cap A_2^c$. The expression on the right-hand side of the last equality can be written in the form

$$\sum_{j=1}^{3} c_j E[S_j^+ - S_{j-1}^+] - c_3 E[(S_3^+ - S_2^+)I_{[A_1 \cup A_2]}] - c_3 E[S_3^+ I_{[A_1^c \cap A_2^c \cap A_3^c]}]$$

$$+ \sum_{j=4}^{n} E[c_j \, S_j^+ I_{[A_j]}]. \qquad (2.5.4)$$

Applying the convexity of the function x^+ again, we get that

$$E[(S_3^+ - S_2^+)I_{[A_1 \cup A_2]}] \geq E[(S_3 - S_2)h(S_2)I_{[A_1 \cup A_2]}].$$

Since the function $h(S_2)I_{[A_1 \cup A_2]}$ is a nonnegative componentwise nondecreasing function of (S_1, S_2), it follows that

$$E[(S_3 - S_2)h(S_2)I_{[A_1 \cup A_2]}] \geq 0$$

by the demisubmartingale property of the sequence $\{S_n, n \geq 1\}$. Hence the quantity defined by (2.5.4) is bounded above by

$$\sum_{j=1}^{3} c_j E[S_j^+ - S_{j-1}^+] - c_3 E[S_3^+ I_{[A_1^c \cap A_2^c \cap A_3^c]}] + \sum_{j=4}^{n} E[c_j \, S_j^+ I_{[A_j]}]. \qquad (2.5.5)$$

Proceeding in this way, we get that

$$\epsilon P(A) \leq \sum_{j=1}^{n} c_j E[S_j^+ - S_{j-1}^+] - c_n E[S_n^+ I_{[A^c]}] \qquad (2.5.6)$$

$$\leq \sum_{j=1}^{n} c_j E[S_j^+ - S_{j-1}^+]$$

since $c_n > 0$. $\qquad\qquad\qquad\qquad\qquad\qquad\qquad\qquad\qquad\qquad\qquad\qquad\qquad\qquad$ $\square$

A corollary to the Chow type maximal inequality is the following Doob type maximal inequality which was derived earlier by other methods.

Theorem 2.5.2. *Suppose the sequence $\{S_n, n \geq 1\}$ is a demisubmartingale. Then, for any $\epsilon > 0$,*

$$\epsilon \, P[\max_{1 \leq k \leq n} S_k \geq \epsilon] \leq \int_{[\max_{1 \leq k \leq n} S_k \geq \epsilon]} S_n dP.$$

Applications of Chow Type Maximal Inequality

As an application of the Chow type maximal inequality, we can obtain the following results.

Theorem 2.5.3. *Let the sequence $\{S_n, \ n \geq 1\}$ be a demimartingale with $S_0 = 0$. Suppose that $\{c_k, \ k \geq 1\}$ is a positive nonincreasing sequence of numbers such that $\lim_{k \to 0} c_k = 0$. Suppose there exists $\nu \geq 1$ such that $E[|S_k|^\nu] < \infty$ for every $k \geq 1$. Assume that*

$$\sum_{k=1}^{\infty} c_k^\nu E(|S_k|^\nu - |S_{k-1}|^\nu) < \infty. \tag{2.5.7}$$

Then

$$c_n S_n \overset{a.s}{\to} 0 \ \ as \ n \to \infty.$$

Proof. Let $\epsilon > 0$. Note that

$$P[\sup_{k \geq n} c_k|S_k| \geq \epsilon] = P[\sup_{k \geq n} c_k^\nu |S_k|^\nu \geq \epsilon^\nu]$$

$$\leq P[\sup_{k \geq n} c_k^\nu (S_k^+)^\nu \geq \epsilon^\nu/2] + P[\sup_{k \geq n} c_k^\nu (S_k^-)^\nu \geq \epsilon^\nu/2].$$

Since the function x^ν, $x \geq 0$ is a nondecreasing convex function for any $\nu \geq 1$, it follows that the sequences $\{(S_k^+)^\nu, \ k \geq 1\}$ and $\{(S_k^-)^\nu, \ k \geq 1\}$ are both demisubmartingales. Applying the Chow type maximal inequality proved earlier, we get that

$$P[\sup_{k \geq n} c_k^\nu (S_k^+)^\nu \geq \epsilon^\nu/2] + P[\sup_{k \geq n} c_k^\nu (S_k^-)^\nu \geq \epsilon^\nu/2] \tag{2.5.8}$$

$$\leq 2\epsilon^{-\nu}(c_n^\nu E[(S_n^+)^\nu] + \sum_{k=n+1}^{\infty} c_k^\nu \, E[(S_k^+)^\nu - (S_{k-1}^+)^\nu]$$

$$+ c_n^\nu E[(S_n^-)^\nu] + \sum_{k=n+1}^{\infty} c_k^\nu \, E[(S_k^-)^\nu - (S_{k-1}^-)^\nu])$$

$$= 2\epsilon^{-\nu}(c_n^\nu E[|S_n|^\nu] + \sum_{k=n+1}^{\infty} c_k^\nu \, E[|S_k|^\nu - |S_{k-1}|^\nu])$$

from the fact that $|S_n|^\nu = (S_n^+)^\nu + (S_n^-)^\nu$. The Kronecker lemma and the condition (2.5.7) imply that $\lim_{n \to \infty} c_n^\nu E|S_n|^\nu = 0$. From the inequality derived above, we get that

$$\lim_{n \to \infty} P[\sup_{k \geq n} c_k|S_k| \geq \epsilon] = 0, \tag{2.5.9}$$

equivalently

$$c_n S_n \overset{a.s}{\to} 0 \ \ as \ n \to \infty. \tag{2.5.10}$$

$\square$

Strong Law of Large Numbers for Associated Sequences

Consider a sequence of mean zero associated random variables. The following Kolmogorov type of strong law of large numbers can be proved for such a sequence as a consequence of Theorem 2.5.3.

Theorem 2.5.4. *Let the sequence* $\{X_n,\ n \geq 1\}$ *be a sequence of* L^2*-mean zero associated random variables. Let* $S_n = X_1 + \ldots + X_n$ *and* $S_0 = 0$. *Suppose that*

$$\sum_{n=1}^{\infty} n^{-2} \operatorname{Cov}(X_n, S_n) < \infty.$$

Then

$$\frac{S_n}{n} \xrightarrow{a.s} 0 \ \text{ as } \ n \to \infty. \tag{2.5.11}$$

Proof. Since the sequence $\{X_n,\ n \geq 1\}$ is mean zero associated, it follows that $\{S_n, n \geq 1\}$ is a demimartingale. Applying Theorem 2.5.3 with $\nu = 2$ and $c_n = \frac{1}{n}$, it follows that

$$\frac{S_n}{n} \xrightarrow{a.s} 0 \ \text{ as } \ n \to \infty$$

provided

$$\sum_{n=1}^{\infty} n^{-2} E(S_n^2 - S_{n-1}^2) < \infty.$$

Observe that

$$E(S_n^2 - S_{n-1}^2) = 2E(X_n S_{n-1}) + E(X_n^2)$$

and

$$2\sum_{n=1}^{\infty} n^{-2} E(X_n S_{n-1}) + \sum_{n=1}^{\infty} n^{-2} E(X_n^2) \leq 2\sum_{n=1}^{\infty} n^{-2} E(X_n S_n) \tag{2.5.12}$$

$$= 2\sum_{n=1}^{\infty} n^{-2} \operatorname{Cov}(X_n, S_n) < \infty. \qquad \square$$

This strong law of large numbers for associated sequences was also proved by Birkel (1989). Wang (2004) generalized Theorem 2.5.1 to obtain a maximal inequality for nonnegative convex functions of a demimartingale.

Theorem 2.5.5. *Let* $S_0 = 0$ *and the sequence* $\{S_n,\ n \geq 1\}$ *be a demimartingale. Let* $g(.)$ *be a nonnegative convex function on* R *with* $g(0) = 0$. *Suppose that the sequence* $\{c_i,\ 1 \leq i \leq n\}$ *is a non-increasing sequence of positive numbers. Let* $S_n^* = \max_{1 \leq i \leq n} c_i g(S_i)$. *Then, for any* $\lambda > 0$,

$$\lambda P(S_n^* \geq \lambda) \leq \sum_{i=1}^{n} c_i E[g(S_i) - g(S_{i-1})] - c_n E[g(S_n) I_{[S_n^* < \lambda]}] \tag{2.5.13}$$

$$\leq \sum_{i=1}^{n} c_i E[g(S_i) - g(S_{i-1})].$$

We now give a sketch of proof of this theorem following the ideas in Hadji-kyriakou (2010) and Wang (2004).

Proof. Define the functions

$$u(x) = g(x)I_{[x \geq 0]} \text{ and } v(x) = g(x)I_{[x < 0]}.$$

Note that the function $u(x)$ is a nonnegative nondecreasing convex function and the function $v(x)$ is a nonnegative nonincreasing convex function. It is obvious that

$$g(x) = u(x) + v(x) = \max\{u(x), v(x)\}.$$

Furthermore

$$P[\max_{1 \leq i \leq n} c_i g(S_i) \geq \lambda) = P[\max_{1 \leq i \leq n} c_i \max\{u(S_i), v(S_i)\} \geq \lambda] \qquad (2.5.14)$$
$$\leq P[\max_{1 \leq i \leq n} c_i u(S_i) \geq \lambda] + P[\max_{1 \leq i \leq n} c_i v(S_i) \geq \lambda].$$

Applying Theorem 2.5.1, we get that

$$\lambda P[\max_{1 \leq i \leq n} c_i u(S_i) \geq \lambda] \leq \sum_{i=1}^{n} c_i E[u(S_i) - u(S_{i-1})]. \qquad (2.5.15)$$

Let

$$A_i = \{c_k v(S_k) < \lambda, \ 1 \leq k < i, \ c_i v(S_i) \geq \lambda\}, \quad i = 1, \dots, n.$$

Following the method in the proof of Theorem 2.5.1, we get that

$$\lambda P[\max_{1 \leq i \leq n} c_i v(S_i) \geq \lambda] \leq \sum_{i=3}^{n} c_i E[v(s_i)I_{A_i}] + c_1 E[v(S_1)] + c_2 E[v(S_2) - v(S_1)]$$
$$- c_2 E[v(S_2)I_{A_1^c \cap A_2^c}] - c_2 E[(v(S_2) - v(S_1))I_{A_1}]. \qquad (2.5.16)$$

Let $h(.)$ be the left derivative of the function $v(.)$ Then $h(.)$ is a non-positive nondecreasing function and

$$v(x) - v(y) \geq (x - y)h(y).$$

Hence

$$v(S_2) - v(S_1) \geq (S_2 - S_1)h(S_1)$$

and

$$E[(v(S_2) - v(S_1))I_{A_1}] \geq E[(S_2 - S_1)h(S_1)I_{A_1}].$$

Since I_{A_1} is a nonincreasing function of S_1 and $h(.)$ is a non-positive nondecreasing function, it follows that $h(S_1)I_{A_1}$ is a nondecreasing function of S_1. By the demimartingale property, we get that

$$E[(S_2 - S_1)h(S_1)I_{A_1}] \geq 0.$$

Applying arguments similar to those in the proof of Theorem 2.5.1, we get that

$$\lambda P[\max_{1 \leq i \leq n} c_i v(S_i) \geq \lambda] \leq \sum_{i=1}^{n} c_i E[v(S_i) - v(S_{i-1})]. \tag{2.5.17}$$

Combining the inequalities (2.5.14), (2.5.15) and (2.5.17), we obtain the inequality (2.5.13). □

Suppose the sequence $\{S_n, n \geq 1\}$ is a nonnegative demimartingale. As a corollary to this theorem, it can be proved that

$$E(\max_{1 \leq i \leq n} S_i) \leq \frac{e}{e-1}[1 + E(S_n \log^+ S_n)].$$

For a proof of this inequality , see Corollary 2.1 in Wang (2004).

2.6 Whittle Type Maximal Inequality

We now discuss a Whittle type maximal inequality for demisubmartingales due to Prakasa Rao (2002). This result generalizes the Kolmogorov inequality and the Hajek-Renyi inequality for independent random variables (Whittle (1969)) and is an extension of the results in Christofides (2000) for demisubmartingales.

Let the sequence $\{S_n, n \geq 1\}$ be a demisubmartingale. Suppose $\phi(.)$ is a nondecreasing convex function. Then the sequence $\{\phi(S_n), n \geq 1\}$ is a demisubmartingale by Theorem 2.1.1 (cf. Christofides (2000)).

Theorem 2.6.1. *Let $S_0 = 0$ and suppose the sequence of random variables $\{S_n, n \geq 1\}$ is a demisubmartingale. Let $\phi(.)$ be nonnegative nondecreasing convex function such that $\phi(0) = 0$. Let $\psi(u)$ be a positive nondecreasing function for $u > 0$. Let A_n be the event that $\phi(S_k) \leq \psi(u_k)$, $1 \leq k \leq n$, where $0 = u_0 < u_1 \leq \ldots \leq u_n$. Then*

$$P(A_n) \geq 1 - \sum_{k=1}^{n} \frac{E[\phi(S_k)] - E[\phi(S_{k-1})]}{\psi(u_k)}. \tag{2.6.1}$$

If, in addition, there exist nonnegative real numbers $\Delta_k, 1 \leq k \leq n$ such that

$$0 \leq E[(\phi(S_k) - \phi(S_{k-1}))f(\phi(S_1), \ldots, \phi(S_{k-1}))] \leq \Delta_k E[f(\phi(S_1), \ldots, \phi(S_{k-1}))]$$

for $1 \leq k \leq n$ for all componentwise nonnegative nondecreasing functions f such that the expectation is defined and

$$\psi(u_k) \geq \psi(u_{k-1}) + \Delta_k, \quad 1 \leq k \leq n,$$

then

$$P(A_n) \geq \prod_{k=1}^{n}(1 - \frac{\Delta_k}{\psi(u_k)}). \qquad (2.6.2)$$

Proof. Since the sequence $\{S_n, n \geq 1\}$ is a demisubmartingale by hypothesis and the function $\phi(.)$ is a nondecreasing convex function, it follows that the sequence $\{\phi(S_n), n \geq 1\}$ forms a demisubmartingale by Theorem 2.1.1. Hence

$$E\{(\phi(S_{n+1}) - \phi(S_n))f(\phi(S_1), \ldots, \phi(S_n))\} \geq 0, \quad n \geq 1 \qquad (2.6.3)$$

for every nonnegative componentwise nondecreasing function f such that the expectation is defined. Let χ_j be the indicator function of the event $[\phi(S_j) \leq \psi(u_j)]$ for $1 \leq j \leq n$. Note that

$$\chi_n \geq (1 - \frac{\phi(S_n)}{\psi(u_n)})$$

and hence

$$P(A_n) = E(\prod_{i=1}^{n}\chi_i) = E(\{\prod_{i=1}^{n-1}\chi_i\}\chi_n)$$

$$\geq E(\{\prod_{i=1}^{n-1}\chi_i\}(1 - \frac{\phi(S_n)}{\psi(u_n)})).$$

Note that

$$E[\{\prod_{i=1}^{n-1}\chi_i\}\{(1 - \frac{\phi(S_n)}{\psi(u_n)}) - (1 - \frac{\phi(S_{n-1})}{\psi(u_n)})\} + \frac{\phi(S_n) - \phi(S_{n-1})}{\psi(u_n)}]$$

$$= E[(1 - \prod_{i=1}^{n-1}\chi_i)(\frac{\phi(S_n) - \phi(S_{n-1})}{\psi(u_n)})] \geq 0$$

since the function $1 - \prod_{i=1}^{n-1}\chi_i$ is a nonnegative componentwise nondecreasing function of $\phi(S_i), 1 \leq i \leq n - 1$. Hence

$$P(A_n) \geq E(\{\prod_{i=1}^{n-1}\chi_i\}(1 - \frac{\phi(S_{n-1})}{\psi(u_n)})) - \frac{E\{\phi(S_n)\} - E\{\phi(S_{n-1})\}}{\psi(u_n)}$$

$$\geq E(\{\prod_{i=1}^{n-2}\chi_i\}(1 - \frac{\phi(S_{n-1})}{\psi(u_{n-1})})) - \frac{E\{\phi(S_n)\} - E\{\phi(S_{n-1})\}}{\psi(u_n)}.$$

The last inequality follows from the observation that the sequence $\psi(u_n), n \geq 1$ is positive and nondecreasing.

Applying this inequality repeatedly, we get that

$$P(A_n) \geq 1 - \sum_{k=1}^{n}\frac{E[\phi(S_k)] - E[\phi(S_{k-1})]}{\psi(u_k)} \qquad (2.6.4)$$

completing the proof of the first part of the theorem. Note that

$$E\{\prod_{i=1}^{n-1} \chi_i(1 - \frac{\phi(S_n)}{\psi(u_n)}) - (1 - \frac{\Delta_n}{\psi(u_n)})(1 - \frac{\phi(S_{n-1})}{\psi(u_{n-1})}) \prod_{i=1}^{n-1} \chi_i\}$$

$$\geq E\{\frac{\phi(S_{n-1})}{\psi(u_n)\psi(u_{n-1})}[\psi(u_n) - \psi(u_{n-1}) - \Delta_n] \prod_{i=1}^{n-1} \chi_i\}$$

and the last term is nonnegative by hypothesis. Hence

$$P(A_n) \geq (1 - \frac{\Delta_n}{\psi(u_n)})E(\{\prod_{i=1}^{n-2} \chi_i\}(1 - \frac{\phi(S_{n-1})}{\psi(u_{n-1})})). \qquad (2.6.5)$$

Applying this inequality repeatedly, we obtain that

$$P(A_n) \geq \prod_{k=1}^{n}(1 - \frac{\Delta_k}{\psi(u_k)}). \qquad (2.6.6)$$

$\square$

Applications

As applications of the Whittle type maximal inequality derived above, the following results can be obtained.

Suppose the sequence $\{S_n, n \geq 1\}$ is a demisubmartingale. Then the sequences $\{(S_n^+)^p, n \geq 1\}$ and $\{(S_n^-)^p, n \geq 1\}$ are demisubmartingales for $p \geq 1$ by Theorem 2.1.1. Furthermore $|S_n|^p = (S_n^+)^p + (S_n^-)^p$ for all $p \geq 1$.

(1) Let $\psi(u) = u^p$, $p \geq 1$ in Theorem 2.6.1. Applying Theorem 2.6.1, we get that

$$P(S_j^+ \leq u_j, 1 \leq j \leq n) \geq 1 - \sum_{j=1}^{n} \frac{E(S_j^+)^p - E(S_{j-1}^+)^p}{u_j^p} \qquad (2.6.7)$$

and

$$P(S_j^- \leq u_j, 1 \leq j \leq n) \geq 1 - \sum_{j=1}^{n} \frac{E(S_j^-)^p - E(S_{j-1}^-)^p}{u_j^p}. \qquad (2.6.8)$$

Hence , for every $\varepsilon > 0$,

$$P(\sup_{1 \leq j \leq n} \frac{|S_j|}{u_j} \geq \varepsilon) = P(\sup_{1 \leq j \leq n} \frac{|S_j|^p}{u_j^p} \geq \varepsilon^p)$$

$$= P(\sup_{1 \leq j \leq n} \frac{(S_j^+)^p + (S_j^-)^p}{u_j^p} \geq \varepsilon^p)$$

$$= P(\sup_{1 \leq j \leq n} \frac{(S_j^+)^p}{u_j^p} \geq \frac{1}{2}\varepsilon^p) + P(\sup_{1 \leq j \leq n} \frac{(S_j^-)^p}{u_j^p} \geq \frac{1}{2}\varepsilon^p)$$

$$\leq 2\varepsilon^{-p} \sum_{j=1}^{n} \frac{E(S_j^+)^p - E(S_{j-1}^+)^p}{u_j^p}$$

$$+ 2\varepsilon^{-p} \sum_{j=1}^{n} \frac{E(S_j^-)^p - E(S_{j-1}^-)^p}{u_j^p}$$

$$= 2\varepsilon^{-p} \sum_{j=1}^{n} \frac{E|S_j|^p - E|S_{j-1}|^p}{u_j^p}.$$

In particular, for $p = 2$, we have

$$P(\sup_{1 \leq j \leq n} \frac{|S_j|}{u_j} \geq \varepsilon) \leq 2\varepsilon^{-2} \sum_{j=1}^{n} \frac{ES_j^2 - ES_{j-1}^2}{u_j^2}. \qquad (2.6.9)$$

which is the Hajek-Renyi type inequality for associated sequences derived in the Corollary 2.3 of Christofides (2000).

Suppose $p = 1$. Let $\phi(x) = \max(0, x)$. Then the function $\phi(x)$ is a nonnegative nondecreasing convex function and it is clear that $S_n \leq S_n^+ = \phi(S_n)$ for every $n \geq 1$. Let $\psi(u) = u$ in Theorem 2.6.1. Then

$$P(\sup_{1 \leq j \leq n} \frac{S_j}{u_j} \geq \varepsilon) \leq P(\sup_{1 \leq j \leq n} \frac{S_j^+}{u_j} \geq \varepsilon) \leq \varepsilon^{-1} \sum_{j=1}^{n} \frac{ES_j^+ - ES_{j-1}^+}{u_j}$$

by Theorem 2.6.1 which is the Chow type maximal inequality derived in Theorem 2.5.1 (cf. Christofides (2000)).

(2) Let $p = 2$ again in the above discussion. If

$$E(S_j^2 - S_{j-1}^2) \leq u_j^2 - u_{j-1}^2,$$

for $1 \leq j \leq n$, then

$$P(A_n) \geq \prod_{j=1}^{n} (1 - \frac{E(S_j^2) - E(S_{j-1}^2)}{u_j^2})$$

which is an analogue of the Dufresnoy type maximal inequality for martingales (cf. Dufresnoy (1967)).

(3) Let the sequence $\{S_n, n \geq 1\}$ be a demisubmartingale and the function $\phi(.)$ be a nonnegative nondecreasing convex function such that $\phi(S_0) = 0$. Let $\psi(u)$ be a positive nondecreasing function for $u > 0$. Then, for any nondecreasing sequence u_n, $n \geq 1$ with $u_0 = 0$,

$$P(\sup_{1 \leq j \leq n} \frac{\phi(S_j)}{\psi(u_j)} \geq \varepsilon) \leq \varepsilon^{-1} \sum_{k=1}^{n} \frac{E[\phi(S_k)] - E[\phi(S_{k-1})]}{\psi(u_k)}. \qquad (2.6.10)$$

In particular, for any fixed $n \geq 1$,

$$P(\sup_{k \geq n} \frac{\phi(S_k)}{\psi(u_k)} \geq \varepsilon) \leq \varepsilon^{-1}[E(\frac{\phi(S_n)}{\psi(u_n)}) + \sum_{k=n+1}^{\infty} \frac{E[\phi(S_k)] - E[\phi(S_{k-1})]}{\psi(u_k)}]. \quad (2.6.11)$$

As a consequence of this inequality, we get the following strong law of large numbers for demisubmartingales (cf. Prakasa Rao (2002)).

Theorem 2.6.2. *Let $S_0 = 0$ and let the sequence $\{S_n, n \geq 1\}$ be a demisubmartingale. Let $\phi(.)$ be a nonnegative nondecreasing convex function such that $\phi(0) = 0$. Let $\psi(u)$ be a positive nondecreasing function for $u > 0$ such that $\psi(u) \to \infty$ as $u \to \infty$. Further suppose that*

$$\sum_{k=1}^{\infty} \frac{E[\phi(S_k)] - E[\phi(S_{k-1})]}{\psi(u_k)} < \infty$$

for a nondecreasing sequence $u_n \to \infty$ as $n \to \infty$. Then

$$\frac{\phi(S_n)}{\psi(u_n)} \stackrel{a.s}{\to} 0 \quad as \quad n \to \infty.$$

2.7   More on Maximal Inequalities

Suppose the sequence $\{S_n, n \geq 1\}$ is a demisubmartingale. Let $S_n^{\max} = \max_{1 \leq i \leq n} S_i$ and $S_n^{\min} = \min_{1 \leq i \leq n} S_i$. As special cases of Theorem 2.3.1, we get that

$$\lambda \ P[S_n^{\max} \geq \lambda] \leq \int_{[S_n^{\max} \geq \lambda]} S_n dP \quad (2.7.1)$$

and

$$\lambda \ P[S_n^{\min} \geq \lambda] \leq \int_{[S_n^{\min} \geq \lambda]} S_n dP \quad (2.7.2)$$

for any $\lambda > 0$.

The inequality (2.7.1) can also be obtained directly without using Theorem 2.3.1 by the standard methods used to prove the Kolomogorov inequality. We now prove a variant of the inequality (2.7.2).

Suppose the sequence $\{S_n, n \geq 1\}$ is a demisubmartingale. Let $\lambda > 0$. Let

$$A = [\min_{1 \leq k \leq n} S_k < \lambda], A_1 = [S_1 < \lambda]$$

and

$$A_k = [S_k < \lambda, S_j \geq \lambda, 1 \leq j \leq k - 1], \quad k > 1.$$

Observe that

$$A = \bigcup_{k=1}^{n} A_k$$

and $A_k \in \mathcal{F}_k = \sigma\{S_1, \ldots, S_k\}$. Furthermore the sets A_k, $1 \le k \le n$ are disjoint and

$$A_k \subset (\bigcup_{i=1}^{k-1} A_i)^c$$

where A^c denotes the complement of the set A in Ω. Note that

$$E(S_1) = \int_{A_1} S_1 dP + \int_{A_1^c} S_1 dP \le \lambda \int_{A_1} dP + \int_{A_1^c} S_2 dP.$$

The last inequality follows by observing that

$$\int_{A_1^c} S_1 dP - \int_{A_1^c} S_2 dP = \int_{A_1^c} (S_1 - S_2) dP = E((S_1 - S_2) I_{[A_1^c]}).$$

Since the indicator function of the set $A_1^c = [S_1 \ge \lambda]$ is a nonnegative nondecreasing function of S_1 and the fact that $\{S_k, 1 \le k \le n\}$ is a demisubmartingale, it follows that

$$E((S_2 - S_1) I_{[A_1^c]}) \ge 0.$$

Therefore

$$E((S_1 - S_2) I_{[A_1^c]}) \le 0$$

which implies that

$$\int_{A_1^c} S_1 dP \le \int_{A_1^c} S_2 dP.$$

This proves the inequality

$$E(S_1) \le \lambda \int_{A_1} dP + \int_{A_1^c} S_2 dP = \lambda P(A_1) + \int_{A_1^c} S_2 dP.$$

Observe that $A_2 \subset A_1^c$. Hence

$$\int_{A_1^c} S_2 dP = \int_{A_2} S_2 dP + \int_{A_2^c \cap A_1^c} S_2 dP$$

$$\le \int_{A_2} S_2 dP + \int_{A_2^c \cap A_1^c} S_3 dP$$

$$\le \lambda \, P(A_2) + \int_{A_2^c \cap A_1^c} S_3 dP.$$

The second inequality in the above chain follows from the observation that the indicator function of the set $A_2^c \cap A_1^c = I_{[S_1 \ge \lambda, S_2 \ge \lambda]}$ is a nonnegative nondecreasing function of S_1, S_2 and the fact that the sequence $\{S_k, 1 \leq k \leq n\}$ is a demisubmartingale. By repeated application of these arguments, we get that

$$E(S_1) \leq \lambda \sum_{i=1}^{n} P(A_i) + \int_{\cap_{i=1}^{n} A_i^c} S_n dP$$

$$= \lambda \ P(A) + \int_{\Omega} S_n dP - \int_{A} S_n dP.$$

Hence

$$\lambda \ P(A) \geq \int_{A} S_n dP - \int_{\Omega} (S_n - S_1) dP$$

and we have the following result.

Theorem 2.7.1 (Wood (1984)). *Suppose that the sequence $\{S_n, n \geq 1\}$ is a demisubmartingale. Let*

$$A = [\min_{1 \leq k \leq n} S_k < \lambda]$$

for any $\lambda > 0$. Then

$$\lambda \ P(A) \geq \int_{A} S_n dP - \int_{\Omega} (S_n - S_1) dP. \qquad (2.7.3)$$

In particular, if the sequence $\{S_n, n \geq 1\}$ is a demimartingale, then it is easy to check that $E(S_n) = E(S_1)$ for all $n \geq 1$ and hence we have the following result as a corollary to Theorem 2.7.1.

Theorem 2.7.2. *Suppose that the sequence $\{S_n, n \geq 1\}$ is a demimartingale. Let $A = [\min_{1 \leq k \leq n} S_k < \lambda]$ for any $\lambda > 0$. Then*

$$\lambda \ P(A) \geq \int_{A} S_n dP. \qquad (2.7.4)$$

We now prove some inequalities for $E(S_n^{\max})$ and $E(S_n^{\min})$ for nonnegative demisubmartingales $\{S_n, n \geq 1\}$. The following results are from Prakasa Rao (2007).

Theorem 2.7.3. *Suppose that the sequence $\{S_n, n \geq 1\}$ is a positive demimartingale with $S_1 = 1$. Let $\gamma(x) = x - 1 - \log x$ for $x > 0$. Then*

$$\gamma(E(S_n^{\max})) \leq E(S_n \log S_n) \qquad (2.7.5)$$

and

$$\gamma(E(S_n^{\min})) \leq E(S_n \log S_n). \qquad (2.7.6)$$

Proof. Note that the function $\gamma(x)$ is a convex function with minimum $\gamma(1) = 0$. Observe that $S_n^{\max} \geq S_1 = 1$ and hence

$$E(S_n^{\max}) - 1 = \int_0^\infty P[S_n^{\max} \geq \lambda]d\lambda - 1$$

$$= \int_0^1 P[S_n^{\max} \geq \lambda]d\lambda + \int_1^\infty P[S_n^{\max} \geq \lambda]d\lambda - 1$$

$$= \int_1^\infty P[S_n^{\max} \geq \lambda]d\lambda \quad (\text{since } S_1 = 1)$$

$$\leq \int_1^\infty \{\frac{1}{\lambda}\int_{[S_n^{\max}\geq\lambda]} S_n dP\}d\lambda \quad (\text{by } (2.7.2))$$

$$= E(\int_1^\infty \frac{S_n I_{[S_n^{\max}\geq\lambda]}}{\lambda}d\lambda)$$

$$= E(S_n \int_1^{S_n^{\max}} \frac{1}{\lambda}d\lambda)$$

$$= E(S_n \log(S_n^{\max})).$$

Using the fact that $\gamma(x) \geq 0$ for all $x > 0$, we get that

$$E(S_n^{\max}) - 1 \leq E[S_n(\log(S_n^{\max}) + \gamma(\frac{S_n^{\max}}{S_n E(S_n^{\max})}))]$$

$$= E[S_n(\log(S_n^{\max}) + \frac{S_n^{\max}}{S_n E(S_n^{\max})} - 1 - \log(\frac{S_n^{\max}}{S_n E(S_n^{\max})}))]$$

$$= 1 - E(S_n) + E(S_n \log S_n) + E(S_n) \log E(S_n^{\max}).$$

Rearranging the terms in the above inequality, we get that

$$\gamma(E(S_n^{\max})) = E(S_n^{\max}) - 1 - \log E(S_n^{\max}) \qquad (2.7.7)$$

$$\leq 1 - E(S_n) + E(S_n \log S_n) + E(S_n) \log E(S_n^{\max}) - \log E(S_n^{\max})$$

$$= E(S_n \log S_n) + (E(S_n) - 1)(\log E(S_n^{(\max)}) - 1)$$

$$= E(S_n \log S_n)$$

since $E(S_n) = E(S_1) = 1$ for all $n \geq 1$. This proves the inequality (2.7.5).
Observe that $0 \leq S_n^{\min} \leq S_1 = 1$ which implies that

$$E(S_n^{\min}) = \int_0^1 P[S_n^{\min} \geq \lambda]d\lambda \qquad (2.7.8)$$

$$= 1 - \int_0^1 P[S_n^{\min} < \lambda]d\lambda \qquad (2.7.9)$$

$$\leq 1 - \int_0^1 \{\frac{1}{\lambda}\int_{[S_n^{\min}<\lambda]} S_n dP\}d\lambda \quad (\text{by Theorem } 2.7.2) \qquad (2.7.10)$$

$$= 1 - E(\int_0^1 \frac{S_n I_{[S_n^{\min}<\lambda]}}{\lambda}d\lambda) \qquad (2.7.11)$$

$$= 1 - E(S_n \int_{S_n^{\min}}^1 \frac{1}{\lambda} d\lambda) \qquad (2.7.12)$$

$$= 1 + E(S_n \log(S_n^{\min})). \qquad (2.7.13)$$

Applying arguments similar to those given above to prove the inequality (2.7.5), we get that

$$\gamma(E(S_n^{\min})) \le E(S_n \log S_n) \qquad (2.7.14)$$

which proves the inequality (2.7.6). $\square$

The above inequalities for positive demimartingales are analogues of maximal inequalities for nonnegative martingales proved in Harremoës (2008).

2.8   Maximal ϕ-Inequalities for Nonnegative Demisubmartingales

Let $\mathcal{C}$ denote the class of *Orlicz functions* that is, unbounded, nondecreasing convex functions $\phi : [0, \infty) \to [0, \infty)$ with $\phi(0) = 0$. If the right derivative ϕ' is unbounded, then the function ϕ is called a *Young function* and we denote the subclass of such functions by $\mathcal{C}'$. Since

$$\phi(x) = \int_0^x \phi'(s)ds \le x\phi'(x)$$

by convexity, it follows that

$$p_\phi = \inf_{x>0} \frac{x\phi'(x)}{\phi(x)}$$

and

$$p_\phi^* = \sup_{x>0} \frac{x\phi'(x)}{\phi(x)}$$

are in $[1, \infty]$. The function ϕ is called *moderate* if $p_\phi^* < \infty$, or equivalently, if for some $\lambda > 1$, there exists a finite constant c_λ such that

$$\phi(\lambda x) \le c_\lambda \phi(x), \quad x \ge 0.$$

An example of such a function is $\phi(x) = x^\alpha$ for $\alpha \in [1, \infty)$. Example of a non-moderate Orlicz function is $\phi(x) = \exp(x^\alpha) - 1$ for $\alpha \ge 1$.

Let $\mathcal{C}^*$ denote the set of all differentiable $\phi \in \mathcal{C}$ whose derivative is concave or convex and $\mathcal{C}'$ denote the set of $\phi \in \mathcal{C}$ such that $\phi'(x)/x$ is integrable at 0, and thus, in particular $\phi'(0) = 0$. Let $\mathcal{C}_0^* = \mathcal{C}' \cap \mathcal{C}^*$.

Given $\phi \in \mathcal{C}$ and $a \ge 0$, define

$$\Phi_a(x) = \int_a^x \int_a^s \frac{\phi'(r)}{r} dr ds, \quad x > 0.$$

It can be seen that the function $\Phi_a I_{[a,\infty)} \in \mathcal{C}$ for any $a > 0$ where I_A denotes the indicator function of the set A. If $\phi \in \mathcal{C}'$, the same holds for $\Phi \equiv \Phi_0$. If $\phi \in \mathcal{C}_0^*$, then $\Phi \in \mathcal{C}_0^*$. Furthermore, if ϕ' is concave or convex, the same holds for

$$\Phi'(x) = \int_0^x \frac{\phi'(r)}{r} dr,$$

and hence $\phi \in \mathcal{C}_0^*$ implies that $\Phi \in \mathcal{C}_0^*$. It can be checked that ϕ and Φ are related through the differential equation

$$x\Phi'(x) - \Phi(x) = \phi(x), \quad x \geq 0$$

under the initial conditions $\phi(0) = \phi'(0) = \Phi(0) = \Phi'(0) = 0$. If $\phi(x) = x^p$ for some $p > 1$, then $\Phi(x) = x^p/(p-1)$. For instance, if $\phi(x) = x^2$, then $\Phi(x) = x^2$. If $\phi(x) = x$, then $\Phi(x) \equiv \infty$ but $\Phi_1(x) = x \log x - x + 1$. It is known that if $\phi \in \mathcal{C}'$ with $p_\phi > 1$, then the function ϕ satisfies the inequalities

$$\Phi(x) \leq \frac{1}{p_\phi - 1} \phi(x), \quad x \geq 0.$$

Furthermore, if ϕ is moderate, that is $p_\phi^* < \infty$, then

$$\Phi(x) \geq \frac{1}{p_\phi^* - 1} \phi(x), \quad x \geq 0.$$

The brief introduction for properties of Orlicz functions given here is based on Alsmeyer and Rosler (2006).

We now prove some maximal ϕ-inequalities for nonnegative demisubmartingales following the techniques in Alsmeyer and Rosler (2006).

Theorem 2.8.1. *Let the sequence $\{S_n, n \geq 1\}$ be a nonnegative demisubmartingale and let $\phi \in \mathcal{C}$. Then*

$$P(S_n^{\max} \geq t) \leq \frac{\lambda}{(1-\lambda)t} \int_t^\infty P(S_n > \lambda s) ds \qquad (2.8.1)$$

$$= \frac{\lambda}{(1-\lambda)t} E(\frac{S_n}{\lambda} - t)^+$$

for all $n \geq 1$, $t > 0$ and $0 < \lambda < 1$. Furthermore,

$$E[\phi(S_n^{\max})] \leq \phi(b) + \frac{\lambda}{1-\lambda} \int_{[S_n > \lambda b]} (\Phi_a(\frac{S_n}{\lambda}) - \Phi_a(b) - \Phi_a'(b)(\frac{S_n}{\lambda} - b)) dP \quad (2.8.2)$$

for all $n \geq 1$, $a > 0$, $b > 0$ and $0 < \lambda < 1$. If $\phi'(x)/x$ is integrable at 0, that is, $\phi \in \mathcal{C}'$, then the inequality (2.8.2) holds for $b = 0$.

Proof. Let $t > 0$ and $0 < \lambda < 1$. The inequality (2.7.1) implies that

$$P(S_n^{\max} \geq t) \leq \frac{1}{t} \int_{[S_n^{\max} \geq t]} S_n dP \tag{2.8.3}$$

$$= \frac{1}{t} \int_0^\infty P[S_n^{\max} \geq t, S_n > s] ds$$

$$\leq \frac{1}{t} \int_0^{\lambda t} P[S_n^{\max} \geq t] ds + \frac{1}{t} \int_{\lambda t}^\infty P[S_n > s] ds$$

$$\leq \lambda P[S_n^{\max} \geq t] ds + \frac{\lambda}{t} \int_t^\infty P[S_n > \lambda s] ds.$$

Rearranging the last inequality, we get that

$$P(S_n^{\max} \geq t) \leq \frac{\lambda}{(1-\lambda)t} \int_t^\infty P(S_n > \lambda s) ds = \frac{\lambda}{(1-\lambda)t} E(\frac{S_n}{\lambda} - t)^+$$

for all $n \geq 1$, $t > 0$ and $0 < \lambda < 1$ proving the inequality (2.8.1). Let $b > 0$. Then

$$E[\phi(S_n^{\max})] = \int_0^\infty \phi'(t) P(S_n^{\max} > t) dt$$

$$= \int_0^b \phi'(t) P(S_n^{\max} > t) dt + \int_b^\infty \phi'(t) P(S_n^{\max} > t) dt$$

$$\leq \phi(b) + \int_b^\infty \phi'(t) P(S_n^{\max} > t) dt$$

$$\leq \phi(b) + \frac{\lambda}{1-\lambda} \int_b^\infty \frac{\phi'(t)}{t} [\int_t^\infty P(S_n > \lambda s) ds] dt \quad \text{(by (2.7.1))}$$

$$= \phi(b) + \frac{\lambda}{1-\lambda} \int_b^\infty (\int_b^s \frac{\phi'(t)}{t} dt) P(S_n > \lambda s) ds$$

$$= \phi(b) + \frac{\lambda}{1-\lambda} \int_b^\infty (\Phi_a'(s) - \Phi_a'(b)) P(S_n > \lambda s) ds$$

$$= \phi(b) + \frac{\lambda}{1-\lambda} \int_{[S_n > \lambda b]} (\Phi_a(\frac{S_n}{\lambda}) - \Phi_a(b) - \Phi_a'(b)(\frac{S_n}{\lambda} - b)) dP$$

for all $n \geq 1$, $b > 0$, $t > 0$, $0 < \lambda < 1$ and $a > 0$. The value of a can be chosen to be 0 if $\phi'(x)/x$ is integrable at 0.   □

As special cases of the above result, we obtain the following inequalities by choosing $b = a$ in (2.8.2). Observe that $\Phi_a(a) = \Phi_a'(a) = 0$.

Theorem 2.8.2. *Let the sequence $\{S_n, n \geq 1\}$ be a nonnegative demisubmartingale and let $\phi \in C$. Then*

$$E[\phi(S_n^{\max})] \leq \phi(a) + \frac{\lambda}{1-\lambda} E[\Phi_a(\frac{S_n}{\lambda})] \tag{2.8.4}$$

for all $a \geq 0$, $0 < \lambda < 1$ and $n \geq 1$. Let $\lambda = \frac{1}{2}$ in (2.8.4). Then

$$E[\phi(S_n^{\max})] \leq \phi(a) + E[\Phi_a(2S_n)] \tag{2.8.5}$$

for all $a \geq 0$ and $n \geq 1$.

The following lemma is due to Alsmeyer and Rosler (2006).

Lemma 2.8.3. *Let X and Y be nonnegative random variables satisfying the inequality*

$$t \ P(Y \geq t) \leq E(X I_{[Y \geq t]})$$

for all $t \geq 0$. Then

$$E[\phi(Y)] \leq E[\phi(q_\phi X)] \tag{2.8.6}$$

for any Orlicz function ϕ where $q_\phi = \frac{p_\phi}{p_\phi - 1}$ and $p_\phi = \inf_{x>0} \frac{x\phi'(x)}{\phi(x)}$.

This lemma follows as an application of the Choquet decomposition

$$\phi(x) = \int_{[0,\infty)} (x - t)^+ \phi'(dt), x \geq 0.$$

In view of the inequality (2.8.2), we can apply the above lemma to the random variables $X = S_n$ and $Y = S_n^{\max}$ to obtain the following result.

Theorem 2.8.4. *Let the sequence $\{S_n, n \geq 1\}$ be a nonnegative demisubmartingale and let $\phi \in C$ with $p_\phi > 1$. Then*

$$E[\phi(S_n^{\max})] \leq E[\phi(q_\phi S_n)] \tag{2.8.7}$$

for all $n \geq 1$.

Theorem 2.8.5. *Let the sequence $\{S_n, n \geq 1\}$ be a nonnegative demisubmartingale. Suppose that the function $\phi \in C$ is moderate. Then*

$$E[\phi(S_n^{\max})] \leq E[\phi(q_\phi S_n)] \leq q_\Phi^{p_\phi^*} E[\phi(S_n)]. \tag{2.8.8}$$

The first part of the inequality (2.8.8) of Theorem 2.8.5 follows from Theorem 2.8.4. The last part of the inequality follows from the observation that if $\phi \in C$ is moderate, that is,

$$p_\phi^* = \sup_{x>0} \frac{x\phi'(x)}{\phi(x)} < \infty,$$

then

$$\phi(\lambda x) \leq \lambda^{p_\phi^*} \phi(x)$$

for all $\lambda > 1$ and $x > 0$ (see equation (1.10) of Alsmeyer and Rosler (2006)).

Theorem 2.8.6. *Let the sequence* $\{S_n, n \geq 1\}$ *be a nonnegative demisubmartingale. Suppose* ϕ *is a nonnegative nondecreasing function on* $[0, \infty)$ *such that* $\phi^{1/\gamma}$ *is also nondecreasing and convex for some* $\gamma > 1$. *Then*

$$E[\phi(S_n^{\max})] \leq (\frac{\gamma}{\gamma - 1})^\gamma E[\phi(S_n)]. \qquad (2.8.9)$$

Proof. The inequality

$$\lambda P(S_n^{\max} \geq \lambda) \leq \int_{[S_n^{\max} \geq \lambda]} S_n dP$$

given in (2.7.1) implies that

$$E[(S_n^{\max})^p] \leq (\frac{p}{p - 1})^p E(S_n^p), \quad p > 1 \qquad (2.8.10)$$

by an application of Hölder's inequality (cf. Chow and Teicher (1997), p. 255). Note that the sequence $\{[\phi(S_n)]^{1/\gamma}, n \geq 1\}$ is a nonnegative demisubmartingale by Theorem 2.1.1. Applying the inequality (2.8.10) for the sequence $\{[\phi(S_n)]^{1/\gamma}, n \geq 1\}$ and choosing $p = \gamma$ in that inequality, we get that

$$E[\phi(S_n^{\max})] \leq (\frac{\gamma}{\gamma - 1})^\gamma E[\phi(S_n)]. \qquad (2.8.11)$$

for all $\gamma > 1$.   □

Examples of functions ϕ satisfying the conditions stated in Theorem 2.8.6 are $\phi(x) = x^p[\log(1 + x)]^r$ for $p > 1$ and $r \geq 0$ and $\phi(x) = e^{rx}$ for $r > 0$. Applying the result in Theorem 2.8.6 for the function $\phi(x) = e^{rx}, r > 0$, we obtain the following inequality.

Theorem 2.8.7. *Let the sequence* $\{S_n, n \geq 1\}$ *be a nonnegative demisubmartingale. Then*

$$E[e^{rS_n^{\max}}] \leq eE[e^{rS_n}], \quad r > 0. \qquad (2.8.12)$$

Proof. Applying the result stated in Theorem 2.8.6 to the function $\phi(x) = e^{rx}$, we get that

$$E[e^{rS_n^{\max}}] \leq (\frac{\gamma}{\gamma - 1})^\gamma E[e^{rS_n}] \qquad (2.8.13)$$

for any $\gamma > 1$. Let $\gamma \to \infty$. Then

$$(\frac{\gamma}{\gamma - 1})^\gamma \downarrow e$$

and we get that

$$E[e^{rS_n^{\max}}] \leq eE[e^{rS_n}], r > 0. \qquad (2.8.14)$$
                                                                                                              □

The next result deals with maximal inequalities for functions $\phi \in \mathcal{C}$ which are k times differentiable with the k-th derivative $\phi^{(k)} \in \mathcal{C}$ for some $k \geq 1$.

Theorem 2.8.8. *Let the sequence $\{S_n, n \geq 1\}$ be a nonnegative demisubmartingale. Let $\phi \in \mathcal{C}$ which is differentiable k times with the k-th derivative $\phi^{(k)} \in \mathcal{C}$ for some $k \geq 1$. Then*

$$E[\phi(S_n^{\max})] \leq (\frac{k+1}{k})^{k+1} E[\phi(S_n)]. \tag{2.8.15}$$

Proof. The proof follows the arguments given in Alsmeyer and Rosler (2006) following the inequality (2.8.9). We present the proof here for completeness. Note that

$$\phi(x) = \int_{[0,\infty)} (x-t)^+ Q_\phi(dt)$$

where

$$Q_\phi(dt) = \phi'(0)\delta_0 + \phi'(dt)$$

and δ_0 is the Kronecker delta function. Hence, if $\phi' \in \mathcal{C}$, then

$$\begin{aligned}
\phi(x) &= \int_0^x \phi'(y)dy \tag{2.8.16} \\
&= \int_0^x \int_{[0,\infty)} (y-t)^+ Q_{\phi'}(dt)dy \\
&= \int_{[0,\infty)} \int_0^x (y-t)^+ dy Q_{\phi'}(dt) \\
&= \int_{[0,\infty)} \frac{((x-t)^+)^2}{2} Q_{\phi'}(dt).
\end{aligned}$$

An inductive argument shows that

$$\phi(x) = \int_{[0,\infty)} \frac{((x-t)^+)^{k+1}}{(k+1)!} Q_{\phi^{(k)}}(dt) \tag{2.8.17}$$

for any $\phi \in \mathcal{C}$ such that $\phi^{(k)} \in \mathcal{C}$. Let

$$\phi_{k,t}(x) = \frac{((x-t)^+)^{k+1}}{(k+1)!}$$

for any $k \geq 1$ and $t \geq 0$. Note that the function $[\phi_{k,t}(x)]^{1/(k+1)}$ is nonnegative, convex and nondecreasing in x for any $k \geq 1$ and $t \geq 0$. Hence the process $\{[\phi_{k,t}(S_n)]^{1/(k+1)}, n \geq 1\}$ is a nonnegative demisubmartingale by Theorem 2.1.1. Following the arguments given to prove (2.8.10), we obtain that

$$E(([\phi_{k,t}(S_n^{\max})]^{1/(k+1)})^{k+1}) \leq (\frac{k+1}{k})^{k+1} E(([\phi_{k,t}(S_n)]^{1/(k+1)})^{k+1})$$

which implies that

$$E[\phi_{k,t}(S_n^{\max})] \leq (\frac{k+1}{k})^{k+1} E[\phi_{k,t}(S_n)]. \quad (2.8.18)$$

Hence

$$E[\phi(S_n^{\max}))] = \int_{[0,\infty)} E[\phi_{k,t}(S_n^{\max})]Q_{\phi(k)}(dt) \quad (\text{by } (2.8.17)) \quad (2.8.19)$$

$$\leq (\frac{k+1}{k})^{k+1} \int_{[0,\infty)} E[\phi_{k,t}(S_n)]Q_{\phi(k)}(dt) \quad (\text{by } (2.8.18))$$

$$= (\frac{k+1}{k})^{k+1} E[\phi(S_n)]$$

which proves the theorem. $\square$

We now consider a special case of the maximal inequality derived in (2.8.2) of Theorem 2.8.1. Let $\phi(x) = x$. Then $\Phi_1(x) = x \log x - x + 1$ and $\Phi_1'(x) = \log x$. The inequality (2.8.2) reduces to

$$E[S_n^{\max}] \leq b + \frac{\lambda}{1-\lambda} \int_{[S_n > \lambda b]} (\frac{S_n}{\lambda} \log \frac{S_n}{\lambda} - \frac{S_n}{\lambda} + b - (\log b)\frac{S_n}{\lambda})dP$$

$$= b + \frac{\lambda}{1-\lambda} \int_{[S_n > \lambda b]} (S_n \log S_n - S_n(\log \lambda + \log b + 1) + \lambda b)dP$$

for all $b > 0$ and $0 < \lambda < 1$. Let $b > 1$ and $\lambda = \frac{1}{b}$. Then we obtain the inequality

$$E(S_n^{\max}) \leq b + \frac{b}{b-1} E(\int_1^{\max(S_n,1)} \log x \ dx), \quad b > 1, \ n \geq 1. \quad (2.8.20)$$

The value of b which minimizes the term on the right-hand side of equation (2.8.20) is

$$b^* = 1 + (E(\int_1^{\max(S_n,1)} \log x \ dx))^{1/2}$$

and hence

$$E(S_n^{\max}) \leq (1 + (E(\int_1^{\max(S_n,1)} \log x \ dx))^{1/2})^2. \quad (2.8.21)$$

Since

$$\int_1^x \log y \ dy = x \log^+ x - (x - 1), \quad x \geq 1,$$

the inequality (2.8.20) can be written in the form

$$E(S_n^{\max}) \leq b + \frac{b}{b-1}(E(S_n \log^+ S_n) - E(S_n - 1)^+), \quad b > 1, \ n \geq 1. \quad (2.8.22)$$

Let $b = E(S_n - 1)^+$ in equation (2.8.22). Then we get the maximal inequality

$$E(S_n^{\max}) \le \frac{1 + E(S_n - 1)^+}{E(S_n - 1)^+} E(S_n \log^+ S_n). \tag{2.8.23}$$

If we choose $b = e$ in equation (2.8.22), then we get the maximal inequality

$$E(S_n^{\max}) \le e + \frac{e}{e-1}(E(S_n \log^+ S_n) - E(S_n - 1)^+), \quad b > 1, \ n \ge 1. \tag{2.8.24}$$

Results discussed in this section are due to Prakasa Rao (2007).

2.9 Maximal Inequalities for Functions of Demisubmartingales

We now derive some maximal inequalities due to Wang and Hu (2009) and Wang et al. (2010) for functions of demimartingales and demisubmartingales.

Theorem 2.9.1. *Let the sequence $\{S_n, n \ge 1\}$ be a demisubmartingale with $S_0 = 0$ and $g(.)$ be a nondecreasing convex function on R with $g(0) = 0$. Suppose that $E|g(S_i)| < \infty$, $i \ge 1$. Let the sequence $\{c_n, n \ge 1\}$ be a nonincreasing sequence of positive numbers. Then, for any $\epsilon > 0$,*

$$\epsilon P[\max_{1 \le k \le n} c_k g(S_k) \ge \epsilon] \le \sum_{k=1}^{n} c_k E[g^+(S_k) - g^+(S_{k-1})]. \tag{2.9.1}$$

Proof. This result follows from the fact that $\{g(S_n), n \ge 1\}$ is a demisubmartingale and applying Theorem 5.1 on Chow type maximal inequality for demisubmartingales (cf. Christofides (2000)). $\qquad \square$

The following theorem is due to Wang (2004). We omit the proof.

Theorem 2.9.2. *Let the sequence $\{S_n, n \ge 1\}$ be a demimartingale, $S_0 = 0$, and $g(.)$ be a nonnegative convex function on R with $g(0) = 0$. Suppose that $E|g(S_i)| < \infty$, $i \ge 1$. Let $\{c_n, n \ge 1\}$ be a nonincreasing sequence of positive numbers. Then, for any $\epsilon > 0$,*

$$\epsilon P[\max_{1 \le k \le n} c_k g(S_k) \ge \epsilon] \le \sum_{k=1}^{n} c_k E[(g(S_k) - g(S_{k-1}))I_{[\max_{1 \le j \le n} c_j g(S_j) \ge \epsilon]}]$$

$$\le \sum_{k=1}^{n} c_k E[g(S_k) - g(S_{k-1})]. \tag{2.9.2}$$

As an application of the above theorem, we obtain the following result by choosing the function $g(x) = |x|$. For any $\epsilon > 0$,

$$\epsilon P[\max_{1 \le k \le n} c_k |S_k| \ge \epsilon] \le \sum_{k=1}^{n} c_k E[|S_k| - |S_{k-1}|]. \tag{2.9.3}$$

The following inequality gives a Doob type maximal inequality for functions of demisubmartingales.

Theorem 2.9.3. *Let the sequence $\{S_n, \ n \geq 1\}$ be a demisubmartingale, $S_0 = 0$, and $g(.)$ be a nondecreasing convex function on R with $g(0) = 0$. Suppose that $E|g(S_i)| < \infty$, $i \geq 1$. Then, for any $\epsilon > 0$,*

$$\epsilon \, P[\max_{1 \leq k \leq n} g(S_k) \geq \epsilon] \leq \int_{[\max_{1 \leq k \leq n} g(S_k) \geq \epsilon]} g(S_n)] \, dP. \qquad (2.9.4)$$

Proof. This result follows from the fact that the sequence $\{g(S_n), \ n \geq 1\}$ is a demisubmartingale and applying Theorem 2.1 on Doob type maximal inequality for demisubmartingales.   □

The following inequality gives a Doob type maximal inequality for nonnegative convex functions of demimartingales.

Theorem 2.9.4. *Let the sequence $\{S_n, \ n \geq 1\}$ be a demimartingale and $g(.)$ be a nonnegative convex function on R with $g(0) = 0$. Suppose that the random variables $g(S_i)$, $i \geq 1$ are integrable. Then, for any $\epsilon > 0$,*

$$\epsilon \, P[\max_{1 \leq k \leq n} g(S_k) \geq \epsilon] \leq \int_{[\max_{1 \leq k \leq n} g(S_k) \geq \epsilon]} g(S_n) \, dP. \qquad (2.9.5)$$

For a proof, see Wang and Hu (2009). In particular, by choosing $g(x) = |x|^r$, $r \geq 1$, we obtain the following result.

Theorem 2.9.5. *Let the sequence $\{S_n, \ n \geq 1\}$ be a demimartingale, $S_0 = 0$, and suppose that $E|S_n|^r < \infty$, $n \geq 1$ for some $r \geq 1$. Then, for any $\epsilon > 0$,*

$$P[\max_{1 \leq k \leq n} |S_k| \geq \epsilon] \leq \frac{1}{\epsilon^r} \int_{[\max_{1 \leq k \leq n} |S_k|^r \geq \epsilon^r]} |S_n|^r \, dP \leq \frac{1}{\epsilon^r} E|S_n|^r. \qquad (2.9.6)$$

Wang et al. (2010) derived some inequalities for expectations of maxima of functions of demisubmartingales. We now discuss some of these results.

Theorem 2.9.6. *Let the sequence $\{S_n, \ n \geq 1\}$ be a demimartingale and $g(.)$ be a nonnegative convex function on R with $g(0) = 0$. Suppose that $E|g(S_k)|^p < \infty$, $k \geq 1$, for some $p > 1$. Let $\{c_n, \ n \geq 1\}$ be a nonincreasing sequence of positive numbers. Then*

$$E[\max_{1 \leq k \leq n} c_k g(S_k)]^p \leq (\frac{p}{p-1})^p E[\sum_{k=1}^{n} c_k(g(S_k) - g(S_{k-1}))]^p. \qquad (2.9.7)$$

If $p = 1$, then

$$E[\max_{1 \leq k \leq n} c_k g(S_k)] \qquad (2.9.8)$$

$$\leq \frac{e}{e-1}(1 + E[(\sum_{k=1}^{n} c_k(g(S_k) - g(S_{k-1})) \log^+ (\sum_{k=1}^{n} c_k(g(S_k) - g(S_{k-1}))))]).$$

Proof. Let $p > 1$. Then, by Theorem 2.9.2 due to Wang (2004) and Hölder's inequality, we have

$$E[\max_{1 \leq k \leq n} c_k g(S_k)]^p \tag{2.9.9}$$

$$= p \int_0^\infty x^{p-1} P[\max_{1 \leq k \leq n} c_k g(S_k) \geq x] \, dx$$

$$\leq p \int_0^\infty x^{p-2} E[\sum_{j=1}^n c_j(g(S_j) - g(S_{j-1})) I_{[\max_{1 \leq k \leq n} c_k g(S_k) \geq x]}] \, dx$$

$$= \frac{p}{p-1} E([\sum_{j=1}^n c_j(g(S_j) - g(S_{j-1}))(\max_{1 \leq k \leq n} c_k g(S_k))^{p-1})$$

$$\leq \frac{p}{p-1} (E([\sum_{j=1}^n c_j(g(S_j) - g(S_{j-1}))]^p)^{1/p} (E(\max_{1 \leq k \leq n} c_k g(S_k))^p)^{1/q}$$

where q is such that $1/p + 1/q = 1$. Rearranging the last inequality, we get that

$$(E[\max_{1 \leq k \leq n} c_k g(S_k)]^p)^{1/p} \leq \frac{p}{p-1} (E[\sum_{j=1}^n c_j(g(S_j) - g(S_{j-1}))]^p)^{1/p} \tag{2.9.10}$$

which implies that

$$E[\max_{1 \leq k \leq n} c_k g(S_k)]^p \leq (\frac{p}{p-1})^p E[\sum_{j=1}^n c_j(g(S_j) - g(S_{j-1}))]^p. \tag{2.9.11}$$

Let us now consider the case $p = 1$. Then

$$E[\max_{1 \leq k \leq n} c_k g(S_k)] \tag{2.9.12}$$

$$\leq 1 + \int_1^\infty P[\max_{1 \leq k \leq n} c_k g(S_k) \geq x] \, dx$$

$$\leq 1 + \int_1^\infty x^{-1} E[\sum_{j=1}^n c_j(g(S_j) - g(S_{j-1})) I_{[\max_{1 \leq k \leq n} c_k g(S_k) \geq x]}] \, dx$$

$$\leq 1 + E[(\sum_{j=1}^n c_j(g(S_j) - g(S_{j-1}))) \log^+(\max_{1 \leq k \leq n} c_k g(S_k))].$$

Note that, for any $a \geq 0$ and $b > 0$,

$$a \log^+ b \leq a \log^+ a + b e^{-1}.$$

Applying this inequality, we get that

$$E[\max_{1\leq k\leq n} c_k g(S_k)] \tag{2.9.13}$$

$$\leq 1 + E[(\sum_{j=1}^{n} c_j(g(S_j) - g(S_{j-1}))) \log^{+}(\sum_{j=1}^{n} c_j(g(S_j) - g(S_{j-1}))]$$

$$+ e^{-1} E[\max_{1\leq k\leq n} c_k g(S_k)].$$

Hence

$$E[\max_{1\leq k\leq n} c_k g(S_k)] \tag{2.9.14}$$

$$\leq \frac{e}{e-1}(1 + E[(\sum_{j=1}^{n} c_j(g(S_j) - g(S_{j-1})) \log^{+}(\sum_{j=1}^{n} c_j(g(S_j) - g(S_{j-1})))]). \quad \square$$

As a corollary to the above result, we get the following bound on the expectations for maximum of a nonnegative convex function of a demimartingale. Choose $c_k = 1$, $k \geq 1$.

Theorem 2.9.7. *Let the sequence $\{S_n,\ n \geq 1\}$ be a demimartingale and $g(.)$ be a nonnegative convex function on R with $g(0) = 0$. Suppose that $E|g(S_k)|^p < \infty$, $k \geq 1$, for some $p \geq 1$. If $p > 1$, then*

$$E[\max_{1\leq k\leq n} g(S_k)]^p \leq (\frac{p}{p-1})^p E[g(S_n)]^p \tag{2.9.15}$$

and

$$E[\max_{1\leq k\leq n} g(S_k)] \leq \frac{e}{e-1}[1 + E(g(S_n) \log^{+} g(S_n))]. \tag{2.9.16}$$

As an additional corollary, we obtain the following Doob type maximal inequality for demimartingales.

Theorem 2.9.8. *Let the sequence $\{S_n,\ n \geq 1\}$ be a demimartingale and $p \geq 1$. Suppose that $E|S_k|^p < \infty$, $k \geq 1$. Then, for every $n \geq 1$,*

$$E[\max_{1\leq k\leq n} |S_k|]^p \leq (\frac{p}{p-1})^p E[|S_n|^p], \quad p > 1 \tag{2.9.17}$$

and

$$E[\max_{1\leq k\leq n} |S_k|] \leq \frac{e}{e-1}[1 + E(|S_n| \log^{+} |S_n|)]. \tag{2.9.18}$$

Remarks. As corollaries to the above results, one can obtain a strong law of large numbers for functions of demimartingales and partial sums of mean zero associated random variables using conditions developed in Fazekas and Klesov (2001) and Hu and Hu (2006). For related results, see Hu et al. (2008). We will not discuss these results.

2.10 Central Limit Theorems

As was mentioned in Chapter 1, Newman (1980,1984) obtained the central limit theorem for partial sums of mean zero stationary associated random variables which form a demimartingale This result also holds for weakly associated random sequences as defined below. This was pointed out by Sethuraman (2000). However it is not known whether the central limit theorem holds for any demimartingale in general.

Let $\{\mathbf{v}(t) = (v_1(t), \ldots, v_m(t)), t \geq 0\}$ be an m-dimensional L^2-process with stationary increments. This process is said to have *weakly positive associated increments* if

$$E[\phi(\mathbf{v}(t+s) - \mathbf{v}(s))\psi(\mathbf{v}(s_1), \ldots, \mathbf{v}(s_n))] \geq E[\phi(\mathbf{v}(t))]E[\psi(\mathbf{v}(s_1), \ldots, \mathbf{v}(s_n)]$$

for all componentwise nondecreasing functions ϕ and ψ, for all $s, t \geq 0$ and for $0 \leq s_1 < \ldots, s_n = s, n \geq 1$.

Sethuraman (2000, 2006) proved the following invariance principle for processes which have weakly positive associated increments. This theorem is a consequence of the central limit theorem due to Newman (1980) and the maximal inequalities for demimartingales discussed earlier in this chapter (cf. Newman and Wright (1982)). Let $\mathbf{d}$ denote a column vector in R^m and $\mathbf{d}'$ the corresponding row vector.

Theorem 2.10.1. *Let $\{\mathbf{v}'(t) = (v_1(t), \ldots, v_m(t)), t \geq 0\}$ be an m-dimensional L^2-process in $C[0, \infty)$ with stationary and weakly positive associated increments such that $E[v_i(t)] = 0$ for $1 \leq i \leq m$ and $t \geq 0$. Further suppose that*

$$\lim_{t \to \infty} t^{-1} E[v_i(t)v_j(t)] = \sigma_{ij} < \infty. \tag{2.10.1}$$

Then

$$\alpha^{-1/2}\mathbf{v}'(\alpha t)\mathbf{d} \to W(\mathbf{d}'\Sigma \mathbf{d}t) \quad as \ \alpha \to \infty \tag{2.10.2}$$

weakly in the uniform topology where $\Sigma = ((\sigma_{ij})_{m \times m}$ is the covariance matrix, $\mathbf{d} \in R^m$ and W is the standard Brownian motion.

Newman (1984) conjectured the following result: Let $S_0 \equiv 0$ and the sequence $\{S_n, n \geq 1\}$ be an L^2- demimartingale whose difference sequence $\{X_n = S_n - S_{n-1}, n \geq 1\}$ is strictly stationary and ergodic with

$$0 < \sigma^2 = Var(X_1) + 2\sum_{j=2}^{\infty} \text{Cov}(X_1, X_j) < \infty.$$

Then

$$n^{-1/2}S_n \overset{\mathcal{L}}{\to} \sigma Z \ as \ n \to \infty$$

where Z is a standard normal random variable. It is not known whether the above conjecture is true. The problem remains open. We will come back to the discussion on Newman's conjecture in Chapter 6.

2.11   Dominated Demisubmartingales

Let $M_0 = N_0 = 0$ and the sequence $\{M_n, n \geq 0\}$ be a sequence of random variables defined on a probability space $(\Omega, \mathcal{F}, P)$. Suppose that

$$E[(M_{n+1} - M_n)f(M_0, \ldots, M_n)|\zeta_n] \geq 0$$

for any nonnegative componentwise nondecreasing function f given a filtration $\{\zeta_n, n \geq 0\}$ contained in $\mathcal{F}$. Then the sequence $\{M_n, n \geq 0\}$ is said to be a *strong demisubmartingale* with respect to the filtration $\{\zeta_n, n \geq 0\}$. It is obvious that a strong demisubmartingale is a demisubmartingale in the sense discussed earlier.

Definition. Let $M_0 = 0 = N_0$. Suppose $\{M_n, n \geq 0\}$ is a strong demisubmartingale with respect to the filtration generated by a demisubmartingale $\{N_n, n \geq 0\}$. The strong demisubmartingale $\{M_n, n \geq 0\}$ is said to be *weakly dominated* by the demisubmartingale $\{N_n, n \geq 0\}$ if for every nondecreasing convex function $\phi : R_+ \to R$, and for any nonnegative componentwise nondecreasing function $f : R^{2n} \to R$,

$$E[(\phi(|e_n|) - \phi(|d_n|))f(M_0, \ldots, M_{n-1}; N_0, \ldots, N_{n-1})|N_0, \ldots, N_{n-1}] \geq 0 \text{ a.s.,}$$
$$(2.11.1)$$

for all $n \geq 1$ where $d_n = M_n - M_{n-1}$ and $e_n = N_n - N_{n-1}$. We write $M \ll N$ in such a case.

In analogy with the inequalities for dominated martingales developed in Osekowski (2007), we will now prove an inequality for domination between a strong demisubmartingale and a demisubmartingale.

Following Osekowski (2007), define the functions $u_{<2}(x, y)$ and $u_{>2}(x, y)$ as given below. Let

$$u_{<2}(x, y) = \begin{cases} 9|y|^2 - 9|x|^2 & \text{if } (x, y) \in D, \\ 2|y| - 1 + 8|y|^2 I_{[|y| \leq 1]} + (16|y| - 8)I_{[|y| > 1]} & \text{if } (x, y) \in D^c \end{cases} \quad (2.11.2)$$

and

$$u_{>2}(x, y) = \begin{cases} 0 & \text{if } (x, y) \in E, \\ 9|y|^2 - (|x| - 1)^2 - 8(|x| - 1)^2 I_{[|x| \geq 1]} & \text{if } (x, y) \in E^c, \end{cases} \quad (2.11.3)$$

where

$$D = \{(x, y) \in R^2 : |y| + 3|x| \leq 1\} \text{ and } E = \{(x, y) \in R^2 : 3|y| + |x| \leq 1\}.$$

We now state a weak-type inequality between dominated demisubmartingales.

Theorem 2.11.1. *Suppose the sequence $\{M_n, n \geq 0\}$ is a strong demisubmartingale with respect to the filtration generated by the sequence $\{N_n, n \geq 0\}$ which is a demisubmartingale . Further suppose that $M \ll N$. Then, for any $\lambda > 0$,*

$$\lambda P(|M_n| \geq \lambda) \leq 6E|N_n|, \quad n \geq 0. \tag{2.11.4}$$

We will first state and sketch the proof of a lemma which will be used to prove Theorem 2.11.1. The method of proof is the same as that in Osekowski (2007).

Lemma 2.11.2. *Suppose the sequence $\{M_n, n \geq 0\}$ is a strong demisubmartingale with respect to the filtration generated by the sequence $\{N_n, n \geq 0\}$ which is a demisubmartingale. Further suppose that $M \ll N$. Then*

$$E[u_{<2}(M_n, N_n)f(M_0, \ldots, M_{n-1}; N_0, \ldots, N_{n-1})] \tag{2.11.5}$$
$$\geq E[u_{<2}(M_{n-1}, N_{n-1})f(M_0, \ldots, M_{n-1}; N_0, \ldots, N_{n-1})]$$

and

$$E[u_{>2}(M_n, N_n)f(M_0, \ldots, M_{n-1}; N_0, \ldots, N_{n-1})] \tag{2.11.6}$$
$$\geq E[u_{>2}(M_{n-1}, N_{n-1})f(M_0, \ldots, M_{n-1}; N_0, \ldots, N_{n-1})]$$

for any nonnegative componentwise nondecreasing function $f : R^{2n} \to R$, $n \geq 1$.

Proof. Define $u(x, y)$ where $u = u_{<2}$ or $u = u_{>2}$. From the arguments given in Osekowski (2007), it follows that there exist a nonnegative function $A(x, y)$ nondecreasing in x and a nonnegative function $B(x, y)$ nondecreasing in y and a convex nondecreasing function $\phi_{x,y}(.) : R_+ \to R$, such that, for any h and k,

$$u(x, y) + A(x, y)h + B(x, y)k + \phi_{x,y}(|k|) - \phi_{x,y}(|h|) \leq u(x + h, y + k). \tag{2.11.7}$$

Let $x = M_{n-1}$, $y = N_{n-1}$, $h = d_n$ and $k = e_n$. Then, it follows that

$$u(M_{n-1}, N_{n-1}) + A(M_{n-1}, N_{n-1})d_n + B(M_{n-1}, N_{n-1})e_n \tag{2.11.8}$$
$$+ \phi_{M_{n-1}, N_{n-1}}(|e_n|) - \phi_{M_{n-1}, N_{n-1}}(|d_n|)$$
$$\leq u(M_{n-1} + d_n, N_{n-1} + e_n) = u(M_n, N_n).$$

Note that,

$$E[A(M_{n-1}, N_{n-1})d_n f(M_0, \ldots, M_{n-1}; N_0, \ldots, N_{n-1})|N_0, \ldots, N_{n-1}] \geq 0 \text{ a.s.}$$

from the fact that $\{M_n, n \geq 0\}$ is a strong demisubmartingale with respect to the filtration generated by the process $\{N_n, n \geq 0\}$ and that the function

$$A(x_{n-1}, y_{n-1})f(x_0, \ldots, x_{n-1}; y_0, \ldots, y_{n-1})$$

is a nonnegative componentwise nondecreasing function in $x_0, \ldots, x_{n-1}$ for any fixed $y_0, \ldots, y_{n-1}$. Taking expectation on both sides of the above inequality, we get that

$$E[A(M_{n-1}, N_{n-1})d_n f(M_0, \ldots, M_{n-1}; N_0, \ldots, N_{n-1})] \geq 0. \tag{2.11.9}$$

Similarly we get that

$$E[B(M_{n-1}, N_{n-1})d_n f(M_0, \ldots, M_{n-1}; N_0, \ldots, N_{n-1})] \geq 0. \qquad (2.11.10)$$

Since the sequence $\{M_n, n \geq 0\}$ is dominated by the sequence $\{N_n, n \geq 0\}$, it follows that

$$E[(\phi_{M_{n-1}, N_{n-1}}(|e_n|) - \phi_{M_{n-1}, N_{n-1}}(|d_n|))f(M_0, \ldots, M_{n-1}; N_0, \ldots, N_{n-1})] \geq 0 \qquad (2.11.11)$$

by taking the expectations on both sides of (2.11.1). Combining the relations (2.11.7) to (2.11.11), we get that

$$E[u(M_n, N_n)f(M_0, \ldots, M_{n-1}; N_0, \ldots, N_{n-1})] \qquad (2.11.12)$$
$$\geq E[u(M_{n-1}, N_{n-1})f(M_0, \ldots, M_{n-1}; N_0, \ldots, N_{n-1})]. \qquad \square$$

Remarks. Let $f \equiv 1$. Repeated application of the inequality obtained in Lemma 2.11.2 shows that

$$E[u(M_n, N_n)] \geq E[u(M_0, N_0)] = 0. \qquad (2.11.13)$$

Proof of Theorem 2.11.1. Let

$$v(x, y) = 18 \ |y| - I[|x| \geq \frac{1}{3}].$$

It can be checked that (cf. Osekowski (2007))

$$v(x, y) \geq u_{<2}(x, y). \qquad (2.11.14)$$

Let $\lambda > 0$. It is easy to see that the strong demisubmartingale $\{\frac{M_n}{3\lambda}, n \geq 0\}$ is weakly dominated by the demisubmartingale $\{\frac{N_n}{3\lambda}, n \geq 0\}$. In view of the inequalities (2.11.7) and (2.11.8), we get that

$$6 \ E|N_n| - \lambda \ P(|M_n| \geq \lambda) = \lambda E[v(\frac{M_n}{3\lambda}, \frac{N_n}{3\lambda})] \geq \lambda E[u_{<2}(\frac{M_n}{3\lambda}, \frac{N_n}{3\lambda})] \geq 0 \qquad (2.11.15)$$

which proves the inequality

$$\lambda \ P(|M_n| \geq \lambda) \leq 6 \ E|N_n|, n \geq 0. \qquad (2.11.16)$$
$$\square$$

Results discussed in this section are from Prakasa Rao (2007).

Chapter 3

N-Demimartingales

3.1 Introduction

We have already introduced the concept of a demimartingale extending the notion of a martingale and studied some of its properties. The class of demimartingales contains the sequences of partial sums of mean zero associated random variables as a special case. Motivated by the theory of demimartingales developed in Chapter 2, we now define (cf. Prakasa Rao (2002a) and Christofides (2003)) the class of N-demimartingales. We will see later that the sequence of partial sums of mean zero negatively associated (NA) random variables form an N-demimartingale . Throughout this chapter, we assume that the expectations of random variables under consideration exist.

Definition. Let the sequence $\{S_n, n \geq 1\}$ be an L^1-sequence of random variables such that

$$E[(S_{j+1} - S_j)f(S_1, \ldots, S_j)] \leq 0, \quad j \geq 1 \tag{3.1.1}$$

for every componentwise nondecreasing function f such that the expectation is defined. Then the sequence $\{S_n, n \geq 1\}$ is called an N-*demimartingale*. If, in addition, f is assumed to be nonnegative, then the sequence $\{S_n, n \geq 1\}$ is called an N-*demisupermartingale*.

Observe that, if the sequence $\{S_n, n \geq 1\}$ is an N-demimartingale, then $E(S_n) = E(S_1)$, $n \geq 1$. This can be seen from equation (3.1.1) by applying the inequality for the functions $f \equiv 1$ and for $f \equiv -1$.

Suppose the sequence $\{S_n, n \geq 1\}$ is an N-demimartingale. It is easy to see that

$$E[(S_{j+k} - S_j)f(S_1, \ldots, S_j)] \leq 0, \quad k \geq 1 \tag{3.1.2}$$

if f is componentwise nondecreasing. If the sequence $\{S_n, n \geq 1\}$ is an N-demisupermartingale, then (3.1.2) holds for any function f which is componentwise nondecreasing and nonnegative.

B.L.S. Prakasa Rao, *Associated Sequences, Demimartingales and Nonparametric Inference*, Probability and its Applications, DOI 10.1007/978-3-0348-0240-6_3, © Springer Basel AG 2012

Theorem 3.1.1. *Suppose the sequence $\{X_i, \ i \geq 1\}$ is a sequence of mean zero negatively associated random variables. Let $S_j = X_1 + \ldots + X_j$, $j \geq 1$ with $S_0 = 0$. Then the sequence $\{S_n, \ n \geq 1\}$ is an N-demimartingale.*

Proof. Let f be a componentwise nondecreasing function. Then

$$E[(S_{j+1} - S_j)f(S_1, \ldots, S_j)] = E[X_{j+1}f(S_1, \ldots, S_j)] \leq 0$$

by the property of negatively associated random variables $\{X_i, \ i \geq 1\}$ (cf. Joag-Dev and Proschan (1983)).   □

We now describe another type of a random sequence which is also an N-demimartingale.

Suppose that $X_1, \ldots, X_n$ are negatively associated random variables. For any fixed integer m such that $1 \leq m \leq n$, let $h(x_1, \ldots, x_m)$ be a kernel mapping R^m into R. Assume that the function $h(.)$ is symmetric. Define the U-statistic

$$U_n = \binom{n}{m}^{-1} \sum_{1 \leq i_1 < \ldots < i_m \leq n} h(X_{i_1}, \ldots, X_{i_m})$$

where $\sum_{1 \leq i_1 < \ldots < i_m \leq n}$ denotes the summation over the $\binom{n}{m}$ combinations of m distinct elements $\{i_1, \ldots, i_m\}$ from $\{1, \ldots, n\}$.

Theorem 3.1.2. *Let U_n be a U-statistic based on a sequence of negatively associated random variables $\{X_n, \ n \geq 1\}$ and on the kernel h where $h(x_1, \ldots, x_m) = \tilde{h}(x_1)\tilde{h}(x_2) \ldots \tilde{h}(x_m)$ for some nondecreasing function $\tilde{h}(.)$ with $E[\tilde{h}(X_1)] = 0$. Then the sequence $\{S_n = \binom{n}{m}U_n, \ n \geq m\}$ is an N-demimartingale.*

Proof. Observe that

$$S_{n+1} - S_n = \sum_{1 \leq i_1 < \ldots < i_m \leq n+1} h(X_{i_1}, \ldots, X_{i_m}) - \sum_{1 \leq i_1 < \ldots < i_m \leq n} h(X_{i_1}, \ldots, X_{i_m})$$

$$= \sum_{1 \leq i_1 < \ldots < i_{m-1} \leq n} h(X_{i_1}, \ldots, X_{i_{m-1}}, X_{n+1}) \tag{3.1.3}$$

Then, for any componentwise nondecreasing function g,

$$E[(S_{n+1} - S_n)g(S_m, \ldots, S_n)] \tag{3.1.4}$$

$$= E[\sum_{1 \leq i_1 < \ldots < i_{m-1} \leq n} h(X_{i_1}, \ldots, X_{i_{m-1}}, X_{n+1})g(S_m, \ldots, S_n)]$$

$$= E[\sum_{1 \leq i_1 < \ldots < i_{m-1} \leq n} \prod_{j=1}^{m-1} \tilde{h}(X_{i_j})\tilde{h}(X_{n+1})g(S_m, \ldots, S_n)]$$

$$= E[\tilde{h}(X_{n+1})v(X_1, \ldots, X_n)]$$

$$\leq 0$$

where the function v is defined as

$$v(x_1, \ldots, x_n) \tag{3.1.5}$$
$$= g(h(x_1, \ldots, x_m), \sum_{1 \leq i_1 < \ldots < i_m \leq m+1} h(x_{i_1}, \ldots, x_{i_m}), \ldots,$$
$$\sum_{1 \leq i_1 < \ldots < i_m \leq n} h(x_{i_1}, \ldots, x_{i_m})) \sum_{1 \leq i_1 < \ldots < i_m \leq n} \prod_{j=1}^{m-1} \tilde{h}(x_{i_j}).$$

Note that the function $v(x_1, \ldots, x_n)$ is componentwise nondecreasing. The last inequality in (3.1.4) follows from the nondecreasing property of the functions v and $\tilde{h}$ and the fact that the sequence $\{X_i, i \geq 1\}$ is a negatively associated random sequence. Hence the sequence $\{S_n, n \geq m\}$ is an N-demimartingale. □

Remarks. It is easy to see that a martingale, with the natural choice of σ-algebras, is a demimartingale as well as an N-demimartingale.

Theorem 3.1.3. *Let the random sequence $\{S_j, j \geq 1\}$ be a supermartingale with respect to the natural choice of σ-algebras $\mathcal{F}_j = \sigma\{S_i, 1 \leq i \leq j\}$. Then the random sequence $\{S_j, j \geq 1\}$ is an N-demisupermartingale.*

Proof. Let f be a nonnegative function defined on R^j.. Then

$$E[(S_{j+1} - S_j)f(S_1, \ldots, S_j)] = E[E((S_{j+1} - S_j)f(S_1, \ldots, S_j)|\mathcal{F}_j)]$$
$$= E[f(S_1, \ldots, S_j)E((S_{j+1} - S_j)|\mathcal{F}_j)]$$
$$\leq 0$$

where the last inequality is a consequence of the nonnegativity of f and the fact that $E((S_{j+1} - S_j)|\mathcal{F}_j)] \leq 0$ since the process $\{S_j, \mathcal{F}_j, j \geq 1\}$ is a supermartingale. □

The following results are easy to check.

Theorem 3.1.4. *Suppose the process $\{S_j, j \geq 1\}$ is an N-demimartingale and $Y_i = aS_i + b, i \geq 1$ where a and b are real numbers. Then the process $\{Y_j, j \geq 1\}$ is also an N-demimartingale.*

Proof. Let f be a componentwise nondecreasing function. Note that

$$E[(Y_{j+1} - Y_j)f(Y_1, \ldots, Y_j)] = E[(S_{j+1} - S_j)af(aS_1 + b, \ldots, aS_j + b)] \leq 0$$

since the sequence $\{S_j, j \geq 1\}$ is an N-demimartingale and the function $g(x_1, \ldots, x_n) = af(ax_1 + b, \ldots, ax_m + b)$ is componentwise nondecreasing. □

Following the same arguments, one can obtain the following result for N-demisupermartingales.

Theorem 3.1.5. *Suppose the process $\{S_j, j \geq 1\}$ is an N-demisupermartingale and $Y_i = aS_i + b$, $i \geq 1$ where $a \geq 0$ and b are real numbers. Then the process $\{Y_j, j \geq 1\}$ is also an N-demisupermartingale.*

The result stated in Theorem 3.1.4 indicates that the linear functions of N-demimartingales are also N-demimartingales. It would be interesting to identify a large class of functions $g(.)$ such that, if the sequence $\{S_j, j \geq 1\}$ is an N-demimartingale, then the sequence $\{g(S_j), j \geq 1\}$ is also an N-demimartingale. Results discussed in Theorems 3.1.2 to 3.1.5 are due to Christofides (2003).

We have seen earlier that the sequence of partial sums of mean zero negatively associated random variables is an N-demimartingale. However it is not necessary that every N-demimartingale is generated by such a process as shown by the following example due to Hadjikyriakou (2010).

Example 3.1.1. Let (X_1, X_2, X_3) be a random vector such that

$$P(X_1 = 5, X_2 = 5, X_3 = -2) = \frac{1}{12} = P(X_1 = 5, X_2 = -3, X_3 = -2),$$

$$P(X_1 = -3, X_2 = 5, X_3 = -2) = \frac{1}{12} = P(X_1 = -3, X_2 = -3, X_3 = -2),$$

$$P(X_1 = 5, X_2 = 5, X_3 = 1) = \frac{1}{24}; \ P(X_1 = 5, X_2 = -3, X_3 = 1) = \frac{4}{24},$$

and

$$P(X_1 = -3, X_2 = 5, X_3 = 1) = \frac{4}{24}; \ P(X_1 = -3, X_2 = -3, X_3 = 1) = \frac{7}{24}.$$

It can be checked that $E(X_1) = E(X_2) = E(X_3) = 0$ and for any nondecreasing function f,
$$E[(X_2 - X_1)f(X_1)] = 2[f(-3) - f(5)] \leq 0.$$

Furthermore, for any componentwise nondecreasing function $g(x_1, x_2)$,

$$E[(X_3 - X_2)g(X_1, X_2)] \leq 0.$$

Therefore the random sequence $\{X_1, X_2, X_3\}$ is an N-demimartingale. Let $h(.)$ be a nondecreasing function such that $h(-7) = h(-4) = h(-3) = 0$, $h(1) = 4$, $h(4) = 8$ and $h(5) = 16$. Check that

$$\text{Cov}(h(X_1), h(X_3 - X_2)) = \frac{2}{3} > 0$$

which shows that the random variables $X_1, X_3 - X_2$ are not negatively associated.

Another remark that was made earlier is that every supermartingale is an N-demisupermartingale. We will now present an example, due to Hadjikyriakou (2010), which shows the converse statement is not true.

Example 3.1.2. Let (X_1, X_2) be a random vector such that

$$P(X_1 = 1, X_2 = 0) = p \quad \text{and} \quad P(X_1 = 0, X_2 = 1) = 1 - p$$

where $\frac{1}{2} \leq p \leq 1$. Then the pair $\{X_1, X_2\}$ is an N-demisupermartingale since, for every nonnegative nondecreasing function f,

$$E[(X_2 - X_1)f(X_1)] = (1 - p)f(0) - pf(1) \leq p(f(0) - f(1)) \leq 0.$$

However the random sequence $\{X_1, X_2\}$ is not a supermartingale since

$$E[X_2|X_1 = 0] = \sum_{x_2=0,1} x_2 P(X_2 = x_2|X_1 = 0) = \frac{P(X_2 = 1, X_1 = 0)}{P(X_1 = 0)} = 1.$$

We will now describe another method of generating N-demimartingales following Hadjikyriakou (2010).

Example 3.1.3. Suppose the sequence $\{X_n, n \geq 1\}$ is a sequence of negatively associated identically distributed random variables with $\psi(t) = E[e^{tX_1}] < \infty$ for some $t \geq 0$. Let $S_n = \sum_{k=1}^{n} X_k$ and

$$Y_n = \frac{e^{tS_n}}{[\psi(t)]^n}, \quad n \geq 1.$$

Then the sequence $\{Y_n, n \geq 1\}$ is an N-demimartingale. This can be seen by the following arguments. Let f be a componentwise nonnegative nondecreasing function on R^n. Note that

$$E[(Y_{n+1} - Y_n)f(Y_1, \ldots, Y_n)] = E[(\frac{e^{tX_{n+1}}}{\psi(t)} - 1)Y_n f(Y_1, \ldots, Y_n)] \qquad (3.1.6)$$

$$= E[g(X_{n+1})Y_n f(Y_1, \ldots, Y_n]$$

$$= E[g(X_{n+1})h(X_1, \ldots, X_n)]$$

where the function $g(x) = \frac{e^{tx}}{\psi(t)}$ is a nondecreasing function of x and $h(x_1, \ldots x_n)$ is a componentwise nondecreasing function of $x_1, \ldots, x_n$. Since the sequence $\{X_n, n \geq 1\}$ is negatively associated, it follows that

$$\text{Cov}[g(X_{n+1}), h(X_1, \ldots, X_n)] \leq 0$$

and hence

$$E[g(X_{n+1})h(X_1, \ldots, X_n)] \leq E[g(X_{n+1})]E(h(X_1, \ldots, X_n)) = 0.$$

The last equality holds since $E[g(X_{n+1}] = 0$. This proves that the sequence $\{Y_n, n \geq 1\}$ is an N-demisupermartingale.

   The following theorem, due to Hu et al. (2010), gives sufficient conditions for a stopped N-demisupermartingale to be a N-demisupermartingale.

Theorem 3.1.6. *Let the sequence $\{S_n, \; n \geq 1\}$ be an N-demisupermartingale and τ be a positive integer-valued random variable. Suppose that the function $I_{[\tau \leq j]} = h_j(S_1, \ldots, S_j)$ is a componentwise nonincreasing function of $S_1, \ldots, S_j$ for $j \geq 1$. Let $S_j^* = S_{\min(\tau, j)}, \; j \geq 1$. Then the sequence $\{S_j^*, \; j \geq 1\}$ is an N-demisupermartingale.*

Proof. Note that

$$S_j^* = S_{\min(\tau, j)} = \sum_{i=1}^{j} (S_i - S_{i-1}) I_{[\tau \geq i]}.$$

We have to show that

$$E[(S_{j+1}^* - S_j^*) f(S_1^*, \ldots, S_j^*)] \leq 0, \quad j \geq 1$$

for any f which is componentwise nondecreasing and nonnegative. Since

$$g_j(S_1, \ldots, S_j) \equiv 1 - I_{[\tau \leq j]} = 1 - h_j(S_1, \ldots, S_j)$$

is a componentwise nondecreasing and nonnegative function, we get that

$$u_j(S_1, \ldots, S_j) \equiv g_j(S_1, \ldots, S_j) f(S_1, \ldots, S_j)$$

is a componentwise nondecreasing and nonnegative function. By the N-demisupermartingale property, we get that

$$
\begin{aligned}
E[(S_{j+1}^* - S_j^*) f(S_1^*, \ldots, S_j^*)] &= E[(S_{j+1} - S_j) I_{[\tau \geq j+1]} f(S_1^*, \ldots, S_j^*)] \quad (3.1.7) \\
&= E[(S_{j+1} - S_j) I_{[\tau \geq j+1]} f(S_1, \ldots, S_j)] \\
&= E[(S_{j+1} - S_j) u_j(S_1, \ldots, S_j)] \leq 0
\end{aligned}
$$

for $j \geq 1$. Hence the sequence $\{S_j^*, j \geq 1\}$ is an N-demisupermartingale. $\square$

   We now obtain some consequences of this theorem.

Theorem 3.1.7. *Let the sequence $\{S_n, \; n \geq 1\}$ be an N-demisupermartingale and τ be a positive integer-valued random variable. Furthermore suppose that the indicator function $I_{[\tau \leq j]} = h_j(S_1, \ldots, S_j)$ is a componentwise nonincreasing function of $S_1, \ldots, S_j$ for $j \geq 1$. Then, for any $1 \leq n \leq m$,*

$$E(S_{\min(\tau, m)}) \leq E(S_{\min(\tau, n)}) \leq E(S_1). \qquad (3.1.8)$$

Suppose the sequence $\{S_n, n \geq 1\}$ is a N-demimartingale and the indicator function $I_{[\tau \leq j]} = h_j(S_1, \ldots, S_j)$ is a componentwise nondecreasing function of $S_1, \ldots, S_j$ for $j \geq 1$. Then, for any $1 \leq n \leq m$,

$$E(S_{\min(\tau, m)}) \geq E(S_{\min(\tau, n)}) \geq E(S_1). \qquad (3.1.9)$$

Proof. Suppose that the random sequence $\{S_n, n \geq 1\}$ is an N-demisupermartingale and the indicator function $I_{[\tau \leq j]} = h_j(S_1, \ldots, S_j)$ is a componentwise non-increasing function of $S_1, \ldots, S_j$ for $j \geq 1$. Then the sequence $\{S_n^*, n \geq 1\}$ is an N-demisupermartingale. The inequalities in equation (3.1.8) follow from the N-demisupermartingale property by choosing the function $f \equiv 1$.

Suppose that the random sequence $\{S_n, n \geq 1\}$ is an N-demimartingale and that the indicator function $I_{[\tau \leq j]} = h_j(S_1, \ldots, S_j)$ is a componentwise non-decreasing function of $S_1, \ldots, S_j$ for $j \geq 1$. Since the sequence $\{S_n, n \geq 1\}$ is an N-demimartingale, we note that

$$
\begin{aligned}
-E(S_{j+1}^* - S_j^*) &= -E[(S_{j+1} - S_j)I_{[\tau \geq j+1]}] \\
&= E[(S_{j+1} - S_j)(h_j(S_1, \ldots, S_j) - 1)] \\
&\leq 0
\end{aligned}
$$

for $j \geq 1$ from the N-demimartingale property. This in turn proves the inequalities given in (3.1.9). □

3.2 Maximal Inequalities

As an application of Theorem 3.1.7, we can obtain the following maximal inequality for N-demisupermartingales due to Christofides (2003).

Theorem 3.2.1. *Let the sequence $\{S_j, j \geq 1\}$ be an N-demimartingale . Then, for any $\lambda > 0$,*

$$\lambda P[\max_{1 \leq k \leq n} S_k \geq \lambda] \leq E(S_1) - E(S_n I_{[\max_{1 \leq k \leq n} S_k < \lambda]}) \qquad (3.2.1)$$

$$\leq E(S_1) + E(S_n^-). \qquad (3.2.2)$$

Proof. Let $\tau = \inf\{k : k \leq n \text{ and } S_k \leq \lambda\}$ and $\tau = n$ if $\min_{1 \leq k \leq n} S_k \geq \lambda$. Then, by the results derived above,

$$
\begin{aligned}
E(S_1) &\geq E(S_{\min(\tau,n)}) \qquad\qquad\qquad\qquad\qquad\qquad (3.2.3)\\
&= E(S_\tau) \\
&= E(S_\tau I_{[\max_{1 \leq k \leq n} S_k \geq \lambda]}) + E(S_\tau I_{[\max_{1 \leq k \leq n} S_k < \lambda]}) \\
&\geq \lambda P(\max_{1 \leq k \leq n} S_k \geq \lambda) + E(S_n I_{[\max_{1 \leq k \leq n} S_k < \lambda]}).
\end{aligned}
$$

Rearranging the above inequality, we get the inequality (3.2.1). The inequality (3.2.2) is an easy consequence of the inequality (3.2.1). □

As a corollary to the above inequality, we get the following result.

Theorem 3.2.2. *Let the sequence* $\{S_j, j \geq 1\}$ *be a nonnegative N-demimartingale. Then, for any* $\lambda > 0$,

$$\text{(i)} \quad \lambda\, P[\max_{1 \leq k \leq n} S_k \geq \lambda] \leq E(S_1)$$

and

$$\text{(ii)} \quad \lambda\, P[\sup_{k \geq n} S_k \geq \lambda] \leq E(S_n).$$

Proof. The first inequality follows immediately from the inequality (3.2.1). We now prove the second inequality. Let $\tau = \inf\{k : k \leq n \text{ and } S_k \geq \lambda\}$ and $\tau = \infty$ if $\sup_{k \geq n} < \lambda$. Let $m \geq n$. Then, from the results derived earlier, it follows that

$$\begin{aligned}
E(S_n) &= E[S_{\min(\tau,n)}] \\
&\geq E[S_{\min(\tau,m)}] \\
&\geq E[S_{\min(\tau,m)} I_{[\tau \leq m]}) \\
&\geq \lambda\, P(\tau \leq m).
\end{aligned}$$

Let $m \to \infty$. Then, we have

$$E(S_n) \geq \lambda\, P(\tau < \infty) = \lambda\, P(\sup_{k \geq n} S_k \geq \lambda). \tag{3.2.4}$$

$\square$

Suppose ϕ is a right continuous nonincreasing function on $(0, \infty)$ satisfying the condition

$$\lim_{t \to \infty} \phi(t) = 0.$$

Further suppose that ϕ is also integrable on any finite interval $(0, x)$. Let

$$\Phi(x) = \int_0^x \phi(t)dt, \quad x \geq 0.$$

Then the function $\Phi(x)$ is a nonnegative nondecreasing function such that $\Phi(0) = 0$. Further suppose that $\Phi(\infty) = \infty$. Such a function is called a *concave Young function*. Properties of such functions are given in Agbeko (1986). An example of such a function is $\Phi(x) = x^p$, $0 < p < 1$. As a consequence of the inequality (3.2.3), Christofides (2003) obtained the following maximal inequality following the arguments in Agbeko (1986) to derive a maximal inequality for nonnegative supermartingales. We omit the details.

Theorem 3.2.3. *Let the sequence* $\{S_n, n \geq 1\}$ *be a nonnegative N-demisupermartingale. Let* $\Phi(x)$ *be a concave Young function and define* $\psi(x) = \Phi(x) - x\phi(x)$. *Then*

$$E[\psi(S_n^{\max})] \leq E[\Phi(S_1)]. \tag{3.2.5}$$

Furthermore, if

$$\limsup_{x\to\infty} \frac{x\phi(x)}{\Phi(x)} < 1,$$

then

$$E[\Phi(S_n^{\max})] \le c_\Phi(1 + E[\Phi(S_1)]) \tag{3.2.6}$$

for some constant c_Φ depending only on the function Φ.

Following the ideas of Agbeko (1986) and Christofides (2003), Wang et al. (2011a) obtained more general maximal inequalities for demimartingales and N-demimartingales based on concave Young functions.

3.3   More on Maximal Inequalities

We now derive some additional maximal inequalities for N-demisupermartingales following the ideas of Newman and Wright (1982) for demisubmartingales as discussed in Chapter 2. The results in this section are due to Prakasa Rao (2002a).

Let the random sequence $\{S_n,\, n \ge 1\}$ with $S_0 = 0$ be an N-demimartingale and

$$S_{nj} = \begin{cases} j\text{-th smallest of } (S_1, \ldots, S_n) & \text{if } j \le n \\ \max(S_1, \ldots, S_n) = S_{nn} & \text{if } j > n. \end{cases}$$

Let

$$S_n^* = \min(S_1, \ldots, S_n).$$

Note that

$$S_{n1} = S_n^*.$$

In other words $(S_{n1}, \ldots, S_{nn})$ is the set of order statistics corresponding to the set of random variables $(S_1, \ldots, S_n)$. Let $m(.)$ be a nondecreasing function with $m(0) = 0$. For fixed n and j, let $Y_k = S_{kj}$ and $Y_0 = 0$. Then

$$S_n m(Y_n) = \sum_{k=0}^{n-1} S_{k+1}(m(Y_{k+1}) - m(Y_k)) + \sum_{k=1}^{n-1} (S_{k+1} - S_k)m(Y_k)$$

and

$$E[S_n m(Y_n)] = E\{\sum_{k=0}^{n-1} S_{k+1}(m(Y_{k+1}) - m(Y_k))\} + \sum_{k=1}^{n-1} E[(S_{k+1} - S_k)m(Y_k)]$$

$$\le E\{\sum_{k=0}^{n-1} S_{k+1}(m(Y_{k+1}) - m(Y_k))\}$$

since the finite sequence $\{S_k,\, 1 \le k \le n\}$ is an N-demimartingale. Note that, for any k,

$$(m(Y_{k+1}) - m(Y_k))S_{k+1} \le Y_k\,[m(Y_{k+1}) - m(Y_k)].$$

This follows from the observation that, for instance, if $\alpha_k = \min(x_1, \ldots, x_k)$ and $\alpha_{k+1} = \min(x_1, \ldots, x_{k+1})$, then either the minimum remains the same or the minimum is decreased, that is either $\alpha_k = \alpha_{k+1}$ or $x_{k+1} < \alpha_k$. Hence

$$E[S_n m(Y_n)] \le E\{\sum_{k=0}^{n-1} Y_k(m(Y_{k+1}) - m(Y_k))\} \tag{3.3.1}$$

$$\le E\{\sum_{k=0}^{n-1} \int_{Y_k}^{Y_{k+1}} u \, dm(u)\}$$

$$= E\{\int_0^{Y_n} u \, dm(u)\}$$

Hence we have the following theorem.

Theorem 3.3.1. *Let the sequence $\{S_n, n \ge 1\}$ be an N-demimartingale and $m(.)$ be a nondecreasing function with $m(0) = 0$. Then, for any n and j,*

$$E\left(\int_0^{S_{nj}} u \, dm(u)\right) \ge E(S_n m(S_{nj})).$$

By analogous arguments, we can prove the following theorem for N-demisupermartingales.

Theorem 3.3.2. *Let the sequence $\{S_n, n \ge 1\}$ be an N-demisupermartingale and $m(.)$ be a nonnegative nondecreasing function with $m(0) = 0$. Then, for any n and j,*

$$E\left(\int_0^{S_{nj}} u \, dm(u)\right) \ge E(S_n m(S_{nj})).$$

Let the sequence $\{S_n, n \ge 1\}$ be an N-demimartingale.

We now give some applications of Theorem 3.3.1.
(i) Let $m(u) = -I_{[u \le \lambda]}$. Note that $Y_n = S_{nj}$. Then, for any $\lambda > 0$,

$$-\lambda \, P(S_{nj} \le \lambda) \ge -\int_{[S_{nj} \le \lambda]} S_n dP$$

or equivalently

$$\lambda \, P(S_{nj} \le \lambda) \le \int_{[S_{nj} \le \lambda]} S_n dP.$$

(ii) (a) In particular, let $j = 1$ in (i). Then, for any $\lambda > 0$,

$$\lambda \, P(\min_{1 \le i \le n} S_i \le \lambda) \le \int_{[\min_{1 \le i \le n} S_i \le \lambda]} S_n dP. \tag{3.3.2}$$

(ii) (b) Let $j = n$ in (i). Then, for any $\lambda > 0$,

$$\lambda P(\max_{1 \leq i \leq n} S_i \leq \lambda) \leq \int_{[\max_{1 \leq i \leq n} S_i \leq \lambda]} S_n dP.$$

(iii) Suppose the sequence $\{S_n, n \geq 1\}$ is an L^2 N-demimartingale. Then

$$E((S_{nj} - S_n)^2) \geq E(S_n^2). \tag{3.3.3}$$

This can be seen by the following arguments. Choose $m(u) = u$. Then

$$E[S_n Y_n] \leq E[\int_0^{Y_n} u \, du] = E\left[\frac{Y_n^2}{2}\right].$$

Hence

$$E[S_n S_{nj}] \leq E\left(\frac{S_{nj}^2}{2}\right) \tag{3.3.4}$$

which holds if and only if

$$E((S_{nj} - S_n)^2) \geq E(S_n^2)$$

since

$$
\begin{aligned}
E[(S_{nj} - S_n)^2] &= E[S_{nj}^2] - 2E[S_n S_{nj}] + E[S_n^2] \\
&\geq E(S_{nj}^2) - E[S_{nj}^2] + E[S_n^2] \\
&= E(S_n^2).
\end{aligned}
$$

(iv) Suppose the sequence $\{X_n, n \geq 1\}$ forms an integrable mean zero negatively associated (NA) sequence of random variables. Let $S_j = X_1 + \ldots + X_j$, $j \geq 1$ and $S_0 = 0$. Let $T_1 = 0$ and

$$T_k = \sum_{i=n-k+2}^{n} X_i, \quad k = 2, 3, \ldots, n+1.$$

Let f be a componentwise nondecreasing function. Then

$$
\begin{aligned}
&E((T_{k+1} - T_k)f(T_2, \ldots, T_k)) \\
&= E[(S_{n-k+1} - S_{n-k})g(X_{n-k+2}, \ldots, X_n)] \leq 0
\end{aligned}
$$

since the sequence $\{X_k, 1 \leq k \leq n\}$ is negatively associated and g is a componentwise nondecreasing function. Hence $\{T_k, 2 \leq k \leq n+1\}$ is an N-demimartingale. For, $j \leq n$, applying the inequality of type (3.3.4) to the sequence $\{T_k\}$, we have

$$E(T_n T_{n,n-j+1}) \leq E\left(\frac{T_{n,n-j+1}^2}{2}\right). \tag{3.3.5}$$

But
$$E(T_n T_{n,n-j+1}) \geq E(T_{n+1} T_{n,n-j+1})$$

since $\{T_n\}$ is an N-demimartingale. Note that $T_{n,n-j+1}$ is the $(n-j+1)$-th order statistic corresponding to $T_2, \ldots, T_n$. Therefore

$$E\left(\frac{T_{n,n-j+1}^2}{2}\right) \geq E(T_{n+1} T_{n,n-j+1})$$

which implies that
$$E[(T_{n+1} - T_{n,n-j+1})^2] \geq E(T_{n+1}^2). \tag{3.3.6}$$

Note that
$$T_{n+1} = S_n$$

and

$$T_{n+1} - T_{n,n-j+1} = j\text{-th smallest of } (T_{n+1} - T_n, \ldots, T_{n+1} - T_2, T_{n+1} - T_1)$$
$$= S_{nj}.$$

Hence (3.3.6) implies that
$$E(S_n^2) \leq E(S_{nj}^2). \tag{3.3.7}$$

(v) Define T_n as given in (iv). Consider $S_n^* = \min(S_1, \ldots, S_n)$ and $S_n - S_n^* = T_{nn}$. Note that S_n^* and T_{nn} are increasing functions of the X_i's which are negatively associated. Hence S_n^* and T_{nn} are themselves negatively associated (cf. Matula (1996)). Note that, if X and Y are negatively associated, then

$$P(X \geq x)P(Y \leq y) - P(X \geq x, \, Y \leq y) = \mathrm{Cov}(I_{[X \geq x]}, -I_{[Y \leq y]}) \leq 0.$$

Hence
$$P(X \geq x)P(Y \leq y) \leq P(X \geq x, \, Y \leq y)$$

for all x and y. Let $Y = S_n^*$ and $X = S_n - S_n^* = T_n$. Then

$$P(S_n - S_n^* \geq x)P(S_n^* \leq y) \leq P(S_n - S_n^* \geq x, \, S_n^* \leq y) \tag{3.3.8}$$

for all x and y. Similarly

$$P(S_n^* \geq x)P(S_n - S_n^* \leq y) \leq P(S_n^* \geq x, \, S_n - S_n^* \leq y)$$

for all x and y. Note that, for $\lambda_2 < \lambda_1$,

$$
\begin{aligned}
P(S_n^* \leq \lambda_2) &= P(S_n^* \leq \lambda_2, \, S_n \geq \lambda_1) + P(S_n^* \leq \lambda_2, \, S_n < \lambda_1) \\
&\geq P(S_n^* \leq \lambda_2, \, S_n^* - S_n \leq \lambda_2 - \lambda_1) + P(S_n \leq \lambda_2) \\
&= P(S_n^* \leq \lambda_2, \, S_n - S_n^* \geq \lambda_1 - \lambda_2) + P(S_n \leq \lambda_2) \\
&= P(S_n^* \leq \lambda_2, \, T_{n,n} \geq \lambda_1 - \lambda_2) + P(S_n \leq \lambda_2)
\end{aligned}
$$

$$\geq P(S_n^* \leq \lambda_2)P(T_{nn} \geq \lambda_1 - \lambda_2) + P(S_n \leq \lambda_2)$$

by (3.3.8). Note that $T_{nn} \geq 0$ and

$$P(T_{nn} \geq \lambda_1 - \lambda_2) \geq \frac{E(T_{nn}^2) - (\lambda_1 - \lambda_2)^2}{a.s. \ \sup T_{nn}^2} \quad \text{(by Loève (1977), p. 159)}.$$

Hence, for $\lambda_2 < \lambda_1$,

$$P(S_n^* \leq \lambda_2) \geq P(S_n \leq \lambda_2) + P(S_n^* \leq \lambda_2) \left\{ \frac{E((S_n^* - S_n)^2) - (\lambda_1 - \lambda_2)^2}{a.s. \ \sup \ T_{nn}^2} \right\}$$

$$\geq P(S_n \leq \lambda_2) + P(S_n^* \leq \lambda_2) \left\{ \frac{E(S_n^2) - (\lambda_1 - \lambda_2)^2}{a.s. \ \sup \ T_{nn}^2} \right\} \quad \text{(by (3.3.7))}.$$

Let $E(S_n^2) = s_n^2$. Then, for all $\lambda_2 < \lambda_1$,

$$\left\{ 1 - \frac{E(S_n^2) - (\lambda_1 - \lambda_2)^2}{a.s. \ \sup \ T_{nn}^2} \right\} P(S_n^* \leq \lambda_2) \geq P(S_n \leq \lambda_2). \qquad (3.3.9)$$

Furthermore

$$P(S_n^* \geq \lambda_2) \leq P(S_n \geq \lambda_2)$$

since $S_n^* \leq S_n$. Hence $P(S_n^* \leq \lambda_2) \geq P(S_n \leq \lambda_2)$ and we have the inequality

$$P(S_n \leq \lambda_2) \leq P(S_n^* \leq \lambda_2) \leq P(S_n \leq \lambda_2) + \frac{E(S_n^2) - (\lambda_1 - \lambda_2)^2}{a.s. \ \sup \ T_{nn}^2}$$

from (3.3.9).

(vi) Note that the random variables $-X_1, -X_2, \ldots, -X_n$ are negatively associated if the random variables $X_1, \ldots, X_n$ are negatively associated (cf. Matula (1996)). Suppose $E(X_i) = 0$, $1 \leq i \leq n$. Let $Y_i = -X_i$, $1 \leq i \leq n$ and $J_n = Y_1 + \ldots + Y_n$ with $J_0 = 0$. Applying the result in (3.3.9), we get that, for all $\lambda_2 < \lambda_1$,

$$\left\{ 1 - \frac{E(T_n^2) - (\lambda_1 - \lambda_2)^2}{a.s. \ \sup(T_n - T_n^*)^2} \right\} P(J_n^* \leq \lambda_2) \geq P(J_n \leq \lambda_2)$$

where $T_n^* = \min(J_1, \ldots, J_n)$. Observe the $J_n = -S_n$, $n \geq 1$ and

$$J_n^* = \min(J_1, \ldots, J_n)$$
$$= \min(-S_1, \ldots, -S_n)$$
$$= -\max(S_1, \ldots, S_n)$$
$$= -S_{nn}.$$

Furthermore $E(J_n^2) = E(S_n^2) = s_n^2$. Note that $J_{nn} = J_n - J_n^* = -S_n + S_{nn}$. Hence

$$P(J_n^* \leq \lambda_2) = P(-S_{nn} \leq \lambda_2)$$

and

$$\left[1 - \left\{\frac{s_n^2 - (\lambda_1 - \lambda_2)^2}{\text{a.s. sup } (-S_n + S_{nn})^2}\right\}\right] P(-S_{nn} \leq \lambda_2) \geq P(-S_n \leq \lambda_2).$$

Therefore

$$\left[1 - \left\{\frac{s_n^2 - (\lambda_1 - \lambda_2)^2}{\text{a.s. sup } (-S_n + S_{nn})^2}\right\}\right] P(S_{nn} \geq -\lambda_2) \geq P(S_n \geq -\lambda_2),$$

or equivalently for any $\nu_1 < \nu_2$,

$$\left[1 - \left\{\frac{s_n^2 - (\nu_1 - \nu_2)^2}{\text{a.s. sup } (S_{nn} - S_n)^2}\right\}\right] P(S_{nn} \geq \nu_2) \geq P(S_n \geq \nu_2). \qquad (3.3.10)$$

(vii) Let the sequence $\{S_n, n \geq 1\}$ be an L^2 N-demimartingale. Then, for $\lambda_1 < \lambda_2$,

$$\lambda_2 \, P(S_n^* \leq \lambda_2) \leq \int_{[S_n^* \leq \lambda_2]} S_n dP \quad \text{(by (3.3.2))}$$

$$= \int_{[S_n^* \leq \lambda_2, S_n > \lambda_1]} S_n dP + \int_{[S_n^* \leq \lambda_2, S_n \leq \lambda_1]} S_n dP$$

$$\leq \int_{[S_n \geq \lambda_1]} S_n dP + \lambda_1 \, P(S_n^* \leq \lambda_2).$$

Hence

$$P(S_n^* \leq \lambda_2) \leq \frac{1}{\lambda_2 - \lambda_1} E[S_n I(S_n > \lambda_1)] \qquad (3.3.11)$$

$$\leq \frac{1}{\lambda_2 - \lambda_1} (E[S_n^2] P(S_n > \lambda_1))^{1/2}.$$

Special Case. Applying the above result to the random variables $-X_1, \ldots, -X_n$, which are NA with mean zero, we have, for $\lambda_1 < \lambda_2$,

$$P(-S_{nn} \leq \lambda_2) \leq \frac{1}{\lambda_2 - \lambda_1} (E[S_n^2] P(-S_n > \lambda_1))^{1/2}$$

or equivalently, for $\nu_1 > \nu_2$.

$$P(S_{nn} \geq \nu_2) \leq \frac{1}{\nu_2 - \nu_1} (E(S_n^2) P(S_n < \nu_1))^{1/2}. \qquad (3.3.12)$$

3.4 Downcrossing Inequality

Let the finite set $\xi_1, \xi_2, \ldots, \xi_n$ be any real numbers and $a < b$. The numbers of downcrossings $D_{a,b}$ of the interval $[a, b]$ by $\xi_1, \ldots, \xi_n$ is defined as the number of times the sequence $\xi_1, \ldots, \xi_n$ passes from above b to below a. Let ξ_{ν_1} be the first

ξ_i, if any, for which $\xi_i \geq b$ and in general let ξ_{ν_j} be the first ξ_i, if any, after $\xi_{\nu_{j-1}}$ for which

$$\xi_i \leq a \quad (\text{if } j \text{ is even})$$

and

$$\xi_i \geq b \quad (\text{if } j \text{ is odd})$$

so that

$$\xi_{\nu_1} \geq b, \quad \xi_{\nu_2} \leq a, \quad \xi_{\nu_3} \geq b, \quad \ldots$$

Then the number of downcrossings is β where 2β is the largest integer j for which ξ_{ν_j} is defined and $\beta = 0$ if ξ_{ν_2} is not defined. Let $\nu_0 \equiv 0$. Here

$$\nu_{2k-1} = \begin{cases} n+1 & \text{if } \{j : \nu_{2k-2} < j \leq n \text{ and } \xi_j \geq b\} \text{ is empty} \\ \min\{j : \nu_{2k-2} < j \leq n \text{ and } \xi_j \geq b\} & \text{otherwise,} \end{cases}$$

and

$$\nu_{2k} = \begin{cases} n+1 & \text{if } \{j : \nu_{2k-1} < j \leq n \text{ and } \xi_j \leq a\} \text{ is empty} \\ \min\{j : \nu_{2k-2} < j \leq n \text{ and } \xi_j \leq a\} & \text{otherwise.} \end{cases}$$

Define

$$\varepsilon_j = \begin{cases} 1 & \text{if for some } k = 1, 2, \ldots, \nu_{2k-1} \leq j < \nu_{2k} \\ 0 & \text{if for some } k = 1, 2, \ldots, \nu_{2k} \leq j < \nu_{2k+1} \end{cases}$$

so that ε_j is the indicator function of the event that the time interval $[j, j+1]$ is a part of a downcrossing (possibly incomplete). Let

$$\Lambda = \{\tilde{\nu} \equiv \nu_{2D_{a,b}+1} < n\}.$$

Note that Λ is the event that the sequence ends with an incomplete downcrossing. Note that

$$\varepsilon_j = \begin{cases} 1 & \text{if either } \xi_i \geq b \text{ for } i = 1, \ldots, j \text{ or else} \\ & \text{for some } i = 1, \ldots, j, \ \xi_i \geq a \text{ and } \xi_k \geq b \text{ for } k = i+1, \ldots, j \\ 0 & \text{otherwise.} \end{cases}$$

Furthermore

$$(b - \xi_n)^+ - (b - \xi_1)^+ = \sum_{j=1}^{n-1} [(b - \xi_{j+1})^+ - (b - \xi_j)^+]$$

$$= H_u + H_d$$

where

$$H_d = \sum_{j=1}^{n-1} (1 - \varepsilon_j)[(b - \xi_{j+1})^+ - (b - \xi_j)^+].$$

Note that

$$(b - \xi_{j+1})^+ \geq b - \xi_{j+1}$$

and

$$(1 - \varepsilon_j)(b - \xi_j)^+ = (1 - \varepsilon_j)(b - \xi_j) \text{ since } \varepsilon_j = 0 \text{ implies } \xi_j < b.$$

Therefore

$$H_d \geq \sum_{j=1}^{n-1}(1 - \varepsilon_j)[(b - \xi_{j+1}) - (b - \xi_j)] \qquad (3.4.1)$$

$$= \sum_{j=1}^{n}(1 - \varepsilon_j)[\xi_j - \xi_{j+1}]$$

and

$$H_u = \sum_{j=1}^{n-1} \varepsilon_j[(b - \xi_{j+1}^+) - (b - \xi_j)^+]$$

$$= \sum_{k=1}^{D_{a,b}} \sum_{j=\nu_{2k-1}}^{\nu_{2k}-1} [(b - \xi_{j+1})^+ - (b - \xi_j)^+] + \sum_{j=\tilde{\nu}}^{n-1}[(b - \xi_{j+1}^+ - (b - \xi_j)^+]$$

$$= \sum_{k=1}^{D_{a,b}} [(b - \xi_{\nu_{2k}})^+ - (b - \xi_{\nu_{2k-1}})^+] + [(b - \xi_n)^+ - (b - \xi_{\tilde{\nu}})^+]I_\Lambda$$

$$= \sum_{k=1}^{D_{a,b}} (b - \xi_{\nu_{2k}})^+ + (b - \xi_n)^+ I_\Lambda$$

$$\geq (b - a)D_{a,b}$$

where $D_{a,b}$ is the number of completed downcrossings. Combining the above inequalities and taking expectations, we get that

$$E[(b - \xi_n)^+ - (b - \xi_1)^+] \geq (b - a)E[D_{a,b}] + \sum_{j=1}^{n-1} E[(1 - \varepsilon_j)(\xi_j - \xi_{j+1})].$$

Suppose the sequence $\{\xi_n\}$ is an N-demisupermartingale. Since $(1 - \varepsilon_j)$ is a nonnegative nondecreasing function of $\xi_1, \ldots, \xi_j$, it follows that

$$E((1 - \varepsilon_j)(\xi_{j+1} - \xi_j)) \leq 0, \quad 1 \leq j \leq n$$

which implies that

$$E(D_{a,b}) \leq \frac{1}{b - a}[E(b - \xi_n)^+ - E(b - \xi_1)^+]$$

and we have the following theorem.

Theorem 3.4.1. *Suppose the sequence $\{\xi_n, n \geq 1\}$ is an N-demisupermartingale. Then, for any $a < b$,*

$$E(D_{a,b}) \leq \frac{1}{b-a}[E(b-\xi_n)^+ - E(b-\xi_1)^+]$$

where $D_{a,b}$ denotes the number of complete downcrossings of the interval $[a, b]$ by the sequence $\{\xi_n, n \geq 1\}$.

Almost Sure Convergence for N-demisupermartingales

If $E|\xi_1| < \infty$, then it follows, from Theorem 3.4.1, that ξ_n converges a.s. to a finite limit by standard arguments and we have the following theorem.

Theorem 3.4.2. *If the sequence $\{\xi_n, n \geq 1\}$ is a N-demisupermartingale with $\sup_n E|\xi_n| < \infty$, then ξ_n converges a.s. to a finite limit as $n \to \infty$.*

As a corollary to the above theorem, we obtain that, if $\{X_n, n \geq 1\}$ is a zero mean sequence of negatively associated random variables such that $\sup_n E|\xi_n| < \infty$ where $\xi_n = X_1 + \ldots + X_n$, then ξ_n converges a.s. to a finite limit as $n \to \infty$.

Results in this section are from Prakasa Rao (2002a).

3.5 Chow Type Maximal Inequality

We now discuss a Chow type maximal inequality for N-demimartingales due to Prakasa Rao (2004).

Theorem 3.5.1. *Let the sequence $\{S_n, n \geq 1\}$ be an N-demimartingale with $S_0 = 0$. Let $m(.)$ be a nonnegative nondecreasing function on R with $m(0) = 0$. Let $g(.)$ be a function such that $g(0) = 0$ and suppose that*

$$g(x) - g(y) \geq (y - x)h(y) \tag{3.5.1}$$

for all x, y where $h(.)$ is a nonnegative nondecreasing function. Further suppose that $\{c_k, 1 \leq k \leq n\}$ is a sequence of positive numbers such that $(c_k - c_{k+1})g(S_k) \geq 0, 1 \leq k \leq n - 1$. Define

$$Y_k = \max\{c_1 g(S_1), \ldots, c_k g(S_k)\}, \quad k \geq 1, \ Y_0 = 0.$$

Then

$$E\left(\int_0^{Y_n} u \, dm(u)\right) \leq \sum_{k=1}^{n} c_k E([g(S_k) - g(S_{k-1})]m(Y_n)). \tag{3.5.2}$$

Proof. Let $Y_0 = 0$. Observe that

$$E\left(\int_0^{Y_n} u \, dm(u)\right) = \sum_{k=1}^{n} E\left(\int_{Y_{k-1}}^{Y_k} u \, dm(u)\right) \tag{3.5.3}$$

$$\leq \sum_{k=1}^{n} E[Y_k(m(Y_k) - m(Y_{k+1}))].$$

From the definition of Y_k, it follows that $m(Y_1) = 0$ for $Y_1 < 0$. Furthermore $Y_k \geq Y_{k-1}$ and either $Y_k = c_k g(S_k)$ or $m(Y_k) = m(Y_{k-1})$. Hence

$$E(\int_0^{Y_n} u \, dm(u)) \leq \sum_{k=1}^{n} c_k E[g(S_k)(m(Y_k) - m(Y_{k-1}))] \qquad (3.5.4)$$

since $m(.)$ is a nondecreasing function and $Y_{k-1} \leq Y_k$. Note that

$$\sum_{k=1}^{n} c_k E[g(S_k)(m(Y_k) - m(Y_{k-1}))] = \sum_{k=1}^{n} c_k E[(g(S_k) - g(S_{k-1}))m(Y_n)] \qquad (3.5.5)$$

$$- \{\sum_{k=1}^{n-1} E[(c_{k+1}g(S_{k+1}) - c_k g(S_k))m(Y_k)]$$

$$+ \sum_{k=1}^{n-1} E[(c_k - c_{k+1})g(S_k)m(Y_n)]\}$$

Let

$$A = \sum_{k=1}^{n-1} E[(c_{k+1}g(S_{k+1}) - c_k g(S_k))m(Y_k)] + \sum_{k=1}^{n-1} E[(c_k - c_{k+1})g(S_k)m(Y_n)].$$
$$(3.5.6)$$

Since $(c_k - c_{k+1})g(S_k) \geq 0$, $1 \leq k \leq n-1$, it follows that

$$A \geq \sum_{k=1}^{n-1} E[(c_{k+1}g(S_{k+1}) - c_k g(S_k))m(Y_k)] + \sum_{k=1}^{n-1} E[(c_k - c_{k+1})g(S_k)m(Y_k)]$$

$$= \sum_{k=1}^{n-1} E[(c_{k+1}g(S_{k+1}) - c_{k+1}g(S_k))m(Y_k)] \qquad (3.5.7)$$

$$= \sum_{k=1}^{n-1} c_{k+1} E[(g(S_{k+1}) - g(S_k))m(Y_k)]$$

$$\geq \sum_{k=1}^{n-1} c_{k+1} E[(S_k - S_{k+1})h(S_k)m(Y_k)]$$

from the property of the function $g(.)$ given by (3.5.1). Note that $h(S_i)m(Y_i)$ is a nondecreasing function of $S_1, \ldots, S_i$. Since $\{S_i, i \geq 1\}$ forms an N-demimartingale, it follows that

$$E[(S_{k+1} - S_k)h(S_k)m(Y_k)] \leq 0, \quad 1 \leq k \leq n-1$$

and hence

$$\sum_{k=1}^{n-1} c_{k+1} E[(S_k - S_{k+1})h(S_k)m(Y_k)] \geq 0 \tag{3.5.8}$$

by the nonnegativity of the sequence c_i, $i \geq 1$. Hence $A \geq 0$. Therefore

$$E\left(\int_0^{Y_n} u \, dm(u)\right) \leq \sum_{k=1}^{n} c_k E([g(S_k) - g(S_{k-1})]m(Y_n)). \tag{3.5.9}$$

$\square$

Remarks. Let $\epsilon > 0$ and define $m(t) = 1$ if $t \geq \epsilon$ and $m(t) = 0$ if $t < \epsilon$. Applying the previous theorem, we get that

$$\epsilon \, P(Y_n \geq \epsilon) \leq \sum_{k=1}^{n} c_k E([g(S_k) - g(S_{k-1})]I_{[Y_n \geq \epsilon]}). \tag{3.5.10}$$

Examples of functions g satisfying (3.5.1) are $g(x) = -\alpha x$ where $\alpha \geq 0$ and $g(x) = -\alpha x^+$ where $\alpha \geq 0$. Here $x^+ = x$ if $x \geq 0$ and $x^+ = 0$ if $x < 0$.

As a corollary to the Chow type maximal inequality derived above, the following result can be obtained (cf. Wang et al. (2011)) as an easy consequence.

Theorem 3.5.2. *Let the sequence $\{S_n, n \geq 1\}$ be an N-demimartingale with $S_0 = 0$. Let $g(.)$ be a nonnegative function such that $g(0) = 0$ and suppose that*

$$g(x) - g(y) \geq (y - x)h(y) \tag{3.5.11}$$

for all x, y where $h(.)$ is a nonnegative nondecreasing function. Further suppose that the sequence $\{c_k, 1 \leq k \leq n\}$ is a nonincreasing sequence of positive numbers. Then, for any $\epsilon > 0$,

$$\epsilon \, P\left(\max_{1 \leq k \leq n} c_k g(S_k) \geq \epsilon\right) \leq \sum_{k=1}^{n} c_k E([g(S_k) - g(S_{k-1})]I_{[\max_{1 \leq k \leq n} c_k g(S_k) \geq \epsilon]}). \tag{3.5.12}$$

Theorem 3.5.3. *Suppose the conditions in Theorem 3.5.1 hold and $E(g(S_k)) < \infty$, for every $k \geq 1$. Then, for any $\epsilon > 0$ and $n \geq 1$,*

$$\epsilon \, P\left(\max_{1 \leq k \leq n} c_k g(S_k) \geq \epsilon\right) \leq c_1 E[g(S_1)] + \sum_{k=2}^{n} c_k E([g(S_k) - g(S_{k-1})] \tag{3.5.13}$$

and, for $1 \leq n < N$,

$$\epsilon \, P\left(\max_{n \leq k \leq n} c_k g(S_k) \geq \epsilon\right) \leq c_n E[g(S_n)] + \sum_{k=n+1}^{n} c_k E([g(S_k) - g(S_{k-1})]). \tag{3.5.14}$$

Proof. Theorem 3.5.1 implies that

$$\epsilon P(\max_{1\le k\le n} c_k g(S_k) \ge \epsilon) \tag{3.5.15}$$

$$\le \sum_{j=2}^{n} c_j E([g(S_j) - g(S_{j-1})I_{[\max_{1\le k\le n} c_k g(S_k)\ge\epsilon]}])$$

$$= \sum_{i=1}^{n-1} (c_i - c_{i+1})E[g(S_i)]I_{[\max_{1\le k\le n} c_k g(S_k)\ge\epsilon]}$$

$$+ c_n E([g(S_n)]I_{[\max_{1\le k\le n} c_k g(S_k)\ge\epsilon]})$$

$$\le \sum_{i=1}^{n-1} (c_i - c_{i+1})E[g(S_i)] + c_n E[g(S_n)]$$

$$= c_1 E[g(S_1)] + \sum_{k=2}^{n} c_k E([g(S_k) - g(S_{k-1})].$$

This proves the inequality (3.5.13). By the definition of N-demimartingale, it is easily seen that the sequence $\{S_k, \ k \ge n\}$ is also an N-demimartingale for any fixed n. Applying the inequality (3.5.13), we can obtain the inequality (3.5.14). $\square$

3.6   Functions of N-Demimartingales

We now obtain bounds on the moments for maxima of functions of N-demimartingales following Wang et al. (2010).

Theorem 3.6.1. *Let the sequence $\{S_n, \ n \ge 1\}$ be an N-demimartingale with $S_0 = 0$. Let $g(.)$ be a nonnegative function such that $g(0) = 0$ and suppose that*

$$g(x) - g(y) \ge (y - x)h(y) \tag{3.6.1}$$

for all x, y where $h(.)$ is a nonnegative nondecreasing function. Further suppose that $\{c_k, \ 1 \le k \le n\}$ is a non-increasing sequence of positive numbers and $E(g(S_k))^p < \infty$, for every $k \ge 1$ for some $p > 1$. Then, for every $n \ge 1$,

$$E[\max_{1\le k\le n} c_k g(S_k)]^p \le (\frac{p}{p-1})^p E[\sum_{j=1}^{n} c_j(g(S_j) - g(S_{j-1}))]^p \tag{3.6.2}$$

and

$$E[\max_{1\le k\le n} c_k g(S_k)] \tag{3.6.3}$$

$$\le (\frac{e}{e-1})(1 + E[(\sum_{j=1}^{n} c_j(g(S_j) - g(S_{j-1})) \log^+ (\sum_{j=1}^{n} c_j(g(S_j) - g(S_{j-1}))]).$$

Proof. Theorem 3.5.2 and Hölder's inequality imply that

$$E[\max_{1 \leq k \leq n} c_k g(S_k)]^p \tag{3.6.4}$$

$$\leq p \int_0^\infty x^{p-1} P[\max_{1 \leq k \leq n} c_k g(S_k) \geq x] \; dx$$

$$\leq p \int_0^\infty x^{p-2} E[\sum_{k=1}^n c_k(g(S_k) - g(S_{k-1})) I_{[\max_{1 \leq k \leq n} c_k g(S_k) \geq x]}] \; dx$$

$$= \frac{p}{p-1} E([\sum_{k=1}^n c_k(g(S_k) - g(S_{k-1}))][\max_{1 \leq k \leq n} c_k g(S_k)]^{p-1})$$

$$\leq \frac{p}{p-1} \{E[\sum_{k=1}^n c_k(g(S_k) - g(S_{k-1}))]^p\}^{1/p} \{E[\max_{1 \leq k \leq n} c_k g(S_k)]^p\}^{1/q}$$

where q is a real number such that $\frac{1}{p} + \frac{1}{q} = 1$. Since $E(g(S_k))^p < \infty$, for every $k \geq 1$, we get that

$$(E[\max_{1 \leq k \leq n} c_k g(S_k)]^p)^{1/p} \leq \frac{p}{p-1} \{E[\sum_{k=1}^n c_k(g(S_k) - g(S_{k-1}))]^p\}^{1/p}, \tag{3.6.5}$$

and therefore

$$E[\max_{1 \leq k \leq n} c_k g(S_k)]^p \leq (\frac{p}{p-1})^p E[\sum_{k=1}^n c_k(g(S_k) - g(S_{k-1}))]^p. \tag{3.6.6}$$

Following arguments similar to those given above, we get that

$$E[\max_{1 \leq k \leq n} c_k g(S_k)] \tag{3.6.7}$$

$$\leq 1 + \int_1^\infty P[\max_{1 \leq k \leq n} c_k g(S_k) \geq x] \; dx$$

$$\leq 1 + \int_1^\infty x^{-1} E[\sum_{k=1}^n c_k(g(S_k) - g(S_{k-1})) I_{[\max_{1 \leq k \leq n} c_k g(S_k) \geq x]}] \; dx$$

$$= 1 + E[(\sum_{k=1}^n c_k(g(S_k) - g(S_{k-1}))) \log^+(\max_{1 \leq k \leq n} c_k g(S_k))].$$

For any $a \geq 0$ and $b > 0$, it is easy to see that

$$a \log^+ b \leq a \log^+ b e^{-1}.$$

Applying the above inequality, we get that

$$E[\max_{1 \leq k \leq n} c_k g(S_k)] \tag{3.6.8}$$

$$\leq (1 + E[(\sum_{j=1}^{n} c_j(g(S_j) - g(S_{j-1})) \log^+(\sum_{j=1}^{n} c_j(g(S_j) - g(S_{j-1})))])$$

$$+ e^{-1} E[\max_{1 \leq k \leq n} c_k g(S_k)].$$

Rearranging the above inequality, we get that

$$E[\max_{1 \leq k \leq n} c_k g(S_k)] \tag{3.6.9}$$

$$\leq (\frac{e}{e-1})(1 + E[(\sum_{j=1}^{n} c_j(g(S_j) - g(S_{j-1})) \log^+(\sum_{j=1}^{n} c_j(g(S_j) - g(S_{j-1})))]). \quad \square$$

As an application of Theorem 3.6.1, choosing $c_k = 1$ for every $k \geq 1$, we get that

$$E[\max_{1 \leq k \leq n} g(S_k)]^p \leq (\frac{p}{p-1})^p E[g(S_n)]^p \tag{3.6.10}$$

and

$$E[\max_{1 \leq k \leq n} c_k g(S_k)] \leq (\frac{e}{e-1})(1 + E[g(S_n) \log^+ g(S_n)]). \tag{3.6.11}$$

Remarks. Additional inequalities of the above type and their applications to derive sufficient conditions for strong laws of large numbers for N-demimartingales and their rates of convergence are given in Wang et al. (2010) following the methods in Fazekas and Klesov (2001) and Hu et al. (2008). We now discuss a result of this type.

3.7   Strong Law of Large Numbers

The following lemma is due to Fazekos and Klesov (2001).

Lemma 3.7.1. *Let the sequence $\{X_n, n \geq 1\}$ be a sequence of random variables and let $S_n = \sum_{i=1}^{n} X_i, n \geq 1$ and b_n be a nondecreasing sequence of positive numbers tending to infinity as $n \to \infty$. Suppose that α_n is a sequence of nonnegative numbers such that, for some $p > 0$ and $c > 0$,*

$$E[\max_{1 \leq i \leq n} |S_i|]^p \leq c \sum_{i=1}^{n} \alpha_i, \tag{3.7.1}$$

and

$$\sum_{i=1}^{\infty} \frac{\alpha_i}{b_i^p} < \infty. \tag{3.7.2}$$

Then

$$\lim_{n \to \infty} \frac{S_n}{b_n} = 0 \quad a.s. \tag{3.7.3}$$

Remarks. Hu et al. (2008) obtained the rate of convergence in (3.7.3) under some additional conditions.

Theorem 3.7.2. *Let the random sequence $\{S_n,\ n \geq 1\}$ be an N-demimartingale and $g(.)$ be a nonnegative function such that $g(0) = 0$ and*

$$g(x) - g(y) \geq (y - x)h(y) \tag{3.7.4}$$

for all x, y where $h(.)$ is a nonnegative nondecreasing function. Further suppose that b_n is a nondecreasing sequence of positive numbers tending to infinity as $n \to \infty$. In addition, suppose there exists $p > 1$ such that $E[g(S_{k-1})]^p \leq E[g(S_k)]^p$ for every $k \geq 1$, and

$$\sum_{n=1}^{\infty} \frac{E[g(S_n)]^p - E[g(S_{n-1})]^p}{b_n^p} < \infty. \tag{3.7.5}$$

Then

$$\lim_{n \to \infty} \frac{g(S_n)}{b_n} = 0 \ a.s. \tag{3.7.6}$$

Proof. Choose

$$\alpha_n = (\frac{p}{p-1})^p [E(g(S_n))^p - E(g(S_{n-1}))^p], \quad n \geq 1. \tag{3.7.7}$$

By hypothesis, the sequence $\alpha_n \geq 0$ for $n \geq 1$. Furthermore

$$E[\max_{1 \leq k \leq n} g(S_k)]^p \leq (\frac{p}{p-1})^p E(g(S_n))^p = \sum_{k=1}^{n} \alpha_k \tag{3.7.8}$$

and

$$\sum_{n=1}^{\infty} \frac{\alpha_n}{b_n^p} = (\frac{p}{p-1})^p \sum_{n=1}^{\infty} \frac{[E(g(S_n))^p - E(g(S_{n-1}))^p}{b_n^p} < \infty. \tag{3.7.9}$$

Applying Lemma 3.7.1 stated above, we get that

$$\lim_{n \to \infty} \frac{g(S_n)}{b_n} = 0 \ a.s. \tag{3.7.10}$$

$\square$

3.8 Azuma Type Inequality

Suppose $\{X_n, n \geq 1\}$ is a sequence of martingale differences such that $\sup_{i \geq 1} |X_i| \leq \alpha < \infty$. Let $S_n = \sum_{i=1}^{n} X_i$. Azuma (1967) proved that

$$P(S_n \geq n\epsilon) \leq \exp(-\frac{n\epsilon^2}{2\alpha^2}).$$

Christofides and Hadjikyriakou (2009) obtained a similar inequality for N-demimartingales. We now discuss their result.

Theorem 3.8.1. *Let the random sequence $\{S_n, \ n \geq 1\}$ with $S_0 = 0$ be an N-demimartingale with $E(S_1) \leq 0$. Further suppose that $|S_i - S_{i-1}| \leq c_i < \infty, \ i \geq 1$ where $c_i > 0, \ i \geq 1$. Then, for every $\epsilon > 0$,*

$$P(S_n \geq n\epsilon) \leq \exp(-\frac{n^2\epsilon^2}{2\sum_{i=1}^n c_i^2}) \tag{3.8.1}$$

and, if in addition $E(S_1) = 0$, then

$$P(|S_n| \geq n\epsilon) \leq 2\exp(-\frac{n^2\epsilon^2}{2\sum_{i=1}^n c_i^2}). \tag{3.8.2}$$

Proof. For any t real and $x \in [-c_i, c_i]$,

$$tx = \frac{1}{2}(1 + \frac{x}{c_i})(c_it) + \frac{1}{2}(1 - \frac{x}{c_i})(-c_it).$$

From the convexity of the function e^{tx}, it follows that

$$e^{tx} \leq \cosh(c_it) + \frac{x}{c_i}\sinh(c_it).$$

Hence

$$E[e^{tS_n}] = E[\prod_{i=1}^n e^{t(S_i - S_{i-1})}] \tag{3.8.3}$$

$$\leq E[\prod_{i=1}^n (\cosh(c_it) + \frac{S_i - S_{i-1}}{c_i}\sinh(c_it))].$$

Observe that, for any $t > 0$,

$$E[e^{tS_2}] = E[e^{tS_1}e^{t(S_2-S_1)}] \tag{3.8.4}$$

$$\leq E[(\cosh(c_1t) + \frac{S_1}{c_1}\sinh(c_1t))(\cosh(c_2t) + \frac{S_2 - S_1}{c_2}\sinh(c_2t))]$$

$$= \cosh(c_1t)\cosh(c_2t) + \frac{\sinh(c_1t)\sinh(c_2t)}{c_1c_2}E[S_1(S_2 - S_1)]$$

$$+ \frac{\cosh(c_1t)\sinh(c_2t)}{c_2}E(S_2 - S_1) + \frac{\cosh(c_2t)\sinh(c_1t)}{c_1}E(S_1)$$

$$= \cosh(c_1t)\cosh(c_2t) + \frac{\sinh(c_1t)\sinh(c_2t)}{c_1c_2}E[S_1(S_2 - S_1)]$$

$$\leq \cosh(c_1t)\cosh(c_2t).$$

The second equality in the above list of inequalities follows from the fact that $E(S_1) \leq 0$ and the fact that $E(S_2) = E(S_1)$ from the N-demimartingale property of the sequence $\{S_n, \ n \geq 1\}$. The last inequality is a consequence of the fact that

$E((S_1)(S_2 - S_1)) \le 0$ again from the N-demimartingale property of the sequence $\{S_n, n \ge 1\}$. Hence

$$E[e^{tS_2}] \le \prod_{i=1}^{2} \cosh(c_i t), \quad t > 0. \tag{3.8.5}$$

We will now show that

$$E[e^{tS_n}] \le \prod_{i=1}^{n} \cosh(c_i t), \quad t > 0 \tag{3.8.6}$$

for all $n \ge 2$ by the method of induction. We have proved that the inequality (3.8.6) holds for $n = 2$. Suppose the inequality (3.8.6) holds for $n = k$. We will show that it holds for $n = k + 1$. Note that

$$E[e^{tS_{k+1}}] = E[e^{t(S_{k+1} - S_k)} e^{tS_k}] \tag{3.8.7}$$

$$\le E[(\cosh(c_{k+1}t) + (S_{k+1} - S_k) \frac{\sinh(c_{k+1}t)}{c_{k+1}}) e^{tS_k}]$$

$$= \cosh(c_{k+1}t) E[e^{tS_k}] + \frac{\sinh(c_{k+1}t)}{c_{k+1}} E[(S_{k+1} - S_k) e^{tS_k}]$$

$$\le \pi_{i=1}^{k+1} \cosh(c_i t).$$

The last inequality follows from the N-demimartingale property and the induction hypothesis.

Since $\cosh(ct) \le \exp(c^2 t^2 / 2)$, the inequality (3.8.6) implies that

$$E[e^{tS_n}] \le \exp(\frac{t^2 \sum_{i=1}^{n} c_i^2}{2}), \quad n \ge 1, \ t > 0. \tag{3.8.8}$$

For any $\epsilon > 0$ and $t > 0$,

$$P(S_n \ge n\epsilon) = P(tS_n \ge tn\epsilon) \tag{3.8.9}$$

$$= P(e^{tS_n} \ge e^{tn\epsilon})$$

$$\le e^{-tn\epsilon} E[e^{tS_n}]$$

$$\le \exp[-tn\epsilon + \frac{t^2 \sum_{i=1}^{n} c_i^2}{2}].$$

The upper bound given above is minimized by choosing $t = n\epsilon / \sum_{i=1}^{n} c_i^2$ and we obtain the inequality (3.8.1).

In addition, suppose that $E(S_1) = 0$. Note that the sequence $\{-S_n, n \ge 1\}$ is an N-demimartingale with $E(-S_1) = 0$. Applying the inequality (3.8.1) derived earlier for the sequence $\{-S_n, n \ge 1\}$, we get that, for any $\epsilon > 0$,

$$P(-S_n \ge n\epsilon) \le \exp[-tn\epsilon + \frac{t^2 \sum_{i=1}^{n} c_i^2}{2}]. \tag{3.8.10}$$

Inequalities (3.8.9) and (3.8.10) show that

$$P(|S_n| \geq n\epsilon) \leq P(S_n \geq n\epsilon) + P(-S_n \geq n\epsilon) \qquad (3.8.11)$$

$$\leq 2\exp[-tn\epsilon + \frac{t^2 \sum_{i=1}^{n} c_i^2}{2}].$$

The upper bound given above is again minimized by choosing $t = n\epsilon/\sum_{i=1}^{n} c_i^2$ and we obtain the inequality (3.8.2). $\square$

As applications of the inequality derived above, the following results can be obtained.

Theorem 3.8.2. *Let the sequence* $\{X_n, \ n \geq 1\}$ *be mean zero negatively associated random variables with* $|X_k| \leq c_k, \ k \geq 1$ *where* $c_k > 0, \ k \geq 1$. *Let* $S_n = \sum_{k=1}^{n} X_k$. *Then, for every* $\epsilon > 0$,

$$P(S_n \geq n\epsilon) \leq \exp(-\frac{n^2\epsilon^2}{2\sum_{i=1}^{n} c_i^2}) \qquad (3.8.12)$$

and

$$P(|S_n| \geq n\epsilon) \leq 2\exp(-\frac{n^2\epsilon^2}{2\sum_{i=1}^{n} c_i^2}). \qquad (3.8.13)$$

The next result gives sufficient conditions for complete convergence for a class of N-demimartingales.

Theorem 3.8.3. *Let the sequence* $\{S_n, \ n \geq 1\}$ *be a mean zero* N-*demimartingale such that* $|S_i - S_{i-1}| \leq \alpha, \ i \geq 1$. *Then, for* $r > \frac{1}{2}$,

$$n^{-r}S_n \to 0 \ completely \ as \ n \to \infty. \qquad (3.8.14)$$

Proof. Observe that

$$\sum_{n=1}^{\infty} P(|S_n| \geq n^r\epsilon) \leq 2\sum_{i=1}^{\infty} \exp[-\frac{n^{2r-1}\epsilon^2}{2\alpha^2}] \qquad (3.8.15)$$

$$= 2\sum_{i=1}^{\infty} \exp[-n^{2r-1}d]$$

where $d = \epsilon^2/2\alpha^2$ and the infinite series in the last equality is convergent for every $\epsilon > 0$. $\square$

3.9   Marcinkiewicz-Zygmund Type inequality

Let the sequence $\{S_n, n \geq 1\}$ be a nonnegative N-demimartingale. We now derive a Marcinkiewicz-Zygmund type inequality for nonnegative N-demimartingales due to Hadjikyriakou (2011). Let $||f||_p$ denote the L_p norm of a function $f \in L_p$. The following result provides an upper bound for $||S_n||_p$ when $p \geq 1$ in terms of the N-demimartingale differences $d_j = S_j - S_{j-1}$.

Theorem 3.9.1. *Let the sequence $\{S_n, \ n \geq 1\}$ be a nonnegative N-demimartingale with $S_0 = 0$. Then*

(i) *for $1 < p \leq 2$,*

$$||S_n||_p^p \leq ||d_1||_p^p + 2^{2-p} \sum_{j=1}^{n-1} ||d_{j+1}||_p^p \leq 2^{2-p} \sum_{j=1}^{n} |d_j||_p^p;$$

(ii) *for $p > 2$,*

$$||S_n||_p^2 \leq ||d_1||_p^2 + (p-1) \sum_{j=1}^{n-1} ||d_{j+1}||_p^2 \leq (p-1) \sum_{j=1}^{n} |d_j||_p^2.$$

We first state a couple of lemmas which will be used in the proof of Theorem 3.9.1.

Lemma 3.9.2. *Let a, b be real and let $p \in (1, 2]$. Then*

$$|a + b|^p \leq |a|^p + p|a|^{p-1} sgn(a)b + 2^{2-p}|b|^p.$$

Lemma 3.9.3. *Let $p > 2$. Then, for $x \geq c > 0$,*

$$[x^{2-\frac{2}{p}} - c^{2-\frac{2}{p}}]^{1/2} \leq x^{1-\frac{2}{p}}[(p-1)(x^{2/p} - c^{2/p})]^{1/2}.$$

The proof of Lemma 3.9.2 follows by standard methods. Lemma 3.9.3 is a consequence of Lemma 2.1 in Rio (2009). We omit the details.

Proof of Theorem 3.9.1. (i) Let $p \in (1, 2]$. Applying Lemma 3.9.2, we get that

$$E[S_{j+1}^p] \leq E[S_j^p] + pE[S_j^{p-1}(S_{j+1} - S_j)] + 2^{2-p}E[d_{j+1}|^p]$$
$$\leq E[S_j^p] + 2^{2-p}E|d_{j+1}|^p]$$

and the last inequality follows by the N-demimartingale property of the sequence $\{S_n, \ n \geq 1\}$. Applying the inequality recursively, we get that

$$E[S_n^p] \leq E[d_1^p] + 2^{2-p} \sum_{j=2}^{n} E[|d_j|^p]$$
$$= ||d_1||_p^p + 2^{2-p} \sum_{j=2}^{n} ||d_j||_p^p$$
$$\leq 2^{2-p} \sum_{j=1}^{n} ||d_j||_p^p.$$

(ii) Suppose $p > 2$. We assume that $||S_j||_p > 0$ for all j as otherwise, the result holds trivially. Let

$$\phi(t) = E([S_j + t(S_{j+1} - S_j)]^p) \tag{3.9.1}$$

be defined on the interval $[0, \infty)$. Applying Taylor's expansion of order 2 with integral form of the remainder for the function $\phi(t)$, we get that

$$||S_j + t(S_{j+1} - S_j)||_p^p = ||S_j||_p^p + pt \, E[(S_{j+1} - S_j)S_j^{p-1}]$$
$$+ p(p-1) \int_0^t (t-s)E[(S_{j+1} - S_j)^2 |S_j + s(S_{j+1} - S_j)|^{p-2}]ds. \tag{3.9.2}$$

Applying Hölder's inequality and the N-demimartingale property, it follows that

$$\phi(t) \le ||S_j||_p^p + p(p-1)||d_{j+1}||_p^p \int_0^t (t-s)[\phi(s)]^{1-\frac{2}{p}}ds$$
$$= \psi(t) \quad \text{(say)}.$$

It can be checked that the first and second derivatives of the function $\psi(t)$ exist and are given by

$$\psi'(t) = p(p-1)||d_{j+1}||_p^2 \int_0^t [\phi(s)]^{1-\frac{2}{p}}ds$$

and

$$\psi''(t) = p(p-1)||d_{j+1}||_p^2[\phi(t)]^{1-\frac{2}{p}} \le p(p-1)||d_{j+1}||_p^2[\psi(t)]^{1-\frac{2}{p}}.$$

Multiplying both sides of the inequality given above by the function $2\psi'$ and integrating between 0 and x, we have

$$\int_0^x 2\psi'(t)\psi''(t)dt \le 2p(p-1)||d_{j+1}||_p^2 \int_0^x \psi'(t)[\psi(t)]^{1-\frac{2}{p}}dt.$$

Therefore

$$\psi'(x) \le p||d_{j+1}||_p[(\psi(x))^{2-\frac{2}{p}} - c^{2-\frac{2}{p}}]^{1/2}$$

where $c = ||S_j||_p^p$. Applying Lemma 3.9.3, we get that

$$\psi'(t) \le p(p-1)^{1/2}||d_{j+1}||_p[\psi(t)]^{1-\frac{2}{p}}[(\psi(t))^{\frac{2}{p}} - c^{\frac{2}{p}}]^{1/2}. \tag{3.9.3}$$

Let $z(t) = [\psi(t)]^{2/p}$. Then the inequality (3.9.3) can be written as

$$z'(t)[z(t) - c^{2/p}]^{-\frac{1}{2}} \le 2(p-1)^{1/2}||d_{j+1}||_p. \tag{3.9.4}$$

Solving this differential inequality for $z(t)$, we get that

$$z(t) \le (p-1)|d_{j+1}||_p^2 t^2 + c^{2/p}. \tag{3.9.5}$$

Since $\phi(t) \leq \psi(t)$, it follows that $[\phi(t)]^{2/p} \leq z(t)$, and we get the inequality

$$[\phi(t)]^{2/p} \leq ||S_j||_p^2 + (p-1)||S_{j+1} - S_j||_p^2 t^2. \tag{3.9.6}$$

Since $\phi(1) = ||S_{j+1}||_p^p$, the result follows by induction. □

As a consequence of Theorem 3.9.1, we obtain the following large deviation inequalities for nonnegative N-demimartingales.

Theorem 3.9.4. *Let the sequence $\{S_n, n \geq 1\}$ be a nonnegative N-demimartingale such that*

$$||S_{j+1} - S_j||_p < M_{j+1} < \infty, \quad j \geq 1 \tag{3.9.7}$$

for some $p > 1$. Let $\epsilon > 0$. Then

$$P(S_n > n\epsilon) \leq \frac{4}{(2n\epsilon)^p} \sum_{j=1}^n M_j^p$$

if $1 < p \leq 2$ and

$$P(S_n > n\epsilon) \leq \frac{(p-1)^{p/2}}{(n\epsilon)^p} \left(\sum_{j=1}^n M_j^2 \right)^{p/2}$$

if $p > 2$.

Proof. Suppose $p \in (1, 2]$. Applying Theorem 3.9.1, we get that

$$E[S_n^p] \leq 2^{2-p} \sum_{j=1}^n ||d_j||_p^p \leq 2^{2-p} \sum_{j=1}^n M_j^p$$

under the condition given in (3.9.7). Hence, for every $\epsilon > 0$,

$$P(S_n > n\epsilon) \leq \frac{E[S_n^p]}{n^p \epsilon^p} \leq \frac{4}{(2n\epsilon)^p} \sum_{j=1}^n M_j^p.$$

Suppose $p > 2$. Applying Theorem 3.9.1 again, we get that

$$E[S_n^p] \leq (p-1)^{p/2} \left(\sum_{j=1}^n M_j^2 \right)^{p/2}.$$

Hence, for every $\epsilon > 0$,

$$P(S_n > n\epsilon) \leq \frac{E[S_n^p]}{n^p \epsilon^p} \leq \frac{(p-1)^{p/2}}{(n\epsilon)^p} \left(\sum_{j=1}^n M_j^2 \right)^{p/2}. \quad □$$

Complete convergence results for nonnegative N-demimartingales can be proved by using the large deviation inequalities derived above. For details, see Hadjikyriakou (2011).

3.10   Comparison Theorem on Moment Inequalities

We now introduce a stronger condition on an N-demimartingale to derive a comparison theorem.

Definition. A random sequence $\{S_n, n \geq 1\}$ is said to be a *strong N-demimartingale* if, for any two componentwise nondecreasing functions f and g,

$$\text{Cov}[g(S_{j+1} - S_j), f(S_1, \ldots, S_j)] \leq 0, \quad j \geq 1 \qquad (3.10.1)$$

whenever the expectation is defined.

It is easy to see that if a sequence $\{X_j, j \geq 1\}$ is a negatively associated sequence of random variables, then the corresponding partial sums form a strong N-demimartingale.

Theorem 3.10.1. *Let the random sequence $\{S_j, 1 \leq j \leq n\}$ be a strong N-demimartingale. Define $X_j = S_j - S_{j-1}, j \geq 1$ with $S_0 = 0$. Let $X_j^*, 1 \leq j \leq n$ be independent random variables such that X_j and X_j^* have the same distribution. Let $S_n^* = X_1^* + \ldots + X_n^*$. Then, for any convex function h,*

$$E(h(S_n)) \leq E(h(S_n^*)) \qquad (3.10.2)$$

whenever the expectations exist. Furthermore, if $h(.)$ is a nondecreasing convex function, then

$$E(\max_{1 \leq j \leq n} h(S_j)) \leq E(\max_{1 \leq j \leq n} h(S_j^*)) \qquad (3.10.3)$$

whenever the expectations exist.

Proof. Let a random vector (Y_1, Y_2) be independent and identically distributed as the random vector (X_1, X_2). It is well known that for any convex function h on the real line, there exists a nondecreasing function $g(.)$ such that for all $a < b$,

$$h(b) - h(a) = \int_a^b g(t)dt.$$

(cf. Roberts and Verberg (1973)). Hence

$$h(X_1 + X_2) + h(Y_1 + Y_2) - h(X_1 + Y_2) - h(Y_1 + X_2)$$

$$= \int_{X_2}^{X_1} [g(Y_1 + t) - g(X_1 + t)] \ dt$$

$$= \int_{-\infty}^{\infty} [g(Y_1 + t) - g(X_1 + t)](I_{[Y_2 > t]} - I_{[X_2 > t]}) \ dt.$$

Therefore

$$2E[h(S_2)] - 2E[h(S_2^*)]$$
$$= 2(E[h(X_1 + X_2) - E[h(X_1^* + X_2^*)]])$$
$$= E[h(X_1 + X_2) + h(Y_1 + Y_2) - h(X_1 + Y_2) - h(Y_1 + X_2)]$$
$$= \int_{-\infty}^{\infty} \text{Cov}(g(X_1 + t), I_{[X_2 > t]}) \, dt.$$

Observe that the functions $g(x+t)$ and $I_{[x>t]}$ are nondecreasing functions in x for each t. Since $S_i, 1 \leq i \leq n$ forms a strong N-demimartingale, it follows that

$$\text{Cov}(g(X_1 + t), I_{[X_2 > t]}) \leq 0$$

for each t which in turn implies that

$$E[h(S_2)] - E[h(S_2^*)] \leq 0. \tag{3.10.4}$$

This proves the theorem for the case $n = 2$. We now prove the result by induction on n. Suppose the result holds for the case of $n - 1$ random variables. Let

$$k(x) = E(h(x + S_{n-1})).$$

Since the function $h(x + .)$ is a convex function for any fixed x, it follows that

$$k(x) \leq E(h(x + S_{n-1}^*)). \tag{3.10.5}$$

Since $\{S_j, 1 \leq j \leq n\}$ forms a strong N-demimartingale, it follows that $\{S_{n-1}, S_n\}$ also forms a strong N-demimartingale which reduces to negative association of the two random variables S_{n-1} and X_n. Therefore

$$E[h(S_n)] \leq E[h(X_n^* + S_{n-1}^*)] \quad \text{(by induction hypothesis)}$$
$$= E(k(X_n^*))$$
$$\leq E(h(X_n^* + S_{n-1}^*)) \quad \text{(by (3.9.5))}$$
$$= E[h(S_n^*)].$$

This proves the result for the case of n random variables completing the proof by induction. The second part of the theorem can be proved by arguments analogous to those given in Shao (2002). We omit the details. □

Remarks. The proofs given above hinge on the fact that two random variables Z_1, Z_2 are negatively associated if and only if $Z_1, Z_1 + Z_2$ forms a strong N-demimartingale. This is obvious from the definition of a strong N-demimartingale. As a consequence of the comparison theorem, the Rosenthal maximal inequality and the Kolomogorov exponential inequality hold for strong N-demimartingales.

Results presented here are from Prakasa Rao (2004).

Chapter 4

Conditional Demimartingales

4.1 Introduction

We will discuss the properties of conditional demimartingales in this chapter. Chow and Teicher (1988) and more recently Majerek et al. (2005) discussed the concept of conditional independence in detail and we introduced the notions of conditional strong mixing and conditional association for sequences of random variables as found in Prakasa Rao (2009). Properties of such random sequences were investigated in Prakasa Rao (2009) and Roussas (2008). Based on these ideas, Hydjikyriakou (2010) introduced the notion of conditional demimartingales.

4.2 Conditional Independence

Let $(\Omega, \mathcal{A}, P)$ be a probability space. Members of a set of events $A_1, A_2, \ldots, A_n$ are said to be *independent* if

$$P(\bigcap_{j=1}^{k} A_{i_j}) = \prod_{j=1}^{k} P(A_{i_j}) \tag{4.2.1}$$

for all $1 \leq i_1 < i_2 < \ldots < i_k \leq n$, $2 \leq k \leq n$.

Definition. The members of the set of events $A_1, A_2, \ldots, A_n$ are said to be *conditionally independent* given an event B with $P(B) > 0$ if

$$P(\bigcap_{j=1}^{k} A_{i_j} | B) = \prod_{j=1}^{k} P(A_{i_j} | B) \tag{4.2.2}$$

for all $1 \leq i_1 < i_2 < \ldots < i_k \leq n$, $2 \leq k \leq n$.

B.L.S. Prakasa Rao, *Associated Sequences, Demimartingales and Nonparametric Inference*, Probability and its Applications, DOI 10.1007/978-3-0348-0240-6_4, © Springer Basel AG 2012

The following examples (cf. Majerak et al. (2005)) show that the independence of events does not imply conditional independence and that the conditional
independence of events does not imply their independence.

Example 4.2.1. Let $\Omega = \{1, 2, 3, 4, 5, 6, 7, 8\}$ and $p_i = 1/8$ be the probability as-
signed to the event $\{i\}$ for $1 \leq i \leq 8$. Let $A_1 = \{1, 2, 3, 4\}$, $A_2 = \{3, 4, 5, 6\}$ and
$B = \{2, 3, 4, 5\}$. It is easy to see that the events A_1 and A_2 are independent but
not conditionally independent given the event B.

Example 4.2.2. Consider an experiment of choosing a coin numbered $\{i\}$ with
probability $p_i = \frac{1}{n}$, $1 \leq i \leq n$ from a set of n coins and suppose it is tossed twice.
Let $p_0^i = \frac{1}{2^i}$ be the probability for tails for the i-th coin. Let A_1 be the event that
tails appears in the first toss, A_2 be the event that tails appears in the second toss
and H_i be the event that the i-th coin is selected. It can be checked that the events
A_1 and A_2 are conditionally independent given H_i but they are not independent
as long as the number of coins $n \geq 2$.

Let $(\Omega, \mathcal{A}, P)$ be a probability space. Let $\mathcal{F}$ be a sub-σ-algebra of $\mathcal{A}$ and let
I_A denote the indicator function of an event A.

Definition. The members of the set of events $A_1, A_2, \ldots, A_n$ are said to be *conditionally independent* given $\mathcal{F}$ or *$\mathcal{F}$-independent* if

$$E(\prod_{j=1}^{k} I_{A_{i_j}} | \mathcal{F}) = \prod_{j=1}^{k} E(I_{A_{i_j}}) | \mathcal{F}) \quad \text{a.s.} \tag{4.2.3}$$

for all $1 \leq i_1 < i_2 < \ldots < i_k \leq n$, $2 \leq k \leq n$.

For the definition of conditional expectation of a measurable function X given
a σ-algebra $\mathcal{F}$, see Doob (1953).

Remarks. If $\mathcal{F} = \{\Phi, \Omega\}$, then the above definition reduces to the usual definition
of stochastic independence for random variables. If $\mathcal{F} = \mathcal{A}$, then equation (4.2.3)
reduces to the product of indicator $\mathcal{A}$-measurable functions on both sides.

Let $\{\zeta_n, n \geq 1\}$ be a sequence of classes of events. The sequence is said to
be conditionally independent given $\mathcal{F}$ if for all choices $A_m \in \zeta_{k_m}$ where $k_i \neq k_j$
for $i \neq j$, $m = 1, 2, \ldots, n$ and $n \geq 2$,

$$E(\prod_{j=1}^{k} I_{A_j} | \mathcal{F}) = \prod_{j=1}^{k} E(I_{A_j} | \mathcal{F}) \quad \text{a.s.} \tag{4.2.4}$$

For any set of real-valued random variables $X_1, X_2, \ldots, X_n$ defined on
$(\Omega, \mathcal{A}, P)$, let

$$\sigma(X_1, X_2, \ldots, X_n)$$

denote the smallest σ-algebra with respect to which it is measurable.

Definition. A sequence of random variables $\{X_n, n \geq 1\}$ defined on a probability space $(\Omega, \mathcal{A}, P)$ is said to be *conditionally independent* given a sub-σ-algebra $\mathcal{F}$ or *$\mathcal{F}$-independent* if the sequence of classes of events $\zeta_n = \sigma(X_n)$, $n \geq 1$ are conditionally independent given $\mathcal{F}$.

It can be checked that a set of random variables $X_1, X_2, \ldots, X_n$ defined on a probability space $(\Omega, \mathcal{A}, P)$ are conditionally independent given a sub-σ-algebra $\mathcal{F}$ if and only if for all $(x_1, x_2, \ldots, x_n) \in R^n$,

$$E(\prod_{i=1}^{n} I_{[X_i \leq x_i]} | \mathcal{F}) = \prod_{i=1}^{n} E(I_{[X_i \leq x_i]} | \mathcal{F}) \text{ a.s.}$$

Remarks. Independent random variables $\{X_n, n \geq 1\}$ may lose their independence under conditioning. For instance, let $\{X_1, X_2\}$ be Bernoulli trials with probability of success p with $0 < p < 1$. Let $S_2 = X_1 + X_2$. Then $P(X_i = 1 | S_2 = 1) > 0$, $i = 1, 2$ but $P(X_1 = 1, X_2 = 1 | S_2 = 1) = 0$. On the other hand, dependent random variables may become independent under conditioning, that is, they become conditionally independent. This can be seen from the following discussion.

Let the sequence $\{X_i, i \geq 1\}$ be independent positive integer-valued random variables. Then the sequence $\{S_n, n \geq 1\}$ is a dependent sequence where $S_n = X_1 + \ldots + X_n$. Let us consider the event $[S_2 = k]$ with positive probability for some positive integer k. Check that

$$P(S_1 = i, S_3 = j | S_2 = k) = P(S_1 = i | S_2 = k) P(S_3 = j | S_2 = k).$$

Hence the random variables S_1 and S_3 are conditionally independent given S_2. If we interpret the subscript n of S_n as "time", "past and future are conditionally independent given the present". This property holds not only for partial sums S_n of independent random variables but also when the random sequence $\{S_n, n \geq 1\}$ forms a time homogeneous Markov chain (cf. Chow and Teicher (1988)).

Remarks. (i) A sequence of random variables $\{X_n, n \geq 1\}$ defined on a probability space $(\Omega, \mathcal{A}, P)$ is said to be *exchangeable* if the joint distribution of every finite subset of k of these random variables depends only upon k and not on the particular subset. It can be proved that the sequence of random variables $\{X_n, n \geq 1\}$ is exchangeable if and only if the random variables are conditionally independent and identically distributed for a suitably chosen sub-σ-algebra $\mathcal{F}$ of $\mathcal{A}$ (cf. Chow and Teicher (1988)). If a random variable Z is independent of a sequence of independent and identically distributed random variables $\{X_n, n \geq 1\}$, then the sequence $\{Y_n, n \geq 1\}$ where $Y_n = Z + X_n$ forms an exchangeable sequence of random variables and hence conditionally independent.

(ii) Another example of a conditionally independent sequence is discussed in Prakasa Rao (1987) (cf. Gyires (1981)). This model can be described as follows. Let $\theta_{h,j}^{(k)}$, $1 \leq h, j \leq p$, $1 \leq k \leq n$ be independent real-valued random variables defined on a probability space $(\Omega, \mathcal{A}, P)$. Let $\{\eta_j, j \geq 0\}$ be a homogeneous Markov chain defined on the same space with state space $\{1, \ldots, p\}$ and a nonsingular transition matrix $A = ((a_{hj}))$. We denote this Markov chain by $\{A\}$. Define $\psi_k = \theta^{(k)}_{\eta_{k-1},\eta_k}$ for $1 \leq k \leq n$. The sequence of random variables $\{\psi_k, 1 \leq k \leq n\}$ is said to be defined on the homogeneous Markov chain $\{A\}$. Let $\mathcal{F}$ be the sub-σ-algebra generated by the sequence $\{\eta_j, j \geq 0\}$. It is easy to check that the random variables $\{\psi_k, 1 \leq k \leq n\}$ are conditionally independent, in fact, $\mathcal{F}$-independent.

4.3   Conditional Association

The concept of conditional association was introduced in Prakasa Rao (2009). Let X and Y be random variables defined on a probability space $(\Omega, \mathcal{A}, P)$ with $E(X^2) < \infty$ and $E(Y^2) < \infty$. Let $\mathcal{F}$- be a sub-σ-algebra of $\mathcal{A}$. We define the *conditional covariance* of X and Y given $\mathcal{F}$ or $\mathcal{F}$-*covariance* as

$$\mathrm{Cov}^{\mathcal{F}}(X, Y) = E^{\mathcal{F}}[(X - E^{\mathcal{F}}X)(Y - E^{\mathcal{F}}Y)]$$

(cf. Prakasa Rao (2009)). It easy to see that the $\mathcal{F}$-covariance reduces to the ordinary concept of covariance when $\mathcal{F} = \{\Phi, \Omega\}$. A set of random variables $\{X_k, 1 \leq k \leq n\}$ is said to be $\mathcal{F}$-*associated* if, for any componentwise nondecreasing functions h, g defined on R^n,

$$\mathrm{Cov}^{\mathcal{F}}(h(X_1, \ldots, X_n), g(X_1, \ldots, X_n)) \geq 0 \text{ a.s.}$$

A sequence of random variables $\{X_n, n \geq 1\}$ is said to be $\mathcal{F}$-associated if every finite subset of the sequence $\{X_n, n \geq 1\}$ is $\mathcal{F}$-associated.

An example of an $\mathcal{F}$-associated sequence $\{X_n, n \geq 1\}$ is obtained by defining $X_n = Z + Y_n$, $n \geq 1$ where Z and Y_n, $n \geq 1$ are $\mathcal{F}$-independent random variables as defined in Section 4.2. It can be shown by standard arguments that

$$\mathrm{Cov}^{\mathcal{F}}(X, Y) = \int_{-\infty}^{\infty} \int_{-\infty}^{\infty} H^{\mathcal{F}}(x, y) \; dx dy \text{ a.s.}$$

where

$$H^{\mathcal{F}}(x, y) = E^{\mathcal{F}}[I_{(X \leq x, Y \leq y)}] - E^{\mathcal{F}}[I_{[X \leq x]}] E^{\mathcal{F}}[I_{[Y \leq y]}].$$

Let X and Y be $\mathcal{F}$-associated random variables. Suppose that f and g are almost everywhere differentiable functions such that $\mathrm{ess\,sup}_x |f'(x)| < \infty$, $\mathrm{ess\,sup}_x |g'(x)| < \infty$, $E^{\mathcal{F}}[f^2(X)] < \infty$ a.s. and $E^{\mathcal{F}}[g^2(Y)] < \infty$ a.s. where f' and g' denote the derivatives of f and g respectively whenever they exist. Then it can be shown that

$$\mathrm{Cov}^{\mathcal{F}}(f(X), g(Y)) = \int_{-\infty}^{\infty} \int_{-\infty}^{\infty} f'(x) g'(y) H^{\mathcal{F}}(x, y) \; dx dy \text{ a.s.} \qquad (4.3.1)$$

and hence

$$|\mathrm{Cov}^{\mathcal{F}}(f(X), g(Y))| \leq \mathrm{ess\,sup}_x |f'(x)| \mathrm{ess\,sup}_y |g'(y)| \, \mathrm{Cov}^{\mathcal{F}}(X, Y)| \text{ a.s..} \qquad (4.3.2)$$

Proofs of these results can be obtained following the methods used for the study of *associated* random variables. As a consequence of these covariance inequalities, one can obtain a central limit theorem for conditionally associated sequences of random variables following the methods in Newman (1984). Note that $\mathcal{F}$-association does not imply association and vice versa (cf. Roussas (2008)). For an extensive discussion on conditionally associated random variables, see Roussas (2008). Yuan et al. (2011) obtained limit theorems for conditionally associated random variables. A related notion of conditionally negatively associated random variables was introduced and their limiting properties were studied in Yuan et al. (2010).

4.4 Conditional Demimartingales

Following the concept of the conditional association discussed above, Hadjikyriakou (2010) introduced the notion of conditional demimartingales and studied their properties. We will now discuss these results.

Definition. Let the random sequence $\{S_n, n \geq 1\}$ be a sequence of random variables defined on a probability space $(\Omega, \mathcal{A}, P)$. Let $\mathcal{F}$ be a sub-σ-algebra of $\mathcal{A}$. The sequence $\{S_n, n \geq 1\}$ is called an $\mathcal{F}$-*demimartingale* if for every componentwise nondecreasing function $f : R^j \to R$,

$$E[(S_j - S_i)f(S_1, S_2, \ldots, S_i)|\mathcal{F}] \geq 0, \quad 1 \leq i < j < \infty.$$

If the above condition holds only for nonnegative and nondecreasing functions f, then the sequence $\{S_n, n \geq 1\}$ is called an $\mathcal{F}$-*demisubmartingale*.

From the property of conditional expectations that $E(E(Z|\mathcal{F})) = E(Z)$ for any random variable Z with $E|Z| < \infty$, it follows that any $\mathcal{F}$-demimartingale defined on a probability space $(\Omega, \mathcal{A}, P)$ is a demimartingale on the probability space $(\Omega, \mathcal{A}, P)$ and any $\mathcal{F}$-demisubmartingale defined on the probability space $(\Omega, \mathcal{A}, P)$ is a demisubmartingale on the probability space $(\Omega, \mathcal{A}, P)$. We will now give an example to show that the converse is not true. This example is due to Hadjikyriakou (2010).

Example 4.4.1. Let (X_1, X_2) be a bivariate random vector such that

$$P(X_1 = 5, X_2 = 7) = \frac{3}{8}, \quad P(X_1 = -3, X_2 = 7) = \frac{1}{8}$$

and

$$P(X_1 = -3, X_2 = -7) = \frac{4}{8}.$$

Let $\mathcal{F}$ be the σ-algebra generated by the event $[|X_1 X_2| = 21]$. It can be checked that the sequence $\{X_1, X_2\}$ is a demimartingale (cf. Example 2.1.2) but it is not an $\mathcal{F}$-demisubmartingale since

$$E[(X_2 - X_1)||X_1 X_2| = 21] = -\frac{6}{8}f(-3) < 0$$

for a nonnegative function f.

For convenience, we denote the conditional probability $P(A|\mathcal{F}) = E(I_A|\mathcal{F})$ of an event A by $P^{\mathcal{F}}(A)$ and the conditional expectation $E(Z|\mathcal{F})$ of a random variable Z with respect to a σ-algebra $\mathcal{F}$ by $E^{\mathcal{F}}(Z)$.

Let $\{X_n, n \geq 1\}$ be a sequence of $\mathcal{F}$-associated random variables such that $E^{\mathcal{F}}(X_n) = 0$ a.s., $n \geq 1$. Let $S_n = \sum_{i=1}^{n} X_i$, $n \geq 1$. Then it is easy to check that the sequence $\{S_n, n \geq 1\}$ is an $\mathcal{F}$-demimartingale. Hadjikyriakou (2010) proved the following result for conditional demimartingales. We omit the proof as it is similar to the corresponding result for demimartingales given in Chapter 2.

Theorem 4.4.1. *Let a random sequence* $\{S_n, n \geq 1\}$ *be an* $\mathcal{F}$-*demimartingale or an* $\mathcal{F}$-*demisubmartingale and* $g(.)$ *be a nondecreasing convex function. Then the random sequence* $\{g(S_n), n \geq 1\}$ *is an* $\mathcal{F}$-*demisubmartingale.*

The following inequality is a Chow type maximal inequality for conditional demimartingales.

Theorem 4.4.2. *Let the random sequence* $\{S_n, n \geq 1\}$ *be an* $\mathcal{F}$-*demimartingale with* $S_0 = 0$ *and let* $g(.)$ *be a nonnegative convex function such that* $g(0) = 0$. *Let*

$$A = [\max_{1 \leq i \leq n} c_i g(S_i) \geq \epsilon]$$

where $\{c_n, n \geq 1\}$ *is a positive nonincreasing sequence of* $\mathcal{F}$-*measurable random variables and* ϵ *is an* $\mathcal{F}$-*measurable random variable such that* $\epsilon > 0$ *a.s. Then*

$$\epsilon P^{\mathcal{F}}(A) \leq \sum_{i=1}^{n} c_i E^{\mathcal{F}}[g(S_i) - g(S_{i-1})] - c_n E^{\mathcal{F}}[g(S_n)I_{A^c}] \quad a.s \qquad (4.4.1)$$

$$\leq \sum_{i=1}^{n} c_i E^{\mathcal{F}}[g(S_i) - g(S_{i-1})] \quad a.s.$$

We omit the proof of this result as it is similar to that of the Chow type maximal inequality for demisubmartingales given in Chapter 2.

As an application of the above results, we can obtain a strong law of large numbers for $\mathcal{F}$-demimartingales. We will first discuss a Kronecker type lemma which will be used to obtain the strong law of large numbers.

Lemma 4.4.3. *Let the sequence* $\{S_n, n \geq 1\}$ *be a random sequence with* $S_0 = 0$ *and* $\{c_n, n \geq 1\}$ *be a nonincreasing sequence of* $\mathcal{F}$-*measurable random variables such that* $c_n \to 0$ *a.s. as* $n \to \infty$. *Further suppose that*

$$\sum_{k=1}^{\infty} c_k E^{\mathcal{F}}[g(S_k) - g(S_{k-1})] < \infty \quad a.s.$$

where $g(.)$ *is a function such that* $g(0) = 0$. *Then*

$$c_n E^{\mathcal{F}}[g(S_n)] \to 0 \quad a.s. \ as \ n \to \infty.$$

Proof. Let

$$A = \{\omega \in \Omega : \sum_{k=1}^{\infty} c_k E^{\mathcal{F}}[g(S_k) - g(S_{k-1})](\omega) < \infty\}.$$

Note that $P(A) = 1$ by hypothesis. Let $\omega_0 \in A$. Then

$$\sum_{k=1}^{\infty} c_k(\omega_0) E^{\mathcal{F}}[g(S_k) - g(S_{k-1})](\omega_0) < \infty.$$

Hence, by a Kronecker type lemma, we get that

$$\sum_{k=1}^{n} c_n(\omega_0) E^{\mathcal{F}}[g(S_k) - g(S_{k-1})](\omega_0) \to 0 \quad \text{as } n \to \infty$$

or equivalently

$$c_n(\omega_0) E^{\mathcal{F}}[g(S_n)](\omega_0) \to 0 \quad \text{as } n \to \infty.$$

Since this holds for every $\omega_0 \in A$ and $P(A) = 1$, we get that

$$c_n E^{\mathcal{F}}[g(S_n)] \to 0 \quad \text{a.s. as } n \to \infty. \qquad \square$$

We now prove a strong law of large numbers for $\mathcal{F}$-demimartingales.

Theorem 4.4.4. *Let the random sequence $\{S_n, n \geq 1\}$ be an $\mathcal{F}$-demimartingale with $S_0 \equiv 0$ and let $g(.)$ be a nonnegative convex function such that $g(0) = 0$. Suppose that the sequence $\{c_n, n \geq 1\}$ is a positive nonincreasing sequence of $\mathcal{F}$-measurable random variables such that $c_n \to 0$ a.s. as $n \to \infty$. Further suppose that*

$$\sum_{k=1}^{\infty} c_k E^{\mathcal{F}}[g(S_k) - g(S_{k-1})] < \infty \quad a.s. \tag{4.4.2}$$

and

$$E^{\mathcal{F}}[g(S_n)] < \infty \quad a.s., \quad n \geq 1.$$

Then, conditionally on $\mathcal{F}$,

$$c_n g(S_n) \to 0 \quad a.s. \text{ as } n \to \infty.$$

Proof. Let ϵ be an $\mathcal{F}$-measurable random variable such that $\epsilon > 0$ a.s. By the Chow type inequality derived earlier, it follows that

$$\epsilon P^{\mathcal{F}}[\sup_{k \geq n} c_k g(S_k) \geq \epsilon] \leq \sum_{k=n}^{\infty} c_k E^{\mathcal{F}}[g(S_k) - g(S_{k-1})] \tag{4.4.3}$$

$$\leq c_n E^{\mathcal{F}}[g(S_n)] + \sum_{k=n+1}^{\infty} c_k E^{\mathcal{F}}[g(S_k) - g(S_{k-1})] \quad \text{a.s.}$$

Condition (4.4.2) implies that

$$\sum_{k=n+1}^{\infty} c_k E^{\mathcal{F}}[g(S_k) - g(S_{k-1})] \to 0 \ \text{ a.s. as } \ n \to \infty. \tag{4.4.4}$$

Applying the conditional version of the Kronecker type lemma derived above, we get that

$$c_n E^{\mathcal{F}}[g(S_n)] \to 0 \ \text{ a.s. as } \ n \to \infty. \tag{4.4.5}$$

Combining equations (4.4.3),(4.4.4) and (4.4.5), we obtain the result.   □

Remarks. Following the notion of conditionally negatively associated random variables introduced in Yuan et al. (2010), it is possible to define conditional N-demimartingales in an obvious manner. An example of such a sequence is the sequence of partial sums of conditionally negatively associated random variables with conditional mean zero. We do not go into the details on results for such random sequences.

Chapter 5

Multidimensionally Indexed Demimartingales and Continuous Parameter Demimartingales

5.1 Introduction

We have studied demimartingales, N-demimartingales and conditional demimartingales in the last three chapters. These are all discrete parameter stochastic processes with index in one-dimension. We now introduce more general demimartingales such as multidimensionally indexed demimartingales and continuous parameter demisubmartingales and discuss maximal inequalities for such processes.

5.2 Multidimensionally Indexed Demimartingales

Christofides and Hadjikyriakou (2010) introduced multidimensionally indexed demimartingales and demisubmartingales. We now describe these random fields.

Let N^d denote the d-dimensional positive integer lattice. For $\mathbf{n}, \mathbf{m}, \in N^d$ with $\mathbf{n} = (n_1, \ldots, n_d)$ and $\mathbf{m} = (m_1, \ldots, m_d)$, we write $\mathbf{n} \leq \mathbf{m}$ if $n_i \leq m_i$, $i = 1, \ldots, d$ and write $\mathbf{n} < \mathbf{m}$ if $n_i \leq m_i$, $i = 1, \ldots, d$ with at least one strict inequality. We say that $\mathbf{k} \to \infty$ if $\min_{1 \leq j \leq d} k_j \to \infty$.

Definition. A collection of multidimensionally indexed random variables $\{X_{\mathbf{i}}, \mathbf{i} \leq \mathbf{n}\}$ is said to be *associated* if for any two componentwise nondecreasing functions

B.L.S. Prakasa Rao, *Associated Sequences, Demimartingales*
and Nonparametric Inference, Probability and its Applications,
DOI 10.1007/978-3-0348-0240-6_5, © Springer Basel AG 2012

f and g,

$$\text{Cov}(f(X_{\mathbf{i}}, \; \mathbf{i} \leq \mathbf{n}), g(X_{\mathbf{i}}, \; \mathbf{i} \leq \mathbf{n})) \geq 0$$

provided that the covariance exists. An infinite collection is said to be *associated* if every finite sub-collection is associated. Such a family of random variables is also called an *associated random field*.

Properties of associated random fields are investigated extensively in Bulinski and Shaskin (2007).

Definition. A collection of multidimensionally indexed random variables $\{X_{\mathbf{n}}, \; \mathbf{n} \in N^d\}$ is called a *multidimensionally indexed demimartingale* if

$$E[(X_{\mathbf{j}} - X_{\mathbf{i}})f(X_{\mathbf{k}}, \; \mathbf{k} \leq \mathbf{i})] \geq 0$$

for all $\mathbf{i}, \mathbf{j} \in N^d$ with $\mathbf{i} \leq \mathbf{j}$ and for all componentwise nondecreasing functions f. If, in addition f is required to be nonnegative, then the collection $\{X_{\mathbf{n}}, \; \mathbf{n} \in N^d\}$ is said to be a *multidimensionally indexed demisubmartingale*.

Remarks. It is easy to see that the partial sums of mean zero associated multidimensionally indexed random variables form a multidimensionally indexed demimartingale.

Hereafter we discuss the case when $d = 2$ for convenience and call such two-dimensional indexed demimartingales as *two-parameter demimartingales*. For some discussion on two-parameter martingales, see Cairoli and Walsh (1975) and more recently in Amirdjanova and Linn (2007) and Prakasa Rao (2010).

5.3   Chow Type Maximal Inequality for Two-Parameter Demimartingales

The following result is a Chow type maximal inequality, due to Christofides and Hadjikyriakou (2010), for the random field $\{g(Y_{\mathbf{n}}), \; \mathbf{n} \in N^2\}$ where $\{Y_{\mathbf{n}}, \; \mathbf{n} \in N^2\}$ is a two-parameter demimartingale with $Y_{\mathbf{k}} = 0$ whenever $k_1 k_2 = 0$.

Theorem 5.3.1. *Let the array* $\{Y_{\mathbf{n}}, \; \mathbf{n} \in N^2\}$ *be a two-parameter demimartingale with* $Y_{\mathbf{k}} = 0$ *whenever* $k_1 k_2 = 0$. *Further suppose that* $\{c_{\mathbf{n}}, \; \mathbf{n} \in N^2\}$ *is a nonincreasing array of positive numbers and* $g(.)$ *be a nonnegative nondecreasing convex function on* R *with* $g(0) = 0$. *Then, for every* $\epsilon > 0$,

$$\epsilon P[\max_{(i,j) \leq (n_1, n_2)} c_{ij} g(Y_{ij}) \geq \epsilon] \tag{5.3.1}$$

$$\leq \min(\sum_{i=1}^{n_1} \sum_{j=1}^{n_2} c_{ij} E[g(Y_{ij}) - g(Y_{i-1,j})], \sum_{i=1}^{n_1} \sum_{j=1}^{n_2} c_{ij} E[g(Y_{ij}) - g(Y_{i,j-1})]).$$

Proof. Let

$$A = [\max_{(i,j) \leq (n_1, n_2)} c_{ij} g(Y_{ij}) \geq \epsilon], \tag{5.3.2}$$

$$B_{1j} = [c_{1j} g(Y_{1j}) \geq \epsilon], \quad 1 \leq j \leq n_2, \tag{5.3.3}$$

and

$$B_{ij} = [c_{rj} g(Y_{rj}) < \epsilon, \ 1 \leq r < i; \ c_{ij} g(Y_{ij}) \geq \epsilon], \quad 2 \leq i \leq n_1, \ 1 \leq j \leq n_2. \tag{5.3.4}$$

From the definition of the sets A and B_{ij}, we note that $A = \cup_{i,j} B_{ij}$. Hence

$$\epsilon P(A) = \epsilon P(\cup_{(i,j) \leq (n_1, n_2)} B_{ij}) \tag{5.3.5}$$

$$\leq \epsilon \sum_{j=1}^{n_2} \sum_{i=1}^{n_1} P(B_{ij})$$

$$= \sum_{j=1}^{n_2} \sum_{i=1}^{n_1} E(\epsilon I_{B_{ij}})$$

$$\leq \sum_{j=1}^{n_2} \sum_{i=1}^{n_1} E(c_{ij} g(Y_{ij}) I_{B_{ij}})$$

$$= \sum_{j=1}^{n_2} E(c_{1j} g(Y_{1j}) I_{B_{1j}}) + \sum_{j=1}^{n_2} \sum_{i=2}^{n_1} E(c_{ij} g(Y_{ij}) I_{B_{ij}})$$

$$= \sum_{j=1}^{n_2} E(c_{1j} g(Y_{1j})) - \sum_{j=1}^{n_2} E(c_{1j} g(Y_{1j}) I_{B_{1j}^c}) + \sum_{j=1}^{n_2} E(c_{2j} g(Y_{2j}) I_{B_{2j}})$$

$$+ \sum_{j=1}^{n_2} \sum_{i=3}^{n_1} E(c_{ij} g(Y_{ij}) I_{B_{ij}})$$

$$\leq \sum_{j=1}^{n_2} E(c_{1j} g(Y_{1j})) + \sum_{j=1}^{n_2} c_{2j} E[g(Y_{2j}) I_{B_{2j}} - g(Y_{1j}) I_{B_{1j}^c}]$$

$$+ \sum_{j=1}^{n_2} \sum_{i=3}^{n_1} E(c_{ij} g(Y_{ij}) I_{B_{ij}}),$$

where the last inequality follows from the nonincreasing property of the array $\{c_{ij}, \ i \geq 1, \ j \geq 1\}$. Since $B_{2j} \subset B_{1j}^c$, it follows that $I_{B_{2j}} = I_{B_{1j}^c} - I_{B_{1j}^c \cap B_{2j}^c}$. Therefore

$$\epsilon P(A) \leq \sum_{j=1}^{n_2} E(c_{1j} g(Y_{1j})) + \sum_{j=1}^{n_2} E[c_{2j} (g(Y_{2j}) - g(Y_{1j})) I_{B_{1j}^c}] \tag{5.3.6}$$

$$- \sum_{j=1}^{n_2} c_{2j} E[g(Y_{2j}) I_{B_{1j}^c \cap B_{2j}^c}] + \sum_{j=1}^{n_2} \sum_{i=3}^{n_1} E(c_{ij} g(Y_{ij}) I_{B_{ij}})$$

$$= \sum_{j=1}^{n_2} E(c_{1j}g(Y_{1j})) + \sum_{j=1}^{n_2} E[c_{2j}(g(Y_{2j}) - g(Y_{1j}))]$$

$$- \sum_{j=1}^{n_2} E[c_{2j}(g(Y_{2j}) - g(Y_{1j}))I_{B_{1j}}] - \sum_{j=1}^{n_2} c_{2j}E[g(Y_{2j})I_{B_{1j}^c \cap B_{2j}^c}]$$

$$+ \sum_{j=1}^{n_2} \sum_{i=3}^{n_1} E(c_{ij}g(Y_{ij})I_{B_{ij}}).$$

Since the function $g(.)$ is nondecreasing and convex, it follows that

$$g(x) - g(y) \geq h(x)$$

where

$$h(y) = \lim_{x \to y-0} \frac{g(x) - g(y)}{x - y}$$

is the left derivative of $g(.)$ Note that the function $I_{B_{1j}}h(Y_{1j})$ is a nonnegative nondecreasing function of Y_{1j}. Hence, by the demimartingale property of the array $\{Y_{ij}, i \geq 1, j \geq 1\}$, we get that

$$E[(g(Y_{2j}) - g(Y_{1j}))I_{B_{1j}}] \geq E[(Y_{2j} - Y_{1j})h(Y_{1j})I_{B_{1j}}] \geq 0, \quad j = 1, 2, \ldots, n_2. \tag{5.3.7}$$

Therefore

$$\epsilon P(A) \leq \sum_{j=1}^{n_2} E(c_{1j}g(Y_{1j})) + \sum_{j=1}^{n_2} E[c_{2j}(g(Y_{2j}) - g(Y_{1j}))] \tag{5.3.8}$$

$$- \sum_{j=1}^{n_2} c_{2j}E[g(Y_{2j})I_{B_{1j}^c \cap B_{2j}^c}] + \sum_{j=1}^{n_2} E[c_{3j}g(Y_{3j})I_{B_{3j}}]$$

$$+ \sum_{j=1}^{n_2} \sum_{i=4}^{n_1} E(c_{ij}g(Y_{ij})I_{B_{ij}})$$

$$\leq \sum_{j=1}^{n_2} E(c_{1j}g(Y_{1j})) + \sum_{j=1}^{n_2} E[c_{2j}(g(Y_{2j}) - g(Y_{1j}))]$$

$$+ \sum_{j=1}^{n_2} c_{3j}E[g(Y_{3j})I_{B_{3j}} - g(Y_{2j})I_{B_{1j}^c \cap B_{2j}^c}]$$

$$+ \sum_{j=1}^{n_2} \sum_{i=4}^{n_1} E(c_{ij}g(Y_{ij})I_{B_{ij}}),$$

where the last inequality follows from the nonincreasing property of the array $\{c_{ij}, i \geq 1, j \geq 1\}$. Since $B_{3j} \subset B_{1j}^c \cap B_{2j}^c$, it follows that $I_{B_{3j}} = I_{B_{1j}^c \cap B_{2j}^c} - I_{B_{1j}^c \cap B_{2j}^c \cap B_{3j}^c}$.

Hence

$$\epsilon\, P(A) \leq \sum_{j=1}^{n_2} E(c_{1j} g(Y_{1j})) + \sum_{j=1}^{n_2} E[c_{2j}(g(Y_{2j}) - g(Y_{1j}))] \qquad (5.3.9)$$

$$+ \sum_{j=1}^{n_2} c_{3j} E[(g(Y_{3j}) - g(Y_{2j})) I_{B_{1j}^c \cap B_{2j}^c}]$$

$$- \sum_{j=1}^{n_2} c_{3j} E[g(Y_{3j}) I_{B_{1j}^c \cap B_{2j}^c \cap B_{3j}^c}] + \sum_{j=1}^{n_2}\sum_{i=4}^{n_1} E(c_{ij} g(Y_{ij}) I_{B_{ij}})$$

$$= \sum_{j=1}^{n_2} E(c_{1j} g(Y_{1j})) + \sum_{j=1}^{n_2} E[c_{2j}(g(Y_{2j}) - g(Y_{1j}))]$$

$$- \sum_{j=1}^{n_2} c_{3j} E[g(Y_{3j}) - g(Y_{2j})] + \sum_{j=1}^{n_2} c_{3j} E[(g(Y_{3j}) - g(Y_{2j})) I_{B_{1j} \cup B_{2j}}]$$

$$- \sum_{j=1}^{n_2} c_{3j} E[g(Y_{3j}) I_{B_{1j}^c \cap B_{2j}^c \cap B_{3j}^c}] + \sum_{j=1}^{n_2}\sum_{i=4}^{n_1} E(c_{ij} g(Y_{ij}) I_{B_{ij}}).$$

Using the demimartingale property of the array $\{Y_{ij},\ i \geq 1,\ j \geq 1\}$, it can be shown that

$$E[(g(Y_{2j}) - g(Y_{1j})) I_{B_{1j} \cup B_{2j}}] \geq 0, \quad j = 1, 2, \ldots, n_2 \qquad (5.3.10)$$

by observing that the function $I_{B_{1j} \cup B_{2j}}$ is a nonnegative nondecreasing function of Y_{1j} and Y_{2j}. Therefore

$$\epsilon\, P(A) \leq \sum_{j=1}^{n_2} E(c_{1j} g(Y_{1j})) + \sum_{j=1}^{n_2} E[c_{2j}(g(Y_{2j}) - g(Y_{1j}))] \qquad (5.3.11)$$

$$+ \sum_{j=1}^{n_2} c_{3j} E[g(Y_{3j}) - g(Y_{2j})] - \sum_{j=1}^{n_2} c_{3j} E[g(Y_{3j}) I_{B_{1j}^c \cap B_{2j}^c \cap B_{3j}^c}]$$

$$+ \sum_{j=1}^{n_2}\sum_{i=4}^{n_1} E(c_{ij} g(Y_{ij}) I_{B_{ij}}).$$

Proceeding in the same way and using the fact that $Y_{0j} = 0$, we get that

$$\epsilon\, P(A) \leq \sum_{i=1}^{n_1}\sum_{j=1}^{n_2} E[c_{ij}(g(Y_{ij}) - g(Y_{i-1,j}))] - \sum_{j=1}^{n_2} c_{n_1 j} E[g(Y_{n_1 j}) I_{\cap_{i=1}^{n_1} B_{ij}^c}]$$

$$\leq \sum_{i=1}^{n_1}\sum_{j=1}^{n_2} E[c_{ij}(g(Y_{ij}) - g(Y_{i-1,j}))]. \qquad (5.3.12)$$

Similarly one can show that

$$\epsilon \, P(A) \le \sum_{i=1}^{n_1} \sum_{j=1}^{n_2} E[c_{ij}(g(Y_{ij}) - g(Y_{i,j-1}))] - \sum_{i=1}^{n_1} c_{in_2} E[g(Y_{in_2}) I_{\cap_{j=1}^{n_2} B_{ij}^c}]$$

$$\le \sum_{i=1}^{n_1} \sum_{j=1}^{n_2} E[c_{ij}(g(Y_{ij}) - g(Y_{i,j-1}))]. \tag{5.3.13}$$

Combining the inequalities (5.3.12) and (5.3.13), we get the result stated in the theorem.   □

The condition that the function g is nondecreasing in the above theorem can be removed and the following theorem can be proved by methods analogous to those given in Chapter 2. We omit the proof. For details, see Christofides and Hadjikyriakou (2010).

Theorem 5.3.2. *Let the array $\{Y_\mathbf{n}, \; \mathbf{n} \in N^2\}$ be a two-parameter demimartingale with $Y_\mathbf{k} = 0$ whenever $k_1 k_2 = 0$. Further suppose that $\{c_\mathbf{n}, \; \mathbf{n} \in N^2\}$ is a nonincreasing array of positive numbers and $g(.)$ be a nonnegative convex function on R with $g(0) = 0$. Then, for every $\epsilon > 0$,*

$$\epsilon P[\max_{(i,j) \le (n_1, n_2)} c_{ij} g(Y_{ij}) \ge \epsilon] \tag{5.3.14}$$

$$\le \min(\sum_{i=1}^{n_1} \sum_{j=1}^{n_2} c_{ij} E[g(Y_{ij}) - g(Y_{i-1,j})], \sum_{i=1}^{n_1} \sum_{j=1}^{n_2} c_{ij} E[g(Y_{ij}) - g(Y_{i,j-1})]).$$

For an application of the above maximal inequality to derive Hajek-Renyi type inequality for multidimensionally indexed associated random variables, see Christofides and Hadjikyriakou (2010). Wang (2004) has also derived a Hajek-Renyi type inequality for multidimensionally indexed associated random variables.

5.4   Continuous Parameter Demisubmartingales

We have studied different classes of discrete parameter demisubmartingales, N-demisupermartingales with index in one-dimension in the previous chapters and multidimensionally indexed demimartingales in the previous sections in this chapter. We will now describe how some of these concepts can be extended to the continuous parameter case. Our discussion here is based on Wood (1984).

Recall that an L^1-sequence of random variables $\{S_n, \; n \ge 1\}$ defined on a complete probability space $(\Omega, \mathcal{F}, P)$ is called a demimartingale if for every componentwise nondecreasing function $f(.)$,

$$E[(S_{n+1} - S_n) f(S_1, \ldots, S_n)] \ge 0.$$

It is called a demisubmartingale if, for every componentwise nonnegative nondecreasing function $f(.)$,

$$E[(S_{n+1} - S_n)f(S_1, \ldots, S_n)] \geq 0.$$

Definition. Let the process $\{S_t, \ t \in [0,T]\}$ be a stochastic process defined on a complete probability space $(\Omega, \mathcal{F}, P)$. It is called a *demisubmartingale* if for every $0 = t_0 < t_1 < \ldots, t_k = T, \ k \geq 1$ the sequence $\{S_{t_j}, \ j = 0, \ldots, k - 1\}$ is a demisubmartingale.

A process $\{S_t, \ t \in [0,T]\}$ is said to be *separable* if there is a measurable set B with $P(B) = 0$ and a countable subset $\tau \subset [0,T]$ such that for every closed interval $A \subset R$ and any open interval $(a, b) \subset [0,T]$, the sets

$$\{\omega : S_t(\omega) \in A, \ t \in (a, b)\}$$

and

$$\{\omega : S_t(\omega) \in A, \ t \in (a, b) \cap \tau\}$$

differ at most by a subset of B. It is known that every real-valued stochastic process indexed by the parameter $t \in [0,T] \subset R$ has a separable version (cf. Doob (1953), p. 57).

5.5   Maximal Inequality for Continuous Parameter Demisubmartingales

We now obtain a maximal inequality for continuous parameter demisubmartingales. Recall the following inequalities derived earlier for discrete parameter demisubmartingales. We now give alternate proofs of these inequalities due to Wood (1984).

Theorem 5.5.1. *Let the sequence $\{S_n, \ n \geq 1\}$ be a discrete parameter demisubmartingale with $M_k = \max_{1 \leq i \leq k} S_i$ and $m_k = \min_{1 \leq i \leq k} S_i$ for $1 \leq i \leq k$. Then, for any $\lambda \in R$,*

$$\lambda \, P(M_k > \lambda) \leq \int_{[M_k > \lambda]} S_k \, dP \qquad (5.5.1)$$

and

$$\lambda \, P(m_k \leq \lambda) \geq E[S_1] - \int_{[m_k > \lambda]} S_k \, dP \geq E[S_1] - E|S_k|. \qquad (5.5.2)$$

Proof. Let i be the smallest index such that $S_i > \lambda$, that is, $i = n$ if $S_k \leq \lambda$, $1 \leq k \leq n - 1$ but $S_n > \lambda$. Let $A_k = [M_k > \lambda]$. Then

$$\int_{A_k} S_k \, dP = \sum_{n=1}^{k} \int_{[i=n]} S_k \, dP \qquad (5.5.3)$$

$$= \sum_{n=1}^{k} [\int_{[i=n]} S_n \, dP + \int_{[i=n]} (S_k - S_n) \, dP]$$

$$= \sum_{n=1}^{k} \int_{[i=n]} S_n \, dP + \sum_{n=1}^{k-1} \int_{[i=n]} (S_k - S_n) \, dP$$

$$\geq \lambda \, P(A_k) + \sum_{n=1}^{k-1} \int_{\Omega} I_{[A_n]}(S_{n+1} - S_n) \, dP.$$

Note that the function $I_{[A_n]}$ is nonnegative and nondecreasing in the variables $S_1, \ldots, S_n$. Hence, by the demisubmartingale property of the sequence $\{S_n, n \geq 1\}$, it follows that

$$\int_{\Omega} I_{[A_n]}(S_{n+1} - S_n) \, dP \geq 0.$$

Applying this inequality, we get that

$$\int_{A_k} S_k \, dP \geq \lambda \, P(M_k > \lambda) \tag{5.5.4}$$

This proves the inequality (5.5.1).

Let $B_k = [m_k > \lambda]$. In order to prove (5.5.2), let i be the smallest index for which $S_i \leq \lambda$. From the arguments given above, it follows that

$$\int_{B_k^c} S_k \, dP \leq \lambda P(B_k^c) + \sum_{n=1}^{k-1} \int_{B_n^c} (S_{n+1} - S_n) \, dP. \tag{5.5.5}$$

Hence

$$\lambda \, P(B_k^c) \geq \int_{B_k^c} S_k \, dP - \sum_{n=1}^{k-1} \int_{B_n^c} (S_{n+1} - S_n) \, dP \tag{5.5.6}$$

$$\geq \int_{B_k^c} S_k \, dP - \sum_{n=1}^{k-1} \int_{B_n^c} (S_{n+1} - S_n) \, dP - \sum_{n=1}^{k-1} \int_{B_n} (S_{n+1} - S_n) \, dP$$

$$= \int_{B_k^c} S_k \, dP - \sum_{n=1}^{k-1} E[S_{n+1} - S_n]$$

$$= \int_{B_k^c} S_k \, dP - E[S_k] + E[S_1]$$

$$= E[S_1] - \int_{B_k} S_k \, dP$$

$$\geq E[S_1] - E|S_k|.$$

Note that the second inequality, in the chain of inequalities given above, is a consequence of the demisubmartingale property of the sequence $\{S_n, n \geq 1\}$. This proves the inequality (5.5.2). $\qquad\square$

Theorem 5.5.2. *Let the process $\{S_t, 0 \leq t \leq T\}$ be a demisubmartingale. Then, $E[S_\alpha] \leq E[S_\beta]$ for any $0 \leq \alpha < \beta \leq T$.*

Proof. Since the function $f \equiv 1$ is positive and nondecreasing and the sequence $\{S_0, S_\alpha, S_\beta, S_T\}$ forms a discrete parameter demisubmartingale by definition, we get that

$$E[1.(S_\beta - S_\alpha)] \geq 0$$

which implies that

$$E[S_\alpha] \leq E[S_\beta], \quad 0 \leq \alpha < \beta \leq T. \qquad \square$$

Theorem 5.5.3. *Let the process $\{S_t, 0 \leq t \leq T\}$ be a separable demisubmartingale. For any $\lambda \in R$, let*

$$A_T = [\omega \in \Omega : \sup_{0 \leq t \leq T} S_t > \lambda]$$

and

$$B_T = [\omega \in \Omega : \inf_{0 \leq t \leq T} S_t \leq \lambda].$$

Then

$$\lambda \, P(A_T) \leq \int_{A_T} S_T \, dP \tag{5.5.7}$$

and

$$\lambda \, P(B_T) \geq E[S_0] - E|S_T|. \tag{5.5.8}$$

Proof. By Theorem 5.5.1 and the fact that $\{S_t, 0 \leq t \leq T\}$ is a demisubmartingale, we observe that the inequalities (5.5.7) and (5.5.8) hold if the supremum and infimum are restricted to a finite subset of parameter values including 0 and T. Since

$$\lim_{n \to \infty} P(\bigcup_{k=1}^{n} D_k) = P(\bigcup_{k=1}^{\infty} D_k)$$

for any sequence $D_k \in \mathcal{F}$, it follows that the inequalities (5.5.7) and (5.5.8) hold if the parameter in the supremum and infimum is allowed to run over the interval $[0, T]$. By the separability of the process, the sets A_T and B_T differ from the corresponding sets where the parameter t takes values in the interval $[0, T]$ by a set of probability measure zero. This proves the theorem. $\qquad \square$

As a corollary to Theorem 5.5.3, we get the following result by methods used in Chapter 2 (cf. Theorem 3.4 in Doob (1953), p. 317). We omit the details.

Theorem 5.5.4. *Let the process $\{S_t, 0 \leq t \leq T\}$ be a nonnegative separable demi-martingale. Suppose that $E|S_t|^p < \infty$, $0 \leq t \leq T$ for some $p \geq 1$. Then, for $p > 1$,*

$$E[\sup_{0 \leq t \leq T} S_t^p] \leq (\frac{p}{p-1})^p E[S_T^p] \tag{5.5.9}$$

and, for $p = 1$,

$$E[\sup_{0 \leq t \leq T} S_t] \leq (\frac{e}{e-1})(1 + E[S_T \log^+ S_T]). \tag{5.5.10}$$

5.6   Upcrossing Inequality

Let the finite random sequence $\{S_j, 1 \leq j \leq m\}$ be a discrete parameter demisub-martingale and $[a, b]$ be an interval. Define a sequence of stopping times τ_k by

$$\tau_1 = \min\{j : S_j \leq a\},$$

$$\tau_k = \min\{j : \tau_{k-1} < j \leq m; \ S_j \geq b\} \ \text{for } k \text{ even}$$

and

$$\tau_k = \min\{j : \tau_{k-1} < j \leq m; \ S_j \leq a\} \ \text{for } k \text{ odd}.$$

We make the convention that the minimum of the empty set is m. The number of *upcrossings* of the interval $[a, b]$ by the sequence $S_1(\omega), \ldots, S_m(\omega)$ is the number of times the sequence crosses from below a to above b.

Theorem 5.6.1. *Let the finite random sequence $\{S_j, 1 \leq j \leq m\}$ be a discrete parameter demisubmartingale. If $a < 0$, then*

$$E[U_m] \leq (b - a)^{-1} E[S_m^+] + 1$$

where $S_m^+ = \max(S_m, 0)$. If $a \geq 0$, then

$$E[U_m] \leq (b - a)^{-1} (b + E[S_m^+]).$$

Proof. Let us first discuss the case $S_1 \equiv 0$, $0 \leq a < b$ and replace S_m with a new random variable (still to be denoted by S_m for convenience), that is, $S_m = S_m^+ + b$. Define

$$Z = (S_{\tau_3} - S_{\tau_2}) + (S_{\tau_5} - S_{\tau_4}) + \ldots. \tag{5.6.1}$$

If $A_k = \{\omega : \tau_{2n} \leq k < \tau_{2n+1}, n = 1, 2, \ldots\}$, then the function $I_{[A_k]}$ is a nondecreasing function of the variables $S_1, \ldots, S_k$ (we use the fact that $S_1 \equiv 0$ here). Hence

$$E[Z] = \sum_{k=2}^{n-1} \int_{A_k} (S_{k+1} - S_k) \, dP = \sum_{k=2}^{n-1} \int_{\Omega} I_{[A_k]}(S_{k+1} - S_k) \, dP \geq 0.$$

Note that

$$S_m - S_1 = S_{\tau_m} - S_{\tau_1} \tag{5.6.2}$$

$$= \sum_{n=1}^{m} [S_{\tau_{n+1}} - S_{\tau_n}]$$

$$= \sum\nolimits^1 [S_{\tau_{n+1}} - S_{\tau_n}] + \sum\nolimits^2 [S_{\tau_{n+1}} - S_{\tau_n}]$$

where $\sum^1$ adds terms over n odd and $\sum^2$ adds terms over n even. Furthermore

$$\sum\nolimits^1 [S_{\tau_{n+1}} - S_{\tau_n}] \geq U_m(b - a)$$

and

$$E[Z] = E(\sum{}^2 [S_{\tau_{n+1}} - S_{\tau_n}]) \geq 0.$$

Therefore

$$E[S_m - S_1] = E[S_m^+ + b - 0] = E[S_m^+ + b] \geq E[U_m](b - a).$$

The last inequality proves the theorem in the special case.

We will now remove the assumption that $S_1 \equiv 0$. Let $Y_1 = S_2$, $Y_2 = S_3, \ldots, Y_m = S_{m+1}$. Then the number of upcrossings $\tilde{U}_m$ of the interval $[a, b]$ by $Y_1, Y_2, \ldots, Y_m$ satisfies

$$\tilde{U}_m(\omega) \leq U_{m+1}(\omega) \leq \tilde{U}_m(\omega) + 1 \tag{5.6.3}$$

and

$$E[\tilde{U}_m] \leq E[U_{m+1}] \leq (b - a)^{-1}(b + E S_{m+1}^+) = (b - a)^{-1}(b + E[Y_m^+]).$$

We will now relax the assumption that $a \geq 0$. Note that, if $a > 0$, then the finite sequence $S_1 - a, S_2, \ldots, S_m$ forms a demisubmartingale with possibly one more upcrossing over the interval $[0, b - a)$ than the sequence $S_1, \ldots, S_m$ makes over $[a, b]$. Finally, we note that changing S_m to $S_m^+ + b$ increases S_m, so the number of upcrossings by $S_1, \ldots, S_{m-1}, S_m$ is less than or equal to the number of upcrossings by $S_1, \ldots, S_{m-1}, S_m^+ + b$. This completes the proof. $\qquad \square$

As a consequence of the upcrossing inequality, the following convergence theorem holds which we have stated in Chapter 2.

Theorem 5.6.2. *Let the sequence $\{S_n, \ n \geq 1\}$ be a demisubmartingale with $\limsup_n E|S_n| < \infty$. Then there exists a random variable X such that $E|X| < \infty$ and $S_n \to X$ a.s. as $n \to \infty$.*

Proof. Define the stopping times τ_k as before but we now define the minimum of the empty set as $+\infty$. Let $B = \cap_{n=1}^\infty [\omega : \tau_n(\omega) < \infty]$. Then

$$P(B) = P(\text{Either } S_n \leq a \text{ infinitely often or } S_n \geq b \text{ infinitely often}).$$

If U_m is the number of upcrossings of the interval $[a, b]$ by the sequence $\{S_j, 1 \leq j \leq m\}$, then, on the event B, the sequence $U_m \to \infty$ as $m \to \infty$. Since $E[U_m]$ is bounded by the inequalities obtained in Theorem 5.6.1 and since $\limsup_{n \to \infty} E|S_n| < \infty$, it follows that $P(B) = 0$. Hence

$$P(\bigcup [\liminf S_n \leq a < b \leq \limsup S_n]) = 0$$

where the union is taken over all a and b rational with $a < b$. Hence, either the sequence S_n converges almost surely, or $|S_n| \to \infty$ with positive probability. However the second case cannot occur, since, by Fatou's lemma,

$$E[\limsup_{n \to \infty} |S_n|] \leq \limsup_{n \to \infty} E|S_n|.$$

Hence the result as stated holds. $\qquad \square$

Remarks. If, in addition to the hypothesis of Theorem 5.6.2, the demisubmartingale is uniformly integrable, then $E|S_n - X| \to 0$ as $n \to \infty$. This follows from the standard results in measure theory.

The upcrossing inequality in Theorem 5.6.1 and the almost sure convergence theorem in Theorem 5.6.2 can be easily extended to the case of a continuous parameter separable demisubmartingale. By taking limit as $\lambda \to \infty$ in (5.5.7) and $\lambda \to -\infty$ in (5.5.8), we get that the sample paths of the process $\{S_t, 0 \leq t \leq T\}$ are almost surely bounded. We will now show that the sample paths of the process have no discontinuities of the second kind with probability 1.

Theorem 5.6.3. *Suppose the process $\{S_t, 0 \leq t \leq T\}$ is a separable demisubmartingale defined on a complete probability space $(|\Omega, \mathcal{F}, P)$. Then there exists a measurable set D with $P(D) = 0$ such that , for any fixed $\omega \in D^c$, the function $S(t, \omega)$ defined over $t \in [0, T]$, is bounded and has no discontinuities of the second kind.*

Proof. It was shown earlier that there exists a measurable set A_1 with $P(A_1) = 0$ such that, if $\omega \in A_1^c$, then the function $f(t) = S(t, \omega)$ is bounded.

From the separability of the process $\{S_t, 0 \leq t \leq T\}$, there exists a countable set $J = \{t_n, n \geq 1\}$ contained in the interval $[0, T]$, and a measurable set A_2 with $P(A_2) = 0$, such that, for $\omega \in A_2^c$,

$$\inf\{S(t, \omega) : t \in J\} = \inf\{S(t, \omega) : t \in [0, T]\}$$

and

$$\sup\{S(t, \omega) : t \in J\} = \sup\{S(t, \omega) : t \in [0, T]\}.$$

Let $0 = t_{i_1} < t_{i_2} < \ldots, < t_{i_n} = T$ be a finite subset of J. Let $[r_1, r_2]$ be an interval in R and $U_n(\omega)$ be the number of upcrossings of the interval $[r_1, r_2]$ by the sequence $S(t_{i_1}, \omega), \ldots, S(t_{i_n}, \omega)$. Let M_{nk} be the event $[U_n \geq k]$. Then, by the upcrossings theorem, we get that

$$P[\cup_n M_{nk}] \leq \frac{c + ES_T^+}{k(r_2 - r_1)} \tag{5.6.4}$$

for some constant $c > 0$. Suppose that, for some $\omega \in A_2^c$, the function $S(t, \omega)$ has a discontinuity of the second kind at some point $s \in [0, T]$ with either

$$\limsup_{t \uparrow s} S(t, \omega) > r_2 > r_1 > \limsup_{t \uparrow s} S(t, \omega)$$

or

$$\limsup_{t \downarrow s} S(t, \omega) > r_2 > r_1 > \limsup_{t \downarrow s} S(t, \omega).$$

Then this inequality also holds if t tends to s in the set J, which implies that the number of upcrossings of the interval $[r_1, r_2]$ by the sequence $S(t_{i_1}, \omega), \ldots, S(t_{i_n}, \omega)$

tends to infinity as n tends to infinity. Observe that the set of $\omega \in A_2^c$ for which this occurs is contained in the set $\cup M_{nk}$ for every k. But

$$P(\bigcap_k \bigcup_n M_{nk}) = \lim_{n \to \infty} P(\bigcup_n M_{nk}) = 0.$$

Let $B(r_1, r_2) = \cap_k \cup_n M_{nk}$ and $A_3 = \cup B(r_1, r_2)$ where the union is over all rational numbers $r_1 < r_2$. Then $P(A_3) = 0$. Therefore, for any $\omega \in [A_1 \cup A_2 \cup A_3]^c$, the sample path $f(t) = S(t, \omega)$ has finite left and right limits at any point of discontinuity and is bounded. Let $D = A_1 \cup A_2 \cup A_3$. Note that $P(D) = 0$ and the theorem is proved. $\qquad \square$

Chapter 6

Limit Theorems for Associated Random Variables

6.1 Introduction

A sequence of partial sums of mean zero associated random variables forms a demimartingale. We have discussed properties of demimartingales in Chapter 2 and some probabilistic properties of associated sequences in Chapter 1. Let $\{\Omega, \mathcal{F}, \mathcal{P}\}$ be a probability space and $\{X_n, n \geq 1\}$ be a sequence of associated random variables defined on it. Recall that a finite collection $\{X_1, X_2, \ldots, X_n\}$ is said to be associated if for every pair of functions $h(\mathbf{x})$ and $g(\mathbf{x})$ from R^n to R, which are nondecreasing componentwise,

$$\mathrm{Cov}(h(\mathbf{X}), \mathbf{g}(\mathbf{X})) \geq \mathbf{0},$$

whenever it is finite, where $\mathbf{X} = (X_1, X_2, \ldots, X_n)$ and an infinite sequence $\{X_n, n \geq 1\}$ is said to be associated if every finite subfamily is associated. As we have mentioned in Chapter 1, associated random variables are of considerable interest in reliability studies (cf. Esary, Proschan and Walkup (1967), Barlow and Proschan (1975)), statistical physics (cf. Newman (1980, 1983)) and percolation theory (cf. Cox and Grimmet (1984)). We have given an extensive review of several probabilistic results for associated sequences in Chapter 1 (cf. Prakasa Rao and Dewan (2001), Roussas (1999)). We now discuss some recent advances in limit theorems for associated random variables. Covariance inequalities of different types play a major role in deriving limit theorems for partial sums of associated random variables. The next section gives some covariance inequalities and their applications.

B.L.S. Prakasa Rao, *Associated Sequences, Demimartingales and Nonparametric Inference*, Probability and its Applications, DOI 10.1007/978-3-0348-0240-6_6, © Springer Basel AG 2012

6.2   Covariance Inequalities

Generalized Hoeffding Identity

We have discussed the Hoeffding identity in Chapter 1. We now discuss a *generalized Hoeffding identity* due to Khoshnevisan and Lewis (1998).

Let $C_b^2(R^2)$ denote the set of all functions $f(x,y)$ from R^2 to R with bounded and continuous mixed second-order partial derivatives $f_{xy}(x,y) = \frac{\partial^2 f}{\partial x \partial y}$. For $f \in C_b^2(R^2)$, let $M(f) = \sup_{(s,t) \in R^2} |f_{xy}(s,t)|$.

Theorem 6.2.1. *Let X, Y, U and Z be random variables with bounded second moments defined on a common probability space. Suppose that X and U are identically distributed and Y and Z are identically distributed. Further suppose that U and Z are independent. Then, for any $f \in C_b^2(R^2)$,*

$$E[f(X,Y)] - E[f(U,Z)] = \int_{-\infty}^{\infty} \int_{-\infty}^{\infty} f_{xy}(u,v) H_{X,Y}(u,v) \, du dv$$

where

$$H_{X,Y}(x,y) = P(X \geq x, \, Y \geq y) - P(X \geq x) P(Y \geq y).$$

If, in addition X and Y are associated, then

$$|E[f(X,Y)] - E[f(U,Z)]| \leq M(f) \operatorname{Cov}(X,Y).$$

Proof. Without loss of generality, we assume that the random vectors (X,Y) and (U,Z) are independent. Let

$$I(u,x) = \begin{cases} 1 & \text{if } u \leq x \\ 0 & \text{if } u > x. \end{cases}$$

Then

$$|X - U||Y - Z| = \int_{-\infty}^{\infty} \int_{-\infty}^{\infty} |I(s,X) - I(s,U)||I(t,Y) - I(t,Z)| \, dsdt$$

and hence

$$E[|X-U||Y-Z|] = E\left[\int_{-\infty}^{\infty} \int_{-\infty}^{\infty} |I(s,X)-I(s,U)||I(t,Y)-I(t,Z)| \, dsdt\right]. \quad (6.2.1)$$

Observe that

$$E[f(X,Y) - f(U,Y) + f(U,Z) - f(X,Z)] \qquad (6.2.2)$$
$$= E\left[\int_{-\infty}^{\infty} \int_{-\infty}^{\infty} f_{xy}(s,t)(I(s,X) - I(s,U))(I(t,Y) - I(t,Z)) \, dsdt\right].$$

The integrand on the right-hand side of the equation is bounded by

$$M(f)|I(s, X) - I(s, U)||I(t, Y) - I(t, Z)|$$

and, by (6.2.1), we can interchange the order of integration by Fubini's theorem. This gives the identity

$$E[f(X, Y)] - E[f(U, Z)] = \int_{-\infty}^{\infty} \int_{-\infty}^{\infty} f_{xy}(u, v) H_{X,Y}(u, v) \, du dv. \qquad (6.2.3)$$

If X and Y are associated, then $H_{X,Y}(x, y) \geq 0$ and it follows that

$$|E[f(X, Y)] - E[f(U, Z)]| \leq M(f) \operatorname{Cov}(X, Y)$$

by using the Hoeffding identity

$$\operatorname{Cov}(X, Y) = \int_{-\infty}^{\infty} \int_{-\infty}^{\infty} H_{X,Y}(u, v) du dv. \qquad \square$$

A slight variation of this result is the following result from Bulinski and Shaskin (2007).

Theorem 6.2.2. *Let $g(x, y)$ be a real-valued function with continuous second partial derivative $g_{xy}(x, y) = \frac{\partial^2 g}{\partial x \partial y}$. Suppose that X, Y and Z are random variables such that the Z is independent of X and the distributions of Y and Z are the same. Then*

$$E[g(X, Y)] - E[g(X, Z)] = \int_{-\infty}^{\infty} \int_{-\infty}^{\infty} \frac{\partial^2 g}{\partial x \partial y} H_{X,Y}(x, y) \, dx dy \qquad (6.2.4)$$

where

$$H_{X,Y}(x, y) = P(X \geq x, \, Y \geq y) - P(X \geq x)P(Y \geq y)$$

provided the expectations and the double integral exist.

Proof. Without loss of generality, we can assume that Z is independent of X as well as Y by enlarging the probability space. Let U be a random variable such that the random vector (U, Z) is independent of the random vector (X, Y) and has the same distribution as that of (X, Y). Note that

$$\int_{R^2} g_{x,y}(x, y) H_{X,Y}(x, y) dx dy$$

$$= \int_{R^2} \int_{R^2} g_{x,y}(t, w) \operatorname{Cov}(I_{[X \geq t]}, I_{[Y \geq w]}) \, dt dw$$

$$= \frac{1}{2} \int_{R^2} g_{x,y}(t, w) E\{[I_{[X \geq t]} - I_{[U \geq t]}][I_{[Y \geq w]} - I_{[Z \geq w]}]\} \, dt dw$$

$$= \frac{1}{2}E\{\int_{R^2} g_{x,y}(t,w)([I_{[X \geq t]} - I_{[U \geq t]}][I_{[Y \geq w]} - I_{[Z \geq w]}])\, dtdw\}$$

$$= \frac{1}{2}E[\int_U^X \int_Z^Y g_{x,y}(t,w)dtdw]$$

$$= \frac{1}{2}E[\int_U^X (\frac{\partial g(t,Y)}{\partial t} - \frac{\partial g(t,Z)}{\partial t})dt]$$

$$= \frac{1}{2}E[g(X,Y) - g(U,Y) - g(X,Z) + g(U,Z)]$$

$$= E[g(X,Y)] - E[g(X,Z)]. \qquad \qquad \square$$

Suppose f and g are real-valued functions with continuous derivatives. Let X and Y be random variables such that $E|f(X)|, E|g(Y)|$ and $E|f(X)G(Y)|$ are finite. As a special case of the above result, it follows that

$$\mathrm{Cov}(f(X), g(Y)) = \int_{R^2} f'(x)g'(y)H_{X,Y}(x,y)\,dxdy.$$

It is easy to see that the above result can be extended to complex-valued functions f and g with obvious interpretations in the notations.

Newman (1980, 1984) proved the following covariance inequality for a random vector (X,Y) which is associated. Note that $\mathrm{Cov}(X,Y) \geq 0$.

Theorem 6.2.3 (Newman's inequality). *Suppose (X,Y) is a random vector which is associated. Let f and g be differentiable functions with $\sup_x |f'(x)| < \infty$ and $\sup_x |g'(x)| < \infty$. Then*

$$|\mathrm{Cov}(f(X), g(Y))| \leq \sup_x |f'(x)| \sup_y |g'(y)| \, \mathrm{Cov}(X,Y). \qquad (6.2.5)$$

This result follows from the Newman's identity discussed in Chapter 1, see equation (1.1.3). Bulinski (1996) generalized this result.

Theorem 6.2.4. *Suppose (X,Y) is a random vector which is associated. Let f and g be Lipschitz functions. Then*

$$|\mathrm{Cov}(f(X), g(Y))| \leq \mathrm{Lip}(f)\,\mathrm{Lip}(g)\,\mathrm{Cov}(X,Y) \qquad (6.2.6)$$

where

$$\mathrm{Lip}(f) = \sup_{x \neq y} \frac{|f(x) - f(y)|}{|x - y|}.$$

A more general version of the above inequality is the following due to Bulinski (1996).

Let $g(.)$ be a continuous function from $R^n \to R$ such that for any $\mathbf{x} \in R^n$ and any $k = 1, \ldots, n$ there exist finite left and right partial derivatives $\frac{\partial^+ g(\mathbf{x})}{\partial x_k}$

and $\frac{\partial^- g(\mathbf{x})}{\partial x_k}$. Further suppose that, for each $k = 1, \ldots, n$, there are at most a finite number of $\mathbf{x}$ at which $\frac{\partial^+ g(\mathbf{x})}{\partial x_k} \neq \frac{\partial^- g(\mathbf{x})}{\partial x_k}$. Let

$$L_k(g) = \max[||\frac{\partial^+ g(\dot{\mathbf{x}})}{\partial x_k}||_\infty, ||\frac{\partial^- g(\mathbf{x})}{\partial x_k}||_\infty].$$

Theorem 6.2.5 (Bulinski's inequality; Bulinski (1996)). *Let* $X_1, \ldots, X_n$ *be associated random variables. Let* $h(x)$ *be a continuous function from* R^n *to* R *such that for any* $x \in R^n$ *and any* $k = 1, \ldots, n$, *there exist finite derivatives* $\frac{\partial^+ h(x)}{\partial x_k}$ *and* $\frac{\partial^- h(x)}{\partial x_k}$. *Further suppose that for each* $k = 1, \ldots, n$, *there are at most a finite number of points* x *at which* $\frac{\partial^+ h(x)}{\partial x_k} \neq \frac{\partial^- h(x)}{\partial x_k}$. *Let*

$$L_k(h) = \max\{||\frac{\partial^+ h(x)}{\partial x_k}||_\infty, ||\frac{\partial^- h(x)}{\partial x_k}||_\infty\}$$

where $||.||$ *stands for the norm in* L^∞. *Further suppose that similar assumptions hold for another function* g. *Then, for any two disjoint subsets* I *and* J *of* $\{1, \ldots, n\}$,

$$|\operatorname{Cov}(h(X_i, i \in I), g(X_j, j \in J))| \leq \sum_{i \in I} \sum_{j \in J} L_i(h) L_j(g) \operatorname{Cov}(X_i, X_j). \quad (6.2.7)$$

The following result is due to Bulinski and Shabanovich (1998).

Theorem 6.2.6. *Let* $\mathbf{X} = \{X_j, j \in Z^d\}$ *be an associated (or negatively associated) random field such that* $E[X_j^2] < \infty$ *for any* $j \in Z^d$. *Let* A *and* B *be two finite subsets of* Z^d. *In addition, suppose that they are disjoint if the random field* $\mathbf{X} = \{X_j, j \in Z^d\}$ *is negatively associated. Let* $|A|$ *denote the cardinality of the set* A. *Let* X_A *denote the random vector with components* X_j *for* $j \in A$. *Then, for any Lipschitzian functions* $f : R^{|A|} \to R$ *and* $g : R^{|B|} \to R$,

$$|\operatorname{Cov}(f(X_A), g(X_B))| \leq \sum_{i \in A, j \in B} \operatorname{Lip}_i(f) \operatorname{Lip}_j(g) |\operatorname{Cov}(X_i, X_j)|.$$

Proof. Define

$$f_+(X_A) = f(X_A) + \sum_{i \in A} \operatorname{Lip}_i(f) X_i$$

and

$$f_-(X_A) = f(X_A) - \sum_{i \in A} \operatorname{Lip}_i(f) X_i.$$

Observe that the function f_+ is componentwise nondecreasing and the function f_- is componentwise nonincreasing. In the associated case, it follows that

$$\operatorname{Cov}(f_+(X_A), g_+(X_B)) \geq 0,$$

$$\text{Cov}(f_-(X_A), g_-(X_B)) \geq 0,$$
$$\text{Cov}(-f_+(X_A), g_-(X_B)) \geq 0,$$

and

$$\text{Cov}(f_-(X_A), -g_+(X_A)) \geq 0.$$

Addition of the first two inequalities gives the upper bound for $\text{Cov}(f(X_A), g(X_B))$ and addition of third and fourth give the lower bound for $\text{Cov}(f(X_A), g(X_B))$. The result in the negatively associated case is proved by reversing the four inequalities given above. $\qquad\square$

Remarks. It is possible to extend the result to the composition of functions of bounded variation instead of Lipschitzian functions. Then the upper bound has to be replaced by the covariance of monotone functions of X_i multiplied by the total variation of f and g (cf. Zhang (2001)).

The following inequality is a consequence of Bulinski's inequality.

Theorem 6.2.7. *Let the set $\{X_i, 1 \leq i \leq n\}$ be a finite sequence of associated random variables. Let g_j, $j = 1, \ldots, n$ be functions as defined above. Then*

$$|E[\exp(it \sum_{j=1}^{n} g_j(X_j))] - \prod_{j=1}^{n} E[\exp(itg_j(X_j))]| \qquad (6.2.8)$$

$$\leq t^2 \sum_{1 \leq j < k \leq n} L_j(g_j) L_k(g_k) \text{Cov}(X_j, X_k).$$

Proof. The proof is by induction. For $n = 2$, using Newman's inequality, it follows that

$$|E[\exp\{it(g_1(X_1) + g_2(X_2))\}] - \prod_{j=1}^{2} E[\exp(itg_j(X_j))]| \qquad (6.2.9)$$

$$= |\text{Cov}(\exp(itg_1(X_1)), \exp(itg_2(X_2))|$$
$$\leq t^2 L_1(g_1) L_2(g_2) \text{Cov}(X_1, X_2).$$

Suppose the result holds for $n = m$. Then, for $n = m+1$, using Bulinski's inequality (1996) and the induction hypothesis, we get that

$$|E[\exp(it \sum_{j=1}^{m+1} g_j(X_j))] - \prod_{j=1}^{m+1} E[\exp(itg_j(X_j))]| \qquad (6.2.10)$$

$$= |\text{Cov}(\exp\{it \sum_{j=1}^{m} g_j(X_j)\}, \exp(itg_{m+1}(X_{m+1})))|$$

$$+ |E[\exp(itg_{m+1}(X_{m+1}))]| |E[\exp(it \sum_{j=1}^{m} g_j(X_j))] - \prod_{j=1}^{m} E[\exp(itg_j(X_j))]|$$

$$\leq t^2 \sum_{j=1}^{m} L_j(g_j) L_{m+1}(g_{m+1}) \operatorname{Cov}(X_j, X_{m+1})$$

$$+ t^2 \sum_{1 \leq j < k \leq m} L_j(g_j) L_k(g_k) \operatorname{Cov}(X_j, X_k)$$

$$= t^2 \sum_{1 \leq j < k \leq m+1} L_j(g_j) L_k(g_k) \operatorname{Cov}(X_j, X_k)$$

which proves the theorem. □

As a special case of the above theorem, we get the following inequality.

Theorem 6.2.8. *Suppose the set of random variables* $\{X_1, \ldots, X_n\}$ *is associated (or negatively associated) such that* $E|X_i|^2 < \infty$, $1 \leq i \leq n$. *Then, for any* $t_i \in R$, $1 \leq i \leq n$,

$$\left| E(e^{it_1 X_1 + \ldots + it_n X_n}) - \prod_{j=1}^{n} E(e^{it_j X_j}) \right| \leq 4 \sum_{1 \leq j \neq k \leq n} |t_j t_k| \left| \operatorname{Cov}(X_j, X_k) \right|.$$

Remarks. Newman (1980) proved the above result with factor "4" replaced by "1". Another important inequality which will be used later in this chapter is the following result proved in Dewan and Prakasa Rao (1999).

Theorem 6.2.9. *Let* $X_1, \ldots, X_n$ *be associated random variables bounded by a constant* M. *Then, for any* $\lambda > 0$,

$$\left| E(e^{\lambda \sum_{i=1}^{n} X_i}) - \prod_{i=1}^{n} E[e^{\lambda X_i}] \right| \leq \lambda^2 e^{n\lambda M} \sum_{1 \leq i < j \leq n} \operatorname{Cov}(X_i, X_j). \qquad (6.2.11)$$

Proof. Using Newman's (1980) inequality, we get that , for $n = 2$, and any $\lambda > 0$,

$$\left| \operatorname{Cov}(e^{\lambda X_1}, e^{\lambda X_2}) \right| \leq \lambda^2 e^{2\lambda M} \left| \operatorname{Cov}(X_1, X_2) \right|. \qquad (6.2.12)$$

The result follows by induction and using the fact that, if X, Y, and Z are associated, so are X and $Y + Z$ as they are increasing functions of associated random variables. □

The following result due to Shaskin (2007) is useful in approximating partial sums of associated random variables by partial sums of independent random variables.

Theorem 6.2.10. *Let the random vector* $\mathbf{Y} = (Y_1, \ldots, Y_m)$ *be a vector with associated components and let* $E[Y_i^2] < \infty$, $i = 1, \ldots, m$. *Then the random vector* $\mathbf{Y}$ *can be redefined on a new probability space on which independent random* variables $Z_1, \ldots, Z_m$ exist such that Z_k and Y_k are identically distributed and $P(|Y_k - Z_k| > a_k) \leq a_k$ for $k = 1, \ldots, m$ where

$$a_k = A \sum_{v=k}^{m} v^{1/3} (\sum_{j=1}^{v-1} \text{Cov}(Y_j, Y_v))^{1/3}$$

and $A > 0$ is an absolute constant.

Another related result is due to Lewis (1998).

Theorem 6.2.11. Let the set of random variables $\{X_1, \ldots, X_n\}$ be associated (or negatively associated) with second moments. Let the set $\{Y_1, \ldots, Y_n\}$ be independent random variables independent of $\{X_1, \ldots, X_n\}$ such that X_i and Y_i are identically distributed for $1 \leq i \leq n$. Suppose that f is a function twice differentiable with a continuous and bounded second derivative f''. Then

$$|E[f(X_1 + \ldots + X_n)] - E[f(Y_1 + \ldots + Y_n)]| \leq \sup_x |f''(x)| \sum_{1 \leq i \neq j \leq n} \text{Cov}(X_i, X_j)|.$$
$$(6.2.13)$$

Proof. The theorem is proved by induction on n. For $n = 1$, the result is obvious since $E[f(X_1)] = E[(f(Y_1)]$. Suppose that the result holds for a set of $n-1$ random variables. Note that

$$\begin{aligned} E[f(X_1 + \ldots + X_n)] &- E[f(Y_1 + \ldots + Y_n)]| & (6.2.14) \\ &\leq |E[f(X_1 + \ldots + X_n)] - E[f(X_1 + \ldots + X_{n-1} + Y_n)]| \\ &+ |E[f(X_1 + \ldots + X_{n-1} + Y_n)] - E[f(Y_1 + \ldots + Y_n)]| \\ &= J_1 + J_2 \quad \text{(say)}. \end{aligned}$$

It can be shown that the function $g(y) \equiv E[f(y + Y_n)]$ is twice differentiable with bounded continuous second derivative. Hence, by the induction hypothesis,

$$J_2 \leq \sup_x |f''(x)| \sum_{1 \leq i \neq j < n} \text{Cov}(X_i, X_j)|. \qquad (6.2.15)$$

Let $S = X_1 + \ldots + X_{n-1}$ and $T = Y_1 + \ldots + Y_{n-1}$. Then the bivariate vector $(S, X_n$ is associated (or negatively associated), the vector (T, Y_n) has independent components independent of (S, X_n) and the random variables X_n and Y_n are identically distributed. Applying the second version of generalized Hoeffding identity, we get that

$$\begin{aligned} J_1 &= |E[f(S + Y_n)] - E[f(T + Y_n)]| & (6.2.16) \\ &= |\int_R^2 f''(t + w) H_{S,T}(t, w) dt dw| \\ &\leq \sup_x |f''(x)||Cov(S, X_n)| \end{aligned}$$

$$\leq \sup_x |f''(x)| \sum_{j=1}^{n-1} ||Cov(X_j, X_n)||.$$

The result follows from equations (6.2.14)–(6.2.16). $\qquad\square$

A more general version of the inequality is given by the following result which follows from results due to Christofides and Vaggelatou (2004).

Theorem 6.2.12. *Let the random vector* $(Y_1, \ldots, Y_m)$ *have associated components with* $E[Y_j^2] < \infty$, $j = 1, \ldots, m$. *Suppose that a random variable* Z *is independent of the vector* $(Y_1, \ldots, Y_{m-1})$ *and the random variables* Z *and* Y_m *are identically distributed. Let* $f : R^m \to R$ *with bounded second-order partial derivatives. Then*

$$|E[f(Y_1, \ldots, Y_{m-1}, Y_m)] - E[f(Y_1, \ldots, Y_{m-1}, Z)]| \qquad (6.2.17)$$

$$\leq \sum_{k=1}^{m-1} \sup_{x_k, x_m} |\frac{\partial^2 f}{\partial x_k \partial x_m}| ||Cov(Y_k, Y_m)|.$$

An important inequality which has been found useful in applications for deriving limit properties of estimators and test statistics for associated sequences of random variables, due to Bagai and Prakasa Rao (1991), is given in Theorem 6.2.14 and it is a consequence of the following probabilistic inequality Sadikova's inequality due to Sadikova (1966).

Theorem 6.2.13 (Sadikova's inequality). *Let* $F(x, y)$ *and* $G(x, y)$ *be two bivariate distribution functions with characteristic functions* $f(s, t)$ *and* $g(s, t)$ *respectively. Define*

$$\hat{f}(s, t) = f(s, t) - f(s, 0) \, f(0, t)$$

and

$$\hat{g}(s, t) = g(s, t) - g(s, 0) \, g(0, t).$$

Suppose that the partial derivatives of G *with respect to* x *and* y *exist. Let*

$$A_1 = \sup_{x, y} \frac{\partial G(x, y)}{\partial x} \quad and \quad A_2 = \sup_{x, y} \frac{\partial G(x, y)}{\partial y}.$$

Suppose A_1 *and* A_2 *are finite. Then, for any* $T > 0$,

$$\sup_{x, y} |F(x, y) - G(x, y)| \qquad (6.2.18)$$

$$\leq \frac{1}{4\pi^2} \int_{-T}^{T} \int_{-T}^{T} |\frac{\hat{f}(s, t) - \hat{g}(s, t)}{st}| \, ds dt + 2 \sup_x |F(x, \infty) - G(x, \infty)|$$

$$+ 2 \sup_y |F(\infty, y) - G(\infty, y)| + 2 \frac{A_1 + A_2}{T} (3\sqrt{2} + 4\sqrt{3}).$$

Theorem 6.2.14 (Bagai and Prakasa Rao (1991)). *Suppose X and Y are associated random variables with bounded continuous densities. Then there exists a constant $C > 0$, such that for any $T > 0$,*

$$\sup_{x,y} |P(X \leq x,\, Y \leq y) - P(X \leq x)\, P(Y \leq y)| \leq C[T^2 \operatorname{Cov}(X,Y) + \frac{1}{T}]. \quad (6.2.19)$$

In particular, choosing the optimum $T = [\operatorname{Cov}(X,Y)]^{-1/3}$, the following inequality holds: there exists a constant $C > 0$, such that

$$\sup_{x,y} |P(X \leq x,\, Y \leq y) - P(X \leq x)\, P(Y \leq y)| \leq C[\operatorname{Cov}(X,Y)]^{1/3}. \quad (6.2.20)$$

Proof. Let $F(x,y) = P(X \leq x,\, Y \leq y)$ and $G(x,y) = P(X \leq x)\, P(Y \leq y)$. It is easy to note that the function $G(x,y)$ satisfies the conditions stated in Theorem 6.2.13. Applying the inequality

$$|f(s,t) - f(s,0)\, f(0,t)| \leq |t||s| \operatorname{Cov}(X,Y)$$

due to Newman (1980), we obtain the result stated in (6.2.20). Note that $\operatorname{Cov}(X,Y) \geq 0$ since the random variables X and Y are associated. $\qquad \square$

A minor variant of the theorem stated above is the following result from Yu (1993).

Theorem 6.2.15. *Let the bivariate random vector (X,Y) be an associated (or negatively associated) random vector such that $E(X^2)$ and $E(Y^2)$ are finite. Further suppose that X and Y have bounded probability density functions bounded by a constant M. Then*

$$\sup_{x,y} |P(X \geq x,\, Y \geq y) - P(X \geq x)P(Y \geq y)| \leq 3.2^{2/3} M^{2/3} |\operatorname{Cov}(X,Y)|^{1/3}.$$

Proof. If $\operatorname{Cov}(X,Y) = 0$, the X and Y are independent since the random vector (X,Y) is associated (or negatively associated). Then the result holds trivially. Suppose that $|\operatorname{Cov}(X,Y)| > 0$. Let $\delta > 0$ and define

$$h_{\delta,x}(s) = \begin{cases} 0 & \text{if } \leq x - \delta \\ \delta^{-1}(s - x) & \text{if } x - \delta < s \leq x \\ 1 & \text{if } s > x. \end{cases}$$

Observe that, for every $s \in R$ and every $\delta > 0$,

$$I_{[s \geq x]} \leq h_{\delta,x}(s)$$

and

$$I_{[s \geq x]}| \geq h_{\delta,x+\delta}(s).$$

Hence

$$
\begin{aligned}
&|P(X \geq x,\, Y \geq y) - P(X \geq x)P(Y \geq y)| \\
&= |\operatorname{Cov}(I_{[X \geq x]}, I_{[Y \geq y]}| \\
&\leq |\operatorname{Cov}(h_{\delta,x}(X), h_{\delta,y}(Y))| + |\operatorname{Cov}(I_{[X \geq x]} - h_{\delta,x}(X), h_{\delta,y}(Y))| \\
&\quad + |\operatorname{Cov}(I_{[X \geq x]}, I_{[Y \geq y]} - h_{\delta,y}(Y))| \\
&\leq \delta^{-2} |\operatorname{Cov}(X,Y)| + 4M\delta
\end{aligned}
$$

from the result given in the theorem stated below. The upper bound stated in the theorem is now a consequence of optimization of the last term in $\delta > 0$. $\qquad\square$

Theorem 6.2.16. *Suppose (X,Y) is an associated random vector. Then, for any $a > 0$,*

$$
\lambda\{(x,y) \in R^2 : P(X \geq x,\, Y \geq y) - P(X \geq x)P(Y \geq y)\} \leq \frac{1}{a}\operatorname{Cov}(X,Y)
$$

where $\lambda(.)$ is the Lebesgue measure on R^2.

This result follows from Chebyshev's inequality and Hoeffding identity.

6.3 Hajek-Renyi Type Inequalities

Prakasa Rao (2002) obtained Hajek-Renyi type inequality for associated random variables.

Theorem 6.3.1. *Let the sequence $\{X_n,\, n \geq 1\}$ be an associated sequence of random variables with $\operatorname{Var}(X_i) = \sigma_i^2 < \infty$ and $\{b_n,\, n \geq 1\}$ be a positive nondecreasing sequence. Then, for any $\varepsilon > 0$,*

$$
P(\max_{1 \leq k \leq n} |\frac{1}{b_k}\sum_{i=1}^{k}(X_i - E[X_i])| \geq \varepsilon) \leq 8\varepsilon^{-2}[\sum_{i=1}^{n}\frac{\operatorname{Var}(X_i)}{b_i^2} + \sum_{1 \leq j \neq k \leq n}\frac{\operatorname{Cov}(X_j, X_k)}{b_j b_k}].
$$

$$(6.3.1)$$

An improved version of this result is due to Sung (2008). Christofides (2000) obtained a slightly different version of the Hajek-Renyi type inequality as a corollary to the Hajek-Renyi inequality for demimartingales.

Theorem 6.3.2. *Let the sequence $\{X_n,\, n \geq 1\}$ be an associated sequence of random variables with $\operatorname{Var}(X_i) = \sigma_i^2 < \infty$ and the sequence $\{b_n,\, n \geq 1\}$ be a positive nondecreasing sequence. Let $S_j = X_1 + \ldots + X_j$, $1 \leq j \leq n$. Then, for any $\varepsilon > 0$,*

$$
P(\max_{1 \leq k \leq n} |\frac{1}{b_k}\sum_{i=1}^{k}(X_i - E[X_i])| \geq \varepsilon) \leq 2\varepsilon^{-2}[\sum_{i=1}^{n}\frac{\operatorname{Var}(X_i)}{b_i^2} + \sum_{1 \leq j \leq n}\frac{\operatorname{Cov}(X_j, S_{j-1})}{b_j^2}].
$$

$$(6.3.2)$$

Esary et al. (1967) proved that monotonic functions of associated random variables are associated. Hence one can easily extend the above inequalities to monotonic functions of associated random variables. We now generalize the above results to some non-monotonic functions of associated random variables following the work in Dewan and Prakasa Rao (2006). As an application, a strong law of large numbers is derived for non-monotonic functions of associated random variables. Let us recall some definitions and results which will be useful in proving our main results.

Definition (Newman (1984)). Let f and f_1 be two real-valued functions defined on R^n. We say that $f \ll f_1$ if $f_1 + f$ and $f_1 - f$ are both nondecreasing componentwise.

In particular, if $f \ll f_1$, then f_1 will be nondecreasing componentwise.

Dewan and Prakasa Rao (2001) observed the following.

Suppose that f is a real-valued function defined on $R.$. Then $f \ll f_1$ for some real-valued function defined f_1 on R if and only if, for $x < y$,

$$f(y) - f(x) \le f_1(y) - f_1(x) \tag{6.3.3}$$

and

$$f(x) - f(y) \le f_1(y) - f_1(x). \tag{6.3.4}$$

It is clear that these relations hold if and only if, for $x < y$,

$$|f(y) - f(x)| \le f_1(y) - f_1(x). \tag{6.3.5}$$

If f is a Lipschitzian function defined on R , that is, there exists a positive constant $c \ll$ such that
$$|f(x) - f(y)| \le c|x - y|,$$

then

$$f \ll \tilde{f}, \text{ with } \tilde{f}(x) = cx. \tag{6.3.6}$$

In general, if f is a Lipschitzian function defined on R^n, then $f \ll \tilde{f}$ where

$$\tilde{f}(x_1, \ldots, x_n) = \mathrm{Lip}(f) \sum_{i=1}^{n} x_i$$

and

$$\mathrm{Lip}(f) = \sup_{x \neq y} \frac{|f(x_1, \ldots, x_n) - f(y_1, \ldots, y_n)|}{\sum_{i=1}^{n} |x_i - y_i|} < \infty.$$

Let the sequence $\{X_n, n \ge 1\}$ be a sequence of associated random variables. Let

(i) $Y_n = f_n(X_1, X_2, \ldots),$

(ii) $\tilde{Y}_n = \tilde{f}_n(X_1, X_2, \ldots),$

(iii) $$f_n \ll \tilde{f}_n, \quad \text{and}$$

(iv) $$E(Y_n^2) < \infty, \quad E(\tilde{Y}_n^2) < \infty, \quad \text{for } n \geq 1. \tag{6.3.7}$$

For convenience, we write that $Y_n \ll \tilde{Y}_n$ if the conditions stated in (i)–(iv) hold. The functions $f_n, \tilde{f}_n$ are assumed to be real-valued and depend only on a finite number of $X_n's$. Let $S_n = \sum_{k=1}^{n} Y_k, \tilde{S}_n = \sum_{k=1}^{n} \tilde{Y}_k$. Matula (2001) proved the following result which will be useful in proving our results. He used them to prove the strong law of large numbers and the central limit theorem for non-monotonic functions of associated random variables.

Lemma 6.3.3 (Matula (2001)). *Suppose the conditions stated above in* (6.3.7) *hold. Then*

(i) $$\mathrm{Var}(f_n) \leq \mathrm{Var}(\tilde{f}_n),$$

(ii) $$|\mathrm{Cov}(f_n, \tilde{f}_n)| \leq \mathrm{Var}(\tilde{f}_n),$$

(iii) $$\mathrm{Var}(S_n) \leq \mathrm{Var}(\tilde{S}_n),$$

(iv) $$f_1 + f_2 + \ldots + f_n \ll \tilde{f}_1 + \tilde{f}_2 + \ldots + \tilde{f}_n,$$

(v) $$\mathrm{Cov}(f_1 + \tilde{f}_1, f_2 + \tilde{f}_2) \leq 4Cov(\tilde{f}_1, \tilde{f}_2), \quad \text{and}$$

(vi) $$\mathrm{Cov}(\tilde{f}_1 - f_1, \tilde{f}_2 - f_2) \leq 4\,\mathrm{Cov}(\tilde{f}_1, \tilde{f}_2). \tag{6.3.8}$$

The following maximal inequality is due to Newman and Wright (1981) for associated random variables.

Lemma 6.3.4 (Newman and Wright (1981)). *Suppose* $X_1, X_2, \ldots, X_m$ *are associated, mean zero, finite variance random variables and* $M_m^* = \max(S_1^*, S_2^*, \ldots, S_m^*)$, *where* $S_n^* = \sum_{i=1}^{n} X_i$. *Then*

$$E((M_m^*)^2) \leq \mathrm{Var}(S_m^*). \tag{6.3.9}$$

Remarks. Note that if $X_1, X_2, \ldots, X_m$ are associated random variables, then $-X_1, -X_2, \ldots, -X_m$ also form a set of associated random variables. Let $M_m^{**} = \max(-S_1^*, -S_2^*, \ldots, -S_m^*)$ and $\tilde{M}_m^* = \max(|S_1^*|, |S_2^*|, \ldots, |S_m^*|)$. Then $\tilde{M}_m^* = \max(M_m^*, M_m^{**})$ and $(\tilde{M}_m^*)^2 \leq (M_m^*)^2 + (M_m^{**})^2$ so that

$$E((\tilde{M}_m^*)^2) \leq 2\,\mathrm{Var}(S_m^*). \tag{6.3.10}$$

We now extend Newman and Wright's (1981) result to non-monotonic functions of associated random variables satisfying conditions (6.3.7).

Theorem 6.3.5. *Let* $Y_1, Y_2, \ldots, Y_m$ *be as defined in* (6.3.7) *with zero mean and finite variances . Let* $M_m = \max(|S_1|, |S_2|, \ldots, |S_m|)$. *Then*

$$E(M_m^2) \leq (20)\,\mathrm{Var}(\tilde{S}_m). \tag{6.3.11}$$

Proof. Observe that

$$\max_{1 \leq k \leq m} |S_k| = \max_{1 \leq k \leq m} |\tilde{S}_k - S_k - E(\tilde{S}_k) - \tilde{S}_k + E(\tilde{S}_k)| \qquad (6.3.12)$$

$$\leq \max_{1 \leq k \leq m} |\tilde{S}_k - S_k - E(\tilde{S}_k)| + \max_{1 \leq k \leq m} |\tilde{S}_k - E(\tilde{S}_k)|.$$

Note that $\tilde{S}_k - E(\tilde{S}_k)$ and $\tilde{S}_k - S_k - E(\tilde{S}_k)$ are partial sums of associated random variables each with mean zero. Hence, using the results of Newman and Wright (1981), we get that

$$E(M_m^2) \leq E(\max_{1 \leq k \leq m} |S_k|)^2 \qquad (6.3.13)$$

$$\leq 2[E(\max_{1 \leq k \leq m} |\tilde{S}_k - S_k - E(\tilde{S}_k)|)^2 + E(\max_{1 \leq k \leq m} |\tilde{S}_k - E(\tilde{S}_k)|)^2]$$

$$\leq 4[\mathrm{Var}(\tilde{S}_m - S_m) + \mathrm{Var}(\tilde{S}_m)]$$

$$\leq 4[\mathrm{Var}(2\tilde{S}_m) + \mathrm{Var}(\tilde{S}_m)]$$

$$= (20)\, \mathrm{Var}(\tilde{S}_m). \qquad \square$$

We have used the fact that

$$\mathrm{Var}(2\tilde{S}_n) = \mathrm{Var}(\tilde{S}_n - S_n + \tilde{S}_n + S_n) \qquad (6.3.14)$$

$$= \mathrm{Var}(\tilde{S}_n - S_n) + \mathrm{Var}(\tilde{S}_n + S_n) + 2\,\mathrm{Cov}(\tilde{S}_n + S_n, \tilde{S}_n - S_n).$$

Since $\tilde{S}_n + S_n$ and $\tilde{S}_n - S_n$ are nondecreasing functions of associated random variables, it follows that $\mathrm{Cov}(\tilde{S}_n + S_n, \tilde{S}_n - S_n) \geq 0$. Hence $\mathrm{Var}(2\tilde{S}_n) \geq Var(\tilde{S}_n - S_n)$.

We now prove a Hajek-Renyi type inequality for some non-monotonic functions of associated random variables satisfying conditions (6.3.7).

Theorem 6.3.6. *Let the sequence $\{Y_n, n \geq 1\}$ be a sequence of non-monotonic functions of associated random variables as defined in (6.3.7). Suppose that $Y_n \ll \tilde{Y}_n$, $n \geq 1$. Let the sequence $\{b_n, n \geq 1\}$ be a positive nondecreasing sequence of real numbers. Then, for any $\epsilon > 0$,*

$$P(\max_{1 \leq k \leq n} |\frac{1}{b_n} \sum_{i=1}^{k} (Y_i - E(Y_i))| \geq \epsilon) \leq 80\epsilon^{-2} [\sum_{j=1}^{n} \frac{\mathrm{Var}(\tilde{Y}_j)}{b_j^2} + \sum_{1 \leq j \neq k \leq n} \frac{\mathrm{Cov}(\tilde{Y}_j, \tilde{Y}_k)}{b_j b_k}].$$

$$(6.3.15)$$

Proof. Let $T_n = \sum_{j=1}^{n} (Y_j - E(Y_j))$. Note that

$$P[\max_{1 \leq k \leq n} |\frac{T_k}{b_k}| \geq \epsilon] \qquad (6.3.16)$$

$$= P[\max_{1 \leq k \leq n} |\frac{\tilde{T}_k - T_k - E(\tilde{T}_k) - \tilde{T}_k + E(\tilde{T}_k)|}{b_k}| \geq \epsilon]$$

$$\leq P[\max_{1\leq k\leq n}|\frac{\tilde{T}_k - T_k - E(\tilde{T}_k)}{b_k}| \geq \frac{\epsilon}{2}] + P[\max_{1\leq k\leq n}|\frac{\tilde{T}_k - E(\tilde{T}_k)|}{b_k}| \geq \frac{\epsilon}{2}]$$

$$\leq 16\epsilon^{-2}[\sum_{j=1}^{n}\frac{\text{Var}(\tilde{Y}_j - Y_j)}{b_j^2} + \sum_{1\leq j\neq k\leq n}\frac{\text{Cov}(\tilde{Y}_j - Y_j, \tilde{Y}_k - Y_k)}{b_j b_k}]$$

$$+ 16\epsilon^{-2}[\sum_{j=1}^{n}\frac{\text{Var}(\tilde{Y}_j)}{b_j^2} + \sum_{1\leq j\neq k\leq n}\frac{\text{Cov}(\tilde{Y}_j, \tilde{Y}_k)}{b_j b_k}].$$

The result follows by applying the inequalities

$$\text{Var}(\tilde{Y}_j - Y_j) \leq 4\,\text{Var}(\tilde{Y}_j)$$
$$\text{Cov}(\tilde{Y}_j - Y_j, \tilde{Y}_k - Y_k) \leq 4\,\text{Cov}(\tilde{Y}_j, \tilde{Y}_k).\qquad\qquad \square$$

Applications

Corollary 6.3.7. *Let the sequence $\{Y_n,\ n \geq 1\}$ be a sequence of non-monotonic functions of associated random variables satisfying the conditions in (6.3.7). Assume that*

$$\sum_{j=1}^{\infty}\text{Var}(\tilde{Y}_j) + \sum_{1\leq j\neq k<\infty}\text{Cov}(\tilde{Y}_j, \tilde{Y}_k) < \infty. \qquad (6.3.17)$$

Then $\sum_{j=1}^{\infty}(Y_j - EY_j)$ converges almost surely.

Proof. Without loss of generality, assume that $EY_j = 0$ for all $j \geq 1$. Let $T_n = \sum_{j=1}^{n} Y_j$ and $\epsilon > 0$. Using Theorem 6.3.2, it is easy to see that there exists a constant $C > 0$, such that

$$P(\sup_{k,m\geq n}|T_k - T_m| \geq \epsilon) \qquad (6.3.18)$$

$$\leq P(\sup_{k\geq n}|T_k - T_n| \geq \frac{\epsilon}{2}) + P(\sup_{m\geq n}|T_m - T_n| \geq \frac{\epsilon}{2})$$

$$\leq C\,\limsup_{N\to\infty} P(\sup_{n\leq k\leq N}|T_k - T_n| \geq \frac{\epsilon}{2})$$

$$\leq C\,\epsilon^{-2}[\sum_{j=n}^{\infty}\text{Var}(\tilde{Y}_j) + \sum_{n\leq j\neq k<\infty}\text{Cov}(\tilde{Y}_j, \tilde{Y}_k)].$$

The last term tends to zero as $n \to \infty$ because of (6.3.17). Hence the sequence of random variables $\{T_n, n \geq 1\}$ is Cauchy almost surely which implies that T_n converges almost surely. $\qquad\qquad \square$

The following corollary proves the strong law of large numbers for non-monotonic functions of associated random variables.

Corollary 6.3.8. *Let the sequence* $\{Y_n, \ n \geq 1\}$ *be a sequence of non-monotonic functions of associated random variables satisfying the conditions in (6.3.7). Suppose that*

$$\sum_{j=1}^{\infty} \frac{\mathrm{Var}(\tilde{Y}_j)}{b_j^2} + \sum_{1 \leq j \neq k < \infty} \frac{\mathrm{Cov}(\tilde{Y}_j, \tilde{Y}_k)}{b_j b_k} < \infty.$$

Then $\frac{1}{b_n} \sum_{j=1}^{n} (Y_j - EY_j)$ *converges to zero almost surely as* $n \to \infty$.

Proof. The proof is an immediate consequence of Theorem 6.3.2 and the Kronecker lemma (Chung (1974)). □

Remarks. Birkel (1989) proved a strong law of large numbers for positively dependent random variables. Prakasa Rao (2002) proved a strong law of large numbers for associated sequences as a consequence of the Hajek-Renyi type inequality. Marcinkiewicz-Zygmund type strong law of large numbers for associated random variables, for which the second moment is not necessarily finite, was studied in Louhichi (2000). A strong law of large numbers for monotone functions of associated sequences follows from these results since monotone functions of associated sequences are associated. However the corollary given above gives sufficient conditions for the strong law of large numbers to hold for possibly non-monotonic functions of associated sequences whose second moments are finite.

For any random variable X and any constant $k > 0$, define

$$X^k = \begin{cases} X & \text{if } |X| \leq k \\ -k & \text{if } X < -k \\ k & \text{if } X > k. \end{cases}$$

The following theorem is an analogue of the three series theorem for non-monotonic functions of associated random variables.

Corollary 6.3.9. *Let the sequence* $\{Y_n, \ n \geq 1\}$ *be a sequence of non-monotonic functions of associated random variables. Further suppose that there exists a constant* $k > 0$ *such that* $Y_n^k \ll \tilde{Y}_n^k$ *satisfying the conditions in (6.3.7) and*

$$\sum_{n=1}^{\infty} P[|Y_n| \geq k] < \infty, \tag{6.3.19}$$

$$\sum_{n=1}^{\infty} E(Y_n^k) < \infty, \tag{6.3.20}$$

$$\sum_{j=1}^{\infty} \mathrm{Var}(\tilde{Y}_j^k) + \sum_{1 \leq j \neq j' < \infty} \mathrm{Cov}(\tilde{Y}_j^k, \tilde{Y}_{j'}^k) < \infty. \tag{6.3.21}$$

Then $\sum_{n=1}^{\infty} Y_n$ *converges almost surely.*

Corollary 6.3.10. *Let the sequence $\{Y_n, n \geq 1\}$ be a sequence of non-monotonic functions of associated random variables satisfying the conditions in (6.3.7). Suppose*

$$\sum_{j=1}^{\infty} \frac{\text{Var}(\tilde{Y}_j)}{b_j^2} + \sum_{1 \leq j \neq k < \infty} \frac{\text{Cov}(\tilde{Y}_j, \tilde{Y}_k)}{b_j b_k} < \infty. \qquad (6.3.22)$$

Let $T_n = \sum_{j=1}^{n}(Y_j - E(Y_j))$. Then, for any $0 < r < 2$,

$$E[\sup_n (\frac{|T_n|}{b_n})^r] < \infty. \qquad (6.3.23)$$

Proof. Note that

$$E[\sup_n (\frac{|T_n|}{b_n})^r] < \infty$$

if and only if

$$\int_1^{\infty} P(\sup_n (\frac{|T_n|}{b_n})^r > t^{1/r}) dt < \infty.$$

The last inequality holds because of Theorem 6.3.2 and the condition (6.3.22). Hence the result stated in equation (6.3.23) holds. $\qquad \square$

6.4 Exponential Inequality

We have discussed an exponential inequality for partial sums of associated random variables in Chapter 1. We will now discuss another variation of the exponential inequality for bounded stationary associated sequences due to Douge (2007).

Theorem 6.4.1. *Let the sequence $\{X_n, n \geq 1\}$ be a stationary associated sequence of random variables such that $|X_n| \leq M < \infty$, $n \geq 1$ and*

$$c_k = \text{Cov}(X_1, X_{k+1}) \leq \alpha_0 \exp\{-\alpha k\}, \quad k \geq 0 \qquad (6.4.1)$$

for some $\alpha > 0$ and $\alpha_0 > 0$. Then, for every $n \geq 2$ and every $\epsilon > \frac{6M}{\sqrt{n}}$,

$$P[\frac{1}{n}|\sum_{i=1}^{n}(X_i - E(X_i))| \geq \epsilon] \leq 8c_0 \exp[-\frac{\min(\alpha, 1)}{12M}\sqrt{n}\epsilon] \qquad (6.4.2)$$

where $c_0 = \exp[\alpha_0/(4M^2(1 - e^{-\alpha}))]$.

We will now sketch a proof of this theorem.

Proof. For $n \geq 2$, let $p = p(n)$ and $r = r(n)$ such that $2pr \leq n$. Let $Y_i = X_i - E(X_i)$, $1 \leq i \leq n$ and $\bar{S}_n = \frac{1}{n}\sum_{i=1}^{n} Y_i$. Define the random variables U_i, V_i, $i = 1, \ldots, n$ and W_n by the equations

$$U_i = Y_{2(i-1)p+1} + \ldots + Y_{(2i-1)p}, \qquad (6.4.3)$$

$$V_i = Y_{(2i-1)p+1} + \ldots + Y_{2ip}, \tag{6.4.4}$$

and

$$W_n = Y_{2pr+1} + \ldots + Y_n. \tag{6.4.5}$$

Let $\bar{U}_n = \frac{1}{n}\sum_{i=1}^r U_i$, $\bar{V}_n = \frac{1}{n}\sum_{i=1}^r V_i$ and $\bar{W}_n = \frac{W_n}{n}$. Let $\epsilon > 0$. Then

$$P(\bar{U}_n \geq \epsilon) \leq e^{-\lambda\epsilon} E[e^{\lambda\bar{U}_n}] \tag{6.4.6}$$

$$\leq e^{-\lambda\epsilon}(|E(e^{\frac{\lambda}{n}\sum_{i=1}^r U_i} - \prod_{i=1}^r E[e^{\frac{\lambda}{n}U_i}]| + \prod_{i=1}^r E[e^{\frac{\lambda}{n}U_i}]).$$

Applying the inequality

$$e^x \leq 1 + x + x^2 \text{ if } |x| \leq \frac{1}{2}$$

for $x = \frac{\lambda}{n}U_n$ where $\lambda = \frac{\min(\alpha,1)}{4M}\frac{n}{p}$, we obtain that

$$\prod_{i=1}^r E[e^{\frac{\lambda}{n}U_i}] \leq \prod_{i=1}^r E(1 + \frac{\lambda}{n}U_i + \frac{\lambda^2}{n^2}U_i^2) \leq \exp(\frac{\alpha_0}{1-e^{-\alpha}}\frac{\lambda^2}{n}). \tag{6.4.7}$$

Note that the random variables $U_1, \ldots, U_n$ are associated. This follows from the properties of associated random variables (cf. Property (4) of Esary et al. (1967)). Applying Theorem 6.2.9, it can be shown that

$$|E(e^{\frac{\lambda}{n}\sum_{i=1}^r U_i} - \prod_{i=1}^r E[e^{\frac{\lambda}{n}U_i}]| \leq \frac{\lambda^2\alpha_0}{2n(1-e^{-\alpha})}e^{\lambda M-\alpha p}. \tag{6.4.8}$$

The inequalities (6.4.7) and (6.4.8) also hold if U_i is replaced by $-U_i$. Choosing $p = [\sqrt{n}]$, and using inequality (6.4.6), we get that

$$P(|\bar{U}_n| \geq \varepsilon) = P(\bar{U}_n \geq \varepsilon) + P(-\bar{U}_n \geq \varepsilon) \leq 4c_0 \exp[-\frac{\min(\alpha,1)}{4M}\sqrt{n}\varepsilon] \tag{6.4.9}$$

where $c_0 = \exp[\alpha_0/(4M^2(1-e^{-\alpha}))]$. It can be seen that $\bar{V}_n$ satisfies the inequality (6.4.9) and furthermore $P(\bar{W}_n \geq \varepsilon) = 0$ for every $\varepsilon > \frac{2M}{\sqrt{n}}$. The result now follows from the fact that $\bar{S}_n = \bar{U}_n + \bar{V}_n + \bar{W}_n$. $\square$

Remarks. As a consequence of the inequality (6.4.2), it follows that

$$\frac{1}{n}|\sum_{i=1}^n (X_i - E(X_i))| = O(\frac{\log n}{\sqrt{n}}) \text{ a.s.} \tag{6.4.10}$$

under the conditions stated in Theorem 6.4.1.

6.5 Non-uniform and Uniform Berry-Esseen Type Bounds

We have stated some results on uniform Berry-Esseen type bounds for partial sums of stationary associated sequences in Chapter 1. We now derive uniform and non-uniform type bounds for such sequences following Dewan and Prakasa Rao (2005a).

Uniform Bound

Let the set of random variables $\{X_i, 1 \le i \le n\}$ be a finite set of stationary associated random variables. We obtain an explicit Berry-Esseen type bound Berry-Esseen type bound for the distribution function of $S_n = X_1 + \ldots + X_n$. The bound is in terms of the moments of random variables $\{X_i, 1 \le i \le n\}$ and bounds on the density function of the partial sums of n independent copies of the random variable X_1 assuming that it exists.

Theorem 6.5.1. *Let the set $\{X_i, 1 \le i \le n\}$ be a finite set of stationary associated random variables with $E(X_1) = 0$, $\mathrm{Var}(X_1) = \sigma^2 > 0$ and $E[|X_1|^3] < \infty$. Suppose the random variable X_1 has an absolutely continuous distribution function. Let $s_n^2 = \mathrm{Var}(S_n)$. Suppose that*

$$\frac{s_n^2}{n} \to \sigma^2 \ \text{as} \ n \to \infty. \tag{6.5.1}$$

Let F_n be the distribution function of $\frac{S_n}{s_n}$ and F_n^* be the distribution function of $\frac{\sum_{i=1}^{n} Z_i}{s_n}$ where $Z_i, 1 \le i \le n$ are independent and identically distributed random variables with the distribution function the same as that of the random variable X_1. Let m_n be a bound on the derivative of F_n^*. Then there exists an absolute constant $C > 0$, such that

$$\sup_x |F_n(x) - \Phi(x)| \le C \left[\frac{d_n^{1/3} m_n^{2/3}}{s_n^{2/3}} + \frac{E|X_1|^3}{\sqrt{n}\sigma^3} + \left(\frac{s_n}{\sigma\sqrt{n}} - 1 \right) \right] \tag{6.5.2}$$

where $\Phi(x)$ is the distribution function of a standard Gaussian random variable and

$$d_n = \sum_{j=2}^{n} (n - j + 1) \, \mathrm{Cov}(X_1, X_j). \tag{6.5.3}$$

Proof. Let $\psi_n(t)$ and $\psi_n^*(t)$ be the characteristic functions corresponding to the distribution functions F_n and F_n^* respectively. Note that, for $T > 0$,

$$\sup_x |F_n(x) - F_n^*(x)| \le \frac{1}{\pi} \int_{-T}^{T} \left| \frac{\psi_n(t) - \psi_n^*(t)}{t} \right| \, dt + 24 \frac{m_n}{\pi T} \tag{6.5.4}$$

by the smoothing inequality (cf. Feller (1977), Vol. II, p. 512). Applying Newman's inequality to the sequence of stationary associated random variables $\{X_i, 1 \le i \le n\}$, we get that

$$|\psi_n(t) - \psi_n^*(t)| \le \sum_{1 \le i < j \le n} \frac{t^2}{s_n^2} \operatorname{Cov}(X_i, X_j) \qquad (6.5.5)$$

$$= \frac{t^2}{s_n^2} \sum_{i=1}^{n-1} \sum_{j=i+1}^{n} \operatorname{Cov}(X_i, X_j)$$

$$= \frac{t^2}{s_n^2} \sum_{j=2}^{n} (n - j + 1) \operatorname{Cov}(X_i, X_j)$$

$$= \frac{t^2}{s_n^2} d_n.$$

From equations (6.5.4) and (6.5.5), it follows that

$$\sup_x |F_n(x) - F_n^*(x)| \le \frac{d_n}{\pi s_n^2} T^2 + 24 \frac{m_n}{\pi T}. \qquad (6.5.6)$$

Choosing $T = \left(\frac{(24)m_n s_n^2}{d_n}\right)^{1/3}$, we get that

$$\sup_x |F_n(x) - F_n^*(x)| \le \frac{2(24)^{2/3}}{\pi} \frac{d_n^{1/3} m_n^{2/3}}{s_n^{2/3}}. \qquad (6.5.7)$$

Applying the classical Berry-Esseen bound for independent and identically distributed random variables $\{Z_i, 1 \le i \le n\}$ (cf. Feller (1977), Vol.II, p. 515), we have

$$\sup_x |F_n^*(x) - \Phi(\frac{s_n x}{\sigma \sqrt{n}})| \le C \frac{E|Z_1|^3}{\sqrt{n}\sigma^3} \qquad (6.5.8)$$

where C is an absolute positive constant such that $\frac{\sqrt{10}+3}{6\sqrt{2\pi}} \le C \le 0.7975$ (cf. Esseen (1956), Van Beek (1972)). Furthermore, applying an inequality from Petrov (1975, p.114), we get that

$$\sup_x |\Phi(\frac{s_n x}{\sigma \sqrt{n}}) - \Phi(x)| \le \frac{(\frac{s_n}{\sigma \sqrt{n}} - 1)}{\sqrt{2\pi e}} \qquad (6.5.9)$$

since $s_n^2 \ge n\sigma^2$ by the associativity property of the sequence $\{X_i, 1 \le i \le n\}$. Combining the inequalities (6.5.7) to (6.5.9), we get the bound in equation (6.5.2). $\qquad\square$

For further comments on the uniform bounds, see Dewan and Prakasa Rao (2005a).

Non-uniform Bound

We now derive non-uniform Berry-Esseen type bounds for sums of stationary associated random variables following the techniques in Petrov (1975) and Hall (1982) for sums of independent random variables. These results are due to Dewan and Prakasa Rao (2005a).

Let G_1 and G_2 be two distribution functions. If the distributions have moments of order $p > 0$, then Petrov (1975, p. 120) proved that

$$|G_1(x) - G_2(x)| \leq \frac{\beta}{|x|^p}, \quad x \neq 0,$$

where $\beta = \max(\beta_1, \beta_2)$, $\beta_k = \int_{-\infty}^{\infty} |x|^p dG_k(x)$, $k = 1, 2$. Furthermore

$$|G_1(x) - G_2(x)| \leq [\frac{(1+\beta)^s \Delta_1^r}{(1+|x|^p)^s}]^{1/(r+s)} \tag{6.5.10}$$

for all x, for $r > 0$, $s > 0$ and $\Delta_1 = \sup_x |G_1(x) - G_2(x)|$.

Lemma 6.5.2 (Petrov (1975)). *Let $F(x)$ be an arbitrary distribution function with finite p-th absolute moment for some $p > 0$. Let*

$$\Delta = \sup_x |F(x) - \Phi(x)|. \tag{6.5.11}$$

Suppose that $0 \leq \Delta \leq e^{-1/2}$. Then

$$|F(x) - \Phi(x)| \leq \frac{C_p \Delta (\log(1/\Delta))^{p/2} + \lambda_p}{1 + |x|^p} \tag{6.5.12}$$

for all x, where C_p is a positive constant depending only on p, and

$$\lambda_p = |\int_{-\infty}^{\infty} |x|^p \, dF(x) - \int_{-\infty}^{\infty} |x|^p \, d\Phi(x)|. \tag{6.5.13}$$

We now derive a non-uniform Berry-Esseen type bound for sums of stationary associated random variables using the lemma stated above. Following the notation used in Theorem 6.5.1, let $F_n(x)$ and $F_n^*(x)$ be the distribution functions of $\frac{S_n}{s_n}$ and $\frac{S_n^*}{s_n}$ respectively where $S_n = \sum_{i=1}^{n} X_i$ and $S_n^* = \sum_{i=1}^{n} Z_i$. Then $\text{Var}[\frac{S_n}{s_n}] = 1$ and $\text{Var}[\frac{S_n^*}{s_n}] \leq 1$ by the associative property of the random variables $X_1, \ldots, X_n$. Hence

$$|F_n(x) - F_n^*(x)| \leq [\frac{2\Delta_{1n}}{1+x^2}]^{1/2} \tag{6.5.14}$$

from equation (6.5.10) by choosing $p = 2, r = 1$ and $s = 1$ where

$$\Delta_{1n} = \sup_x |F_n(x) - F_n^*(x)| \leq C \frac{d_n^{1/3} m_n^{2/3}}{s_n^{2/3}} \tag{6.5.15}$$

from equation (6.5.7). Let

$$\Delta_{2n} = \sup_x |F_n^*(x) - \Phi(x)|. \tag{6.5.16}$$

Suppose that $0 < \Delta_{2n} < e^{-1/2}$. Then, it follows from the above lemma that there exists an absolute constant $c_2 > 0$ such that

$$|F_n^*(x) - \Phi(x)| \leq \frac{c_2\Delta_{2n} \log(1/\Delta_{2n}) + \lambda_{2n}}{1 + x^2} \tag{6.5.17}$$

where

$$\lambda_{2n} = |\int_{-\infty}^{\infty} x^2 dF_n^*(x) - 1| = |Var(S_n^*/s_n) - 1| = |(n\sigma^2/s_n^2) - 1|. \tag{6.5.18}$$

Note that Δ_{2n} is the Berry-Esseen bound obtained from equations (6.5.8) and (6.5.9).

The bounds derived above lead to the following result giving a non-uniform Berry-Esseen type bound for sums of stationary associated sequence of random variables.

Theorem 6.5.3. *Let the sequence $\{X_i, 1 \leq i \leq n\}$ be a sequence of stationary associated random variables satisfying the conditions stated in Theorem 6.5.1. Suppose that $0 < \Delta_{2n} < e^{-1/2}$. Then there exists absolute positive constants c_1 and c_2 such that*

$$|F_n(x) - \Phi(x)| \leq c_1[\frac{2\Delta_{1n}}{1 + x^2}]^{1/2} + \frac{c_2\Delta_{2n} \log(1/\Delta_{2n}) + \lambda_{2n}}{1 + x^2}. \tag{6.5.19}$$

Hall (1982) proved the following result for real-valued functions.

Theorem 6.5.4. *Let $F(x)$ be a nondecreasing function and $G(x)$ be a differentiable function of bounded variation. Suppose that $F(-\infty) = G(-\infty)$ and $F(\infty) = G(\infty)$. Assume that*

$$\int_{-\infty}^{\infty} x^2 |d\{F(x) - G(x)\}| < \infty \tag{6.5.20}$$

and

$$\sup_x (1 + x^2)|G'(x)| \leq k < \infty. \tag{6.5.21}$$

Let

$$\chi(t) = \int_{-\infty}^{\infty} \exp(itx)d[x^2\{F(x) - G(x)\}] \tag{6.5.22}$$

and $f(t)$ and $g(t)$ denote the Fourier-Stieltjes transforms of F and G respectively. Then there exists an absolute constant $C > 0$ such that

$$\sup_x (1 + x^2)|F(x) - G(x)| \leq C[\int_0^T \frac{|f(t) - g(t)|}{t} dt + \int_0^T \frac{|\chi(t)|}{t} dt + \frac{k}{T}] \tag{6.5.23}$$

for $T \geq 1$.

As an application of this result, Dewan and Prakasa Rao (2005a) derived another class of non-uniform bounds for partial sums of random variables which can be dependent. We do not go into the details here.

Newman's Conjecture

Recall the central limit theorem for stationary associated sequences stated in Theorem 1.2.19. It is not known whether the central limit theorem holds for stationary demimartingales. Newman (1980) conjectured that, for a strictly stationary associated random field $\mathbf{X} = \{X_j, j \in R^d\}$, it is sufficient, for the central limit theorem to hold (when partial sums are taken over growing sets), if the function

$$u_{\mathbf{X}}(n) = \sum_{|j| \leq n} \mathrm{Cov}(X_0, X_j), \quad n \geq 1$$

is slowly varying. Herrndorf (1984) showed that this conjecture is not true, even in the case $d = 1$, by constructing an example. He showed that there exists a strictly stationary associated sequence of real-valued random variables $\{X_n, n \geq 1\}$ with $u_X(n) \simeq \log n$ as $n \to \infty$ such that the sequence of normalized partial sums

$$\frac{X_1 + \ldots + X_n}{\sqrt{n u_X(n)}}$$

does not have any non-degenerate limiting distribution as $n \to \infty$. Shaskin (2005) obtained a result generalizing the example of Herrndorf (1984).

Recall that a function $L : R_+ \to R - \{0\}$ is called *slowly varying in the Karamata sense* at infinity, if for any $a > 0$,

$$\frac{L(ax)}{L(x)} \to 1 \text{ as } x \to \infty.$$

A sequence $L : N \to R - \{0\}$ is *slowly varying* at infinity if the above relation holds for any $a \in N$ as $x \to \infty$, $x \in N$.

Let L be a nondecreasing slowly varying function such that $L(n) \to \infty$ as $n \to \infty$. Shaskin (2005) proved that, for any positive unbounded sequence $b(n)$, $n \geq 1$, there exists a strictly stationary associated sequence $\{X_n, n \geq 1\}$ such that $E(X_1) = 0$, $E(X_1^2) = 1$, $u_X(n) \simeq L(n)$ as $n \to \infty$, and the normalized partial sums

$$\frac{X_1 + \ldots + X_n}{\sqrt{n b(n)}}$$

do not have any non-degenerate limit distribution as $n \to \infty$.

6.6   Limit Theorems for U-Statistics

Some limit theorems for U-statistics were discussed earlier in Chapter 2 as corollaries to inequalities for demimartingales. The concept of U-statistic was introduced by Hoeffding (1948). It generalizes the notion of sample mean and includes many important test statistics as special cases. Limit theorems for U-statistics of independent random variables are presented in Serfling (1980) and Lee (1990). The study of U-statistics for dependent random variables is made in Sen (1963), Nandi and Sen (1963), Serfling (1968), Denker and Keller (1983) and Becker and Utev (2001) among others. We now discuss results on limit theorems for U-statistics for associated random variables from Dewan and Prakasa Rao (2001, 2002) and Christofides (2004).

Suppose the random sequences $\{X_j,\ j \geq 1\}$ and $\{Y_j,\ j \geq 1\}$ are independent stationary sequences of associated random variables with one-dimensional marginal distribution functions F and G respectively and the problem is to test the hypothesis $H_0 : F = G$ based on the two samples $\{X_i,\ 1 \leq i \leq n\}$ and $\{Y_j,\ 1 \leq j \leq m\}$. A test statistic for testing the hypothesis H_0 is the two-sample U-statistic

$$U_{nm} = \frac{1}{nm} \sum_{i=1}^{n} \sum_{j=1}^{m} I_{[X_i \leq Y_j]}.$$

Hence it is of interest to study the asymptotic behaviour of the statistic U_{nm} under the hypothesis H_0.

Let

$$U_n = \frac{1}{\binom{n}{m}} \sum_{1 \leq i_1 < \ldots < i_m \leq n} h(X_{i_1}, \ldots, X_{i_m}) \tag{6.6.1}$$

be a U-statistic based on associated random variables and the kernel $h(x_1, \ldots, x_m)$ of degree m. Without loss of generality, we assume that h is symmetric in its arguments. Further suppose that h is componentwise nondecreasing and $E(h) = 0$. Then the process $\{S_n = \binom{n}{m}U_n,\ n \geq m\}$ is a demimartingale as shown in Chapter 2. Christofides (2000) proved the following result for demimartingales.

Theorem 6.6.1. *Let the sequence $\{S_n,\ n \geq 0\}$ be a demimartingale and the sequence $\{c_k,\ k \geq 1\}$ be a nonincreasing sequence of positive numbers such that $\lim_{k \to \infty} c_k = 0$. Suppose that $E|S_k|^\nu < \infty,\ k \geq 1$ for some $\nu \geq 1$. If*

$$\sum_{k=1}^{\infty} c_k^\nu E(|S_k|^\nu - |S_{k-1}|^\nu) < \infty, \tag{6.6.2}$$

then

$$c_n S_n \overset{a.s.}{\to} 0 \ as \ n \to \infty. \tag{6.6.3}$$

Applying the above lemma, we get the following strong law of large numbers for U-statistics based on associated random variables.

Strong Law of Large Numbers

Theorem 6.6.2 (Christofides (2004)). *Let U_n be a U-statistic based on a family of associated random variables and on the kernel h of degree m. Suppose that, for some $\nu \geq 1$, $E|S_k|^\nu < \infty$ for all $k \geq m$. Further suppose that the function h is componentwise nondecreasing. If*

$$\sum_{k=m}^{\infty} (k+1)^{-1} E|U_k|^\nu < \infty, \tag{6.6.4}$$

then

$$U_n - E(U_n) \overset{a.s}{\to} 0 \ as \ \to \infty. \tag{6.6.5}$$

Proof. Let $S_n = \binom{n}{m} U_n$ for $n \geq m$ and $S_n = 0$ for $n < m$. Since h is nondecreasing, it follows that the sequence $\{S_n, n \geq m\}$ is a demimartingale. Furthermore the sequence $c_n = \binom{n}{m}^{-1}$, $n \geq m$ is a decreasing sequence of positive numbers. Applying Theorem 6.6.1, we get that

$$\binom{n}{m}^{-1} (S_n - E(S_n)) \overset{a.s}{\to} 0 \text{ as } \to \infty \tag{6.6.6}$$

provided

$$\sum_{k=m}^{\infty} \binom{k}{m}^{-\nu} E(|S_k|^\nu - |S_{k-1}|^\nu) < \infty. \tag{6.6.7}$$

Note that

$$\sum_{k=m}^{\infty} \binom{k}{m}^{-\nu} E(|S_k|^\nu - |S_{k-1}|^\nu) \tag{6.6.8}$$

$$= \sum_{k=m}^{\infty} [\binom{k}{m}^{-\nu} - \binom{k+1}{m}^{-\nu}] E|S_k|^\nu$$

$$< \nu \sum_{k=m}^{\infty} \binom{k}{m}^{-\nu+1} [\binom{k}{m}^{-1} - \binom{k+1}{m}^{-1}] E|S_k|^\nu$$

$$= \nu m \sum_{k=m}^{\infty} \binom{k}{m}^{-\nu+1} \binom{k}{m}^{-1} (k+1)^{-1} E|S_k|^\nu$$

$$= \nu m \sum_{k=m}^{\infty} (k+1)^{-1} E|U_k|^\nu$$

$$< \infty.$$

We have used the inequality

$$x^r - y^r < r x^{r-1} (x - y)$$

for $x \geq y \geq 0$ and $r \geq 1$ in the above estimates. $\qquad \square$

The following lemma is useful in checking the condition (6.6.4).

Lemma 6.6.3. *Let*

$$u(n) \tag{6.6.9}$$

$$= \sup_{k; i_1, \ldots, i_{m-1}} \sum_{\substack{j:|k-j| \geq n \\ k, j \neq i_1, \ldots, i_{m-1}}} \mathrm{Cov}(h(X_{i_1}, \ldots, X_{i_{m-1}}, X_j), h(X_{i_1}, \ldots, X_{i_{m-1}}, X_k)).$$

Suppose that $E(h) = 0$ and for some $\nu > 2, \delta > 0$,

$$\sup_{i_1, \ldots, i_m} E|h(X_{i_1}, \ldots, X_{i_m})|^{2+\delta} < \infty \tag{6.6.10}$$

and

$$u(n) = O(n^{-(\nu-2)(\nu+\delta)}). \tag{6.6.11}$$

Then

$$E|U_n|^\nu = O(n^{-\nu/2}). \tag{6.6.12}$$

For a detailed proof of this lemma, see Christofides (2004). The lemma stated above is a consequence of the inequality that, for arbitrary random variables $X_1, \ldots, X_k$, with finite r-th absolute moment, $r \geq 1$,

$$E|X_1 + \ldots + X_k|^r \leq k^{r-1} \sum_{i=1}^k E|X_i|^r.$$

Since nondecreasing functions of associated random variables are associated, it follows, by results from Birkel (1988) (see Chapter 1), that

$$E\left| \sum_{i_m=i_{m-1}+1}^n h(X_{i_1}, \ldots, X_{i_m}) \right|^\nu \leq cn^{\nu/2} \tag{6.6.13}$$

where c is a constant not depending on n.

Note that a U-statistic based on associated random variables is a demimartingale if the kernel h is componentwise nondecreasing. However there are kernels such as the function $h(x, y) = |x - y|$ which does not satisfy this requirement and the corresponding U-statistic for associated random variables is not a demimartingale. Christofides (2004) considers kernels h of bounded variation in R^2 and obtains the strong law of large numbers for the corresponding U-statistics.

Central Limit Theorem

Let the sequence $\{X_n, n \geq 1\}$ be a stationary sequence of associated random variables. Let F be the distribution function of X_1 and f be its density function assuming that the random variable X_1 has a probability density function.

Let $\psi(x, y)$ be a real-valued function symmetric in its arguments. Define the U-statistic by

$$U_n = \binom{n}{2}^{-1} \sum_{1 \le i < j \le n} \psi(X_i, X_j). \tag{6.6.14}$$

Let

$$\theta = \int_{-\infty}^{\infty} \int_{-\infty}^{\infty} \psi(x, y) dF(x) dF(y), \tag{6.6.15}$$

$$\psi_1(x_1) = E\psi(x_1, X_2)$$

$$= \int_{-\infty}^{\infty} \psi(x_1, x_2) dF(x_2), \tag{6.6.16}$$

$$h^{(1)}(x_1) = \psi_1(x_1) - \theta, \tag{6.6.17}$$

$$h^{(2)}(x_1, x_2) = \psi(x_1, x_2) - \psi_1(x_1) - \psi_1(x_2) + \theta. \tag{6.6.18}$$

Then, the Hoeffding-decomposition (H-decomposition) for U_n is given by (see Lee (1990))

$$U_n = \theta + 2H_n^{(1)} + H_n^{(2)} \tag{6.6.19}$$

where $H_n^{(j)}$ is the U-statistic of degree j based on the kernel $h^{(j)}$, $j = 1, 2$, that is,

$$H_n^{(j)} = \frac{1}{\binom{n}{j}} \sum h^{(j)} \tag{6.6.20}$$

where summation is taken over all subsets $1 \le i_1 < \ldots < i_j \le n$ of $\{1, \ldots, n\}$.

Many authors have studied the limiting properties of U_n under various dependence conditions. Sen (1972) and Qiying (1995) have studied U-statistics for mixing sequences and Yoshihara (1976, 1984) has studied U-statistics for absolutely regular processes. Dewan and Prakasa Rao (2001) established a central limit theorem for U_n using an orthogonal expansion for the kernel associated with U_n. We now establish the central limit theorem for U-statistics based on associated random variables using Hoeffding's decomposition (cf. Dewan and Prakasa Rao (2002)).

We use the following central limit theorem for stationary associated sequences due to Newman (1980) discussed in Chapter 1.

Lemma 6.6.4 (Newman (1980)). *Let the sequence $\{X_n, n \ge 1\}$ be a strictly stationary sequence of associated random variables . Let*

$$\sigma^2 = \mathrm{Var}(X_1) + 2 \sum_{j=2}^{\infty} \mathrm{Cov}(X_1, X_j)$$

with $0 < \sigma^2 < \infty$. Then

$$n^{-1/2} \frac{\sum_{j=1}^{n}(X_j - E(X_j))}{\sigma} \xrightarrow{\mathcal{L}} N(0, 1) \text{ as } n \to \infty.$$

The next lemma is due to Serfling (1968).

Lemma 6.6.5 (Serfling (1968)). *Let* $X_1, \ldots, X_m$ *and* $Y_1, \ldots, Y_n$ *be independent samples. Let*

$$\Delta = \sum_{i=1}^{m} \sum_{j=1}^{n} \sum_{k=1}^{m} \sum_{\ell=1}^{n} \Delta(i, j, k, \ell), \qquad (6.6.21)$$

where

$$\Delta(i, j, k, \ell) = \mathrm{Cov}(h^{(2)}(X_i, Y_j), h^{(2)}(X_k, Y_\ell)). \qquad (6.6.22)$$

Suppose that for some nonnegative function $r(k)$ *satisfying*

$$\sum_{k=0}^{\infty} r(k) < \infty, \qquad (6.6.23)$$

we have, for all (i, j, k, ℓ),

$$|\Delta(i, j, k, \ell)| \le r(\max[|i - k|, |j - \ell|]). \qquad (6.6.24)$$

Then $\Delta = o(mn^2)$, *as* m *and* $n \to \infty$ *such that* $\frac{m}{n} \to c > 0$.

Let

$$\sigma_1^2 = \mathrm{Var}(\psi_1(X_1)) , \qquad (6.6.25)$$

$$\sigma_{1j}^2 = \mathrm{Cov}(\psi_1(X_1), \psi_1(X_{1+j})) , \qquad (6.6.26)$$

and

$$\sigma_U^2 = \sigma_1^2 + 2\sum_{j=1}^{\infty} \sigma_{1j}^2 . \qquad (6.6.27)$$

We now state and prove a result concerning $\mathrm{Var}(U_n)$.

Lemma 6.6.6. *Let the sequence* $\{X_j,\ 1 \le j \le n\}$ *be a sequence of stationary associated random variables. Suppose the function* $\psi(x_1, x_2)$ *is continuous in* x_1 *and satisfies the conditions in Theorem 6.2.5. Further suppose that*

$$\max(L_1(\psi), L_2(\psi)) < \infty. \qquad (6.6.28)$$

Further assume that

$$\sum_{j=1}^{\infty} \sigma_{1j}^2 < \infty. \qquad (6.6.29)$$

Then

$$\mathrm{Var}(U_n) = 4\sigma_U^2 + o(\frac{1}{n}). \qquad (6.6.30)$$

Proof. Let c denote a generic positive constant in the sequel. In view of the H-decomposition, we have

$$\text{Var}(U_n) = 4\,\text{Var}(H_n^{(1)}) + \text{Var}(H_n^{(2)}) + 4\,\text{Cov}(H_n^{(1)}, H_n^{(2)}). \tag{6.6.31}$$

Since $H_n^{(1)} = \frac{1}{n}\sum_{j=1}^{n} h^{(1)}(X_j)$, we get

$$\text{Var}(H_n^{(1)}) = \frac{\sigma_1^2}{n} + \frac{2}{n^2}\sum_{j=1}^{n-1}(n-j)\,\text{Cov}(h^{(1)}(X_1), h^{(1)}(X_{1+j}))$$

$$= \frac{1}{n}\left(\sigma_1^2 + 2\sum_{j=1}^{\infty}\sigma_{1j}^2\right) - \frac{2}{n}\sum_{j=n}^{\infty}\sigma_{1j}^2 - \frac{2}{n^2}\sum_{j=1}^{n-1}j\sigma_{1j}^2. \tag{6.6.32}$$

Note that

$$\frac{2}{n}\sum_{j=n}^{\infty}\sigma_{1j}^2 = o\left(\frac{1}{n}\right) \tag{6.6.33}$$

since $\sum_{j=1}^{\infty}\sigma_{1j}^2 < \infty$. Take an integer-valued function $g(n)$ which satisfies $g(n) \to \infty$ and $g(n) = o(n)$ as $n \to \infty$. Then

$$\sum_{j=1}^{n-1}j\sigma_{1j}^2 = \sum_{j=1}^{g}j\sigma_{1j}^2 + \sum_{j=g+1}^{n-1}j\sigma_{1j}^2$$

$$\leq g\sum_{j=1}^{\infty}\sigma_{1j}^2 + (n-1)\sum_{j=g+1}^{\infty}\sigma_{1j}^2$$

$$= O(g) + (n-1)o(1)$$

$$= o(n). \tag{6.6.34}$$

Therefore, we get

$$\text{Var}(H_n^{(1)}) = \frac{1}{n}\left(\sigma_1^2 + 2\sum_{j=1}^{\infty}\sigma_{1j}^2\right) + o\left(\frac{1}{n}\right). \tag{6.6.35}$$

Furthermore

$$E(H_n^{(2)}) = 0, \tag{6.6.36}$$

and

$$E(H_n^{(2)})^2 = \binom{n}{2}^{-2}\sum_{1\leq i<j\leq n}\sum_{1\leq k<\ell\leq n}E\{h^{(2)}(X_i, X_j)h^{(2)}(X_k, X_\ell)\}$$

$$= \binom{n}{2}^{-2}\sum_{1\leq i<j\leq n}\sum_{1\leq k<l\leq n}\Delta(i, j, k, \ell) \quad (say). \tag{6.6.37}$$

Using Bulinski's inequality (see Theorem 6.2.5), we get that

$$\Delta(i,j,k,\ell) \leq C[\text{Cov}(X_i, X_k) + \text{Cov}(X_i, X_\ell) + \text{Cov}(X_j, X_k) + \text{Cov}(X_j, X_\ell)]. \tag{6.6.38}$$

From Serfling (1968) (see Lemma 6.6.5), it follows that

$$\sum_{1 \leq i < j \leq n} \sum_{1 \leq k < l \leq n} \Delta(i,j,k,\ell) = o(n^3). \tag{6.6.39}$$

Hence, from (6.6.37) we get that

$$E(H_n^{(2)^2}) = o\left(\frac{1}{n}\right) \tag{6.6.40}$$

and, from (6.6.36) , it follows that

$$\text{Var}(H_n^{(2)}) = o\left(\frac{1}{n}\right). \tag{6.6.41}$$

Finally

$$|\text{Cov}(H_n^{(1)}, H_n^{(2)})| \leq (\text{Var}(H_n^{(1)}) \, \text{Var}(H_n^{(2)}))^{1/2}$$
$$= (O(n^{-1})o(\frac{1}{n}))^{\frac{1}{2}}$$
$$= o(n^{-1}). \tag{6.6.42}$$

The main result follows from equations (6.6.35), (6.6.41) and (6.6.42).   □

The following theorem is due to Dewan and Prakasa Rao (2002).

Theorem 6.6.7. *Let the sequence $\{X_n, n \geq 1\}$ be a stationary associated sequence. Let U_n be a U-statistic based on a symmetric kernel $\psi(x,y)$ of degree 2. Suppose the conditions of Theorem 6.6.6 hold. Further assume that $\psi(x,y)$ is monotonic in x with $0 < \sigma_U^2 < \infty$. Then*

$$\frac{n^{1/2}(U_n - \theta)}{2\sigma_U} \overset{\mathcal{L}}{\to} N(0,1) \text{ as } n \to \infty$$

where σ_U^2 is as defined by (6.6.27).

Proof. From (6.6.19), we know that

$$U_n = \theta + 2H_n^{(1)} + H_n^{(2)} .$$

Then

$$\frac{n^{1/2}(U_n - \theta)}{2\sigma_U} = n^{-1/2} \sum_{j=1}^{n} \frac{h^{(1)}(X_j)}{\sigma_U} + n^{1/2}\frac{H_n^{(2)}}{2\sigma_U}. \tag{6.6.43}$$

In addition,

$$E(n^{1/2}H_n^{(2)}) = 0, \qquad (6.6.44)$$

and

$$n \ \mathrm{Var} \ H_n^{(2)} \to 0 \text{ as } n \to \infty \qquad (6.6.45)$$

from (6.6.41). Hence

$$n^{1/2}\frac{H_n^{(2)}}{2\sigma_U} \xrightarrow{p} 0 \text{ as } n \to \infty. \qquad (6.6.46)$$

Since ψ is monotonic in its arguments, the sequence $\{h^{(1)}(X_j), \ j \geq 1\}$ constitutes a stationary associated sequence. Then, using Lemma 6.6.4, it follows that

$$n^{-1/2}\sum_{j=1}^{n}\frac{h^{(1)}(X_j)}{\sigma_U} \xrightarrow{\mathcal{L}} N(0,1) \text{ as } n \to \infty. \qquad (6.6.47)$$

Relations (6.6.43), (6.6.46) and (6.6.47) prove the theorem following the representation (6.6.19). □

6.7 More Limit Theorems for U-Statistics

Let the sequence $\{X_n, \ n \geq 1\}$ be a stationary sequence of associated random variables and let F be the distribution function of X_1. Let $\psi(x, y)$ be a real-valued function symmetric in its arguments. Define a U-statistic by

$$U_n = \frac{\displaystyle\sum_{1 \leq i < j \leq n} \psi(X_i, X_j)}{\binom{n}{2}}. \qquad (6.7.1)$$

Suppose that

$$\int_{-\infty}^{\infty}\int_{-\infty}^{\infty}\psi^2(x, y)dF(x)dF(y) < \infty. \qquad (6.7.2)$$

Let us consider an orthonormal basis $\{e_k(x), \ k \geq 0\}$, with $e_0(x) = 1$, such that

$$\psi(x, y) = \sum_{k=0}^{\infty}\lambda_k e_k(x)e_k(y). \qquad (6.7.3)$$

Then

$$\int_{-\infty}^{\infty}e_k(x)\psi(x, y)dF(x) = \lambda_k e_k(y) \qquad (6.7.4)$$

and

$$\sum_{k=0}^{\infty}\lambda_k^2 < \infty. \qquad (6.7.5)$$

Definition (Gregory (1977)). The U-statistic U_n and its kernel ψ are called *degenerate* if

$$\int_{-\infty}^{\infty} \psi(x,y) dF(y) = 0 \tag{6.7.6}$$

for all x.

Gregory (1977) discussed the central limit theorem for degenerate U-statistics based on i.i.d. sequences and its applications to the Cramer-Von Mises test. Hall (1979) made a unified study of the invariance principle of degenerate and non-degenerate U-statistics in the i.i.d. case. Eagleson (1979) extended the above methods to derive the limiting distribution of U-statistics based on stationary mixing samples. We now obtain obtain similar results for stationary associated random sequences.

We now recall a few definitions and results (cf. Section 3) from Newman (1984) which will be used later in this section.

Let f and f_1 be two complex-valued functions on R^n. Then we say that $f \ll f_1$ if $f_1 - Re(e^{i\alpha} f)$ is componentwise nondecreasing for all real α. If f and f_1 are two real-valued functions, then $f \ll f_1$ if and only if $f_1 + f$ and $f_1 - f$ are both nondecreasing. In particular, if $f \ll f_1$ and f, f_1 are functions of a single variable, then f_1 will be nondecreasing. We say that $f \overset{A}{\ll} f_1$ if $f \ll f_1$ and both f_1 and f depend only on $x'_j s$ with $j \in A$.

Suppose that f is a real-valued function. Then $f \ll f_1$ for f_1 real if and only if for $x < y$,

$$f(y) - f(x) \le f_1(y) - f_1(x) \tag{6.7.7}$$

and

$$f(x) - f(y) \le f_1(y) - f_1(x). \tag{6.7.8}$$

These relations hold if and only if , for $x < y$,

$$|f(y) - f(x)| \le f_1(y) - f_1(x). \tag{6.7.9}$$

Theorem 6.7.1 (Newman (1984)). *Let the sequence $\{X_n, \ n \ge 1\}$ be a stationary sequence of associated random variables. For each j, let $Y_j = f_j(X_1, X_2, \ldots)$, and $\tilde{Y}_j = \tilde{f}_j(X_1, X_2, \ldots)$. Suppose that $f_j \overset{A}{\ll} \tilde{f}_j$ for each j where $A = \{k : k \ge 1\}$. Then*

$$\left| \phi - \prod_{j=1}^{n} \phi_j \right| \le 2 \sum_{1 \le k < \ell \le n} |r_k r_\ell \operatorname{Cov}(\tilde{Y}_k, \tilde{Y}_\ell)|, \tag{6.7.10}$$

where ϕ and ϕ_j are given by

$$\phi = E(\exp[i \sum_{j=1}^{n} r_j Y_j]) \ \ and \ \ \phi_j = E(\exp[i r_j Y_j]).$$

Theorem 6.7.2 (Newman (1984)). *Let the sequence $\{X_n, n \geq 1\}$ be a stationary sequence of associated random variables. For each j, let $Y_j = f(X_j)$ and $\tilde{Y}_j = \tilde{f}(X_j)$. Suppose that $f \ll \tilde{f}$. Let*

$$\sum_{j=2}^{\infty} \mathrm{Cov}(\tilde{Y}_1, \tilde{Y}_j) < \infty. \qquad (6.7.11)$$

Then

$$n^{-\frac{1}{2}} \sum_{j=1}^{n} (Y_j - EY_j) \xrightarrow{\mathcal{L}} \sigma Z \text{ as } n \to \infty \qquad (6.7.12)$$

where Z is a standard normal random variable and

$$\sigma^2 = \mathrm{Var}(Y_1) + 2 \sum_{j=2}^{\infty} \mathrm{Cov}(Y_1, Y_j). \qquad (6.7.13)$$

Remarks. Note that if $\tilde{f}$ is differentiable and $\sup_x |f'(x)|$ is finite, then, using Newman's inequality implies that the inequality in (6.7.11) holds provided

$$\sum_{j=2}^{\infty} \mathrm{Cov}(X_1, X_j) < \infty. \qquad (6.7.14)$$

Suppose $f \ll \tilde{f}$. Following equations (6.7.7) to (6.7.9), we get that

$$|f(y) - f(x)| \leq \tilde{f}(y) - \tilde{f}(x) \qquad (6.7.15)$$

for $x < y$. Let $\tilde{f}(x) = cx$ for some constant $c > 0$. Then $f \ll \tilde{f}$ if and only if, for $x < y$,

$$|f(y) - f(x)| \leq c(y - x) \qquad (6.7.16)$$

which indicates that f is Lipschitzian. A sufficient condition for (6.7.16) is that

$$\sup_x |f'(x)| < \infty. \qquad (6.7.17)$$

Limit Theorem for U-Statistics for Kernels of Degree 2

Let C denote a generic positive constant in the following discussion.

Theorem 6.7.3. *Let the sequence $\{X_n, n \geq 1\}$ be a stationary sequence of associated random variables. Let U_n be a degenerate U-statistic where the kernel $\psi(.,.)$ satisfies (6.7.2). Assume that the eigenfunctions $e_k(x)$ given by (6.7.4) are differentiable and*

$$\sup_j \sup_x |e_j'(x)| < \infty. \qquad (6.7.18)$$

Furthermore, assume that,

$$\sum_{j=1}^{\infty} \mathrm{Cov}(X_1, X_j) < \infty \tag{6.7.19}$$

and

$$\sum_{k=1}^{\infty} |\lambda_k| < \infty. \tag{6.7.20}$$

Then

$$nU_n \stackrel{\mathcal{L}}{\to} \sum_{k=1}^{\infty} \lambda_k(\tilde{Z}_k^2 - 1) \ as \ n \to \infty \tag{6.7.21}$$

where $\{\tilde{Z}_k\}$ is a sequence of correlated jointly normal random variables with mean zero and

$$\mathrm{Cov}(\tilde{Z}_k, \tilde{Z}_j) = \mathrm{Cov}(e_k(X_1), e_j(X_1)) + 2\sum_{i=1}^{\infty} \mathrm{Cov}(e_k(X_1), e_j(X_{1+i}))$$

for $k \neq j$.

Proof. Since the kernel ψ satisfies (6.7.2) and it is degenerate, we have

$$\psi(x, y) = \sum_{k=1}^{\infty} \lambda_k \ e_k(x)e_k(y). \tag{6.7.22}$$

Given $\epsilon > 0$, there exists a positive integer N such that if

$$\psi_N(x, y) = \sum_{k=1}^{N} \lambda_k \ e_k(x)e_k(y), \tag{6.7.23}$$

then

$$\int_{-\infty}^{\infty} \int_{-\infty}^{\infty} |\psi(x, y) - \psi_N(x, y)|^2 dF(x)dF(y) = \sum_{k=N+1}^{\infty} \lambda_k^2 < \epsilon. \tag{6.7.24}$$

Let $U_{n,N}$ be the U-statistic based on the kernel $\psi_N(x, y)$. Then

$$U_{n,N} = \frac{1}{\binom{n}{2}} \sum_{1 \leq i < j \leq n} \psi_N(X_i, X_j)$$

$$= \frac{1}{\binom{n}{2}} \sum_{k=1}^{N} \left(\sum_{1 \leq i < j \leq n} \lambda_k e_k(X_i)e_k(X_j) \right)$$

$$= \frac{1}{n(n-1)} \sum_{k=1}^{N} \lambda_k \left(\sum_{i=1}^{n} \sum_{j=1}^{n} e_k(X_i)e_k(X_j) - \sum_{i=1}^{n} e_k^2(X_i) \right)$$

$$= \frac{1}{n(n-1)} \sum_{k=1}^{N} \lambda_k \left(\left(\sum_{i=1}^{n} e_k(X_i) \right)^2 - \sum_{i=1}^{n} e_k^2(X_i) \right) \tag{6.7.25}$$

and

$$nU_{n,N} = \frac{n}{n-1} \sum_{k=1}^{N} \lambda_k \left(\left(\frac{\sum_{i=1}^{n} e_k(X_i)}{\sqrt{n}} \right)^2 - \frac{1}{n} \sum_{i=1}^{n} e_k^2(X_i) \right). \tag{6.7.26}$$

Because of (6.7.18), (6.7.19) and using the strong law of large numbers for differentiable functions of associated random variables (Bagai and Prakasa Rao (1995)), we get that

$$\frac{1}{n} \sum_{i=1}^{n} e_k^2(X_i) \to 1 \text{ a.s. as } n \to \infty. \tag{6.7.27}$$

Next we consider the joint distribution of

$$\left(\frac{1}{\sqrt{n}} \sum_{i=1}^{n} e_1(X_i), \dots, \frac{1}{\sqrt{n}} \sum_{i=1}^{n} e_N(X_i) \right). \tag{6.7.28}$$

Consider a linear combination

$$T = \sum_{k=1}^{N} \frac{a_k}{\sqrt{n}} \sum_{i=1}^{n} e_k(X_i)$$

$$= \sum_{i=1}^{n} \frac{1}{\sqrt{n}} \sum_{k=1}^{N} a_k e_k(X_i)$$

$$= \sum_{i=1}^{n} \frac{1}{\sqrt{n}} B_N(X_i) \tag{6.7.29}$$

where

$$B_N(X_i) = \sum_{k=1}^{N} a_k e_k(X_i). \tag{6.7.30}$$

Then, under the condition (6.7.18), B_N satisfies the conditions of Theorem 6.7.2 for every vector $(a_1, \dots, a_N) \in R^N$. Note that

$$ET = 0,$$
$$\text{Var } T = \sigma_N^2$$
$$= \sum_{k=1}^{N} a_k^2 + \frac{2}{n} \sum_{j=1}^{n-1} (n-j) \text{Cov}(B_N(X_1), B_N(X_{1+j})). \tag{6.7.31}$$

Hence, by Theorem 6.7.2,

$$n^{-\frac{1}{2}} \sum_{i=1}^{n} B_N(X_i) \xrightarrow{\mathcal{L}} N(0, \sigma_N^2) \text{ as } n \to \infty. \tag{6.7.32}$$

Therefore, using (6.7.26), (6.7.27) and (6.7.32), we get that

$$nU_{n,N} \overset{\mathcal{L}}{\to} \sum_{k=1}^{N} \lambda_k(\tilde{Z}_k^2 - 1) \text{ as } n \to \infty \tag{6.7.33}$$

where $\{\tilde{Z}_k, 1 \leq k \leq N\}$ are jointly normal random variables with mean zero and

$$\text{Cov}(\tilde{Z}_k, \tilde{Z}_j) = \text{Cov}(e_k(X_1), e_j(X_1)) + 2\sum_{i=1}^{\infty} \text{Cov}(e_k(X_1), e_j(X_{1+i})) \tag{6.7.34}$$

for $k \neq j$. Note that

$$E(\tilde{Z}_k^2) = \text{Var}(e_k(X_1)) + 2\sum_{i=1}^{\infty} \text{Cov}(e_k(X_1), e_k(X_{1+i}))$$

$$\leq 1 + C\sum_{i=1}^{\infty} \text{Cov}(X_1, X_{1+i}) \text{ (by using (6.7.18) and Newman's inequality)}$$

$$\leq C \text{ (by using (6.7.19))} \tag{6.7.35}$$

uniformly in $k \geq 1$.

Let $\eta_n(t)$, $\phi(t)$, $\phi_N(t)$ and $\eta_{n,N}(t)$ be the characteristic functions of nU_n, $\sum_{k=1}^{\infty} \lambda_k(\tilde{Z}_k^2 - 1)$, $\sum_{k=1}^{N} \lambda_k(\tilde{Z}_k^2 - 1)$ and $nU_{n,N}$ respectively. Then

$$|\eta_n(t) - \phi(t)| \leq |\eta_n(t) - \eta_{n,N}(t)| + |\eta_{n,N}(t) - \phi_N(t)| + |\phi_N(t) - \phi(t)|. \tag{6.7.36}$$

Relation (6.7.33) implies that, given $\epsilon > 0$, for large n depending on N, ϵ and t,

$$|\eta_{n,N}(t) - \phi_N(t)| \leq \frac{\epsilon}{3}. \tag{6.7.37}$$

Since

$$E|\sum_{k=N+1}^{\infty} \lambda_k(\tilde{Z}_k^2 - 1)| \leq E\{\sum_{k=N+1}^{\infty} |\lambda_k||\tilde{Z}_k^2 - 1|\} \leq (C+1)\sum_{k=N+1}^{\infty} |\lambda_k| < \epsilon, \tag{6.7.38}$$

it follows that, for large N , depending on t,

$$|\phi_N(t) - \phi(t)| = |E(e^{it\sum_{k=1}^{N} \lambda_k(\tilde{Z}_k^2-1)} - e^{it\sum_{k=1}^{\infty} \lambda_k(\tilde{Z}_k^2-1)})|$$

$$\leq E|e^{it\sum_{k=N+1}^{\infty} \lambda_k(\tilde{Z}_k^2-1)} - 1|$$

$$\leq |t|E|\sum_{k=N+1}^{\infty} \lambda_k(\tilde{Z}_k^2 - 1)|$$

$$\leq |t|\sum_{k=N+1}^{\infty} |\lambda_k|(E\tilde{Z}_k^2 + 1)$$

$$\leq C|t| \sum_{k=N+1}^{\infty} |\lambda_k|$$

$$\leq \frac{\epsilon}{3} \quad \text{(by (6.7.20) and (6.7.35))}. \tag{6.7.39}$$

Observe that

$$E(n^{-\frac{1}{2}} \sum_{i=1}^{n} e_k(X_i))^2 = E\frac{1}{n}(\sum_{i=1}^{n} \sum_{j=1}^{n} e_k(X_i)e_k(X_j))$$

$$= \frac{1}{n}E(\sum_{i=1}^{n} e_k^2(X_i) + \sum_{1\leq i\neq j\leq n} e_k(X_i)e_k(X_j))$$

$$= 1 + \frac{1}{n} \sum_{1\leq i\neq j\leq n} E(e_k(X_i)e_k(X_j))$$

$$\leq 1 + \frac{1}{n} \sum_{1\leq i\neq j\leq n} (\sup_x |e_k'(x)|)^2 \operatorname{Cov}(X_i, X_j)$$

$$\leq 1 + \frac{C}{n} \sum_{j=2}^{n} (n-j) \operatorname{Cov}(X_1, X_j) \quad \text{(using stationarity)}$$

$$\leq C \quad \text{(using (6.7.19))}. \tag{6.7.40}$$

Hence,

$$E|nU_n - nU_{n,N}|$$

$$= E|\frac{n}{n-1} \sum_{k=N+1}^{\infty} \lambda_k\{(\frac{\sum_{i=1}^{n} e_k(X_i)}{\sqrt{n}})^2 - \frac{1}{n}\sum_{i=1}^{n} e_k^2(X_i)\}|$$

$$\leq \frac{n}{n-1} \sum_{k=N+1}^{\infty} |\lambda_k| E|\{(\frac{\sum_{i=1}^{n} e_k(X_i)}{\sqrt{n}})^2 - \frac{1}{n}\sum_{i=1}^{n} e_k^2(X_i)\}|$$

$$\leq \frac{n}{n-1} \sum_{k=N+1}^{\infty} |\lambda_k|\{E(\frac{\sum_{i=1}^{n} e_k(X_i)}{\sqrt{n}})^2 + \frac{1}{n}\sum_{i=1}^{n} E(e_k^2(X_i))\}$$

$$\leq \frac{n}{n-1}(C+1) \sum_{k=N+1}^{\infty} |\lambda_k|. \tag{6.7.41}$$

Hence, we have

$$|\eta_n(t) - \eta_{n,N}(t)| = |E(e^{itnU_n} - e^{itnU_{n,N}})|$$

$$\leq E|e^{itn(U_n-U_{n,N})} - 1|$$

$$\leq |t|E|nU_n - nU_{n,N}|$$

$$< \frac{\epsilon}{3} \tag{6.7.42}$$

for large n, N depending on t. The result now follows by combining (6.7.36), (6.7.37), (6.7.39) and (6.7.42). $\qquad\qquad\qquad\qquad\qquad\qquad\qquad\qquad\qquad$ $\square$

Theorem 6.7.4. *Let the sequence* $\{X_n, n \geq 1\}$ *be a stationary sequence of associated random variables. Suppose* U_n *is a non-degenerate U-statistic corresponding to the kernel* ψ *with the eigenfunction expansion (6.7.3). Assume that*

$$\sup_{j} \sup_{x} |e'_j(x)| < \infty, \tag{6.7.43}$$

and

$$\sum_{j=1}^{\infty} \mathrm{Cov}(X_1, X_j) < \infty. \tag{6.7.44}$$

Further assume that

$$\sum_{k=0}^{\infty} |\lambda_k| < \infty. \tag{6.7.45}$$

If U_n *has a finite variance, then*

$$n^{\frac{1}{2}}(U_n - EU_n) \xrightarrow{\mathcal{L}} N(0, 4\sigma^2) \text{ as } n \to \infty. \tag{6.7.46}$$

where

$$\sigma^2 = \mathrm{Var}(g(X_1)) + 2\sum_{j=1}^{\infty} \mathrm{Cov}(g(X_1), g(X_{j+1})), \tag{6.7.47}$$

and

$$g(x) = \int_{-\infty}^{\infty} \psi(x, y)dF(y), \tag{6.7.48}$$

provided $g(.)$ *is monotone or* g *is Lipschitzian, that is,*

$$|g(x) - g(y)| \leq C|x - y|, \quad x, y \in R \tag{6.7.49}$$

and $\sum_{k=0}^{\infty} |a_k^*| < \infty$, *where* $a_k^* = E(e_k(X_1))$.

Proof. Note that

$$\begin{aligned}
g(x) &= \int_{-\infty}^{\infty} \psi(x, y)dF(y) \\
&= \int_{-\infty}^{\infty} \{\sum_{k=0}^{\infty} \lambda_k \, e_k(x)e_k(y)\}dF(y) \\
&= \sum_{k=0}^{\infty} \lambda_k \, e_k(x)a_k^*.
\end{aligned} \tag{6.7.50}$$

Then

$$|a_k^*| \leq E|e_k(X_1)| \leq \{E(e_k^2(X_1))\}^{\frac{1}{2}} = 1, \tag{6.7.51}$$

and

$$Eg(X_1) = \sum_{k=0}^{\infty} \lambda_k a_k^{*2}. \tag{6.7.52}$$

Using (6.7.52) and the fact that the functions $\{e_k(x), k \geq 0\}$ form an orthonormal basis, we get

$$E(g^2(X_1)) = \sum_{k=0}^{\infty} \lambda_k^2 a_k^{*2} \leq \sum_{k=0}^{\infty} \lambda_k^2 < \infty. \tag{6.7.53}$$

Define

$$\hat{\psi}(x, y) = \psi(x, y) - g(x) - g(y) + Eg(X_1). \tag{6.7.54}$$

It is easy to see that $\int_{-\infty}^{\infty} \hat{\psi}(x, y) dF(y)$ is zero. Thus $\hat{\psi}$ is symmetric, square integrable and degenerate. Let $\hat{U}_n$ be the U-statistic corresponding to $\hat{\psi}$. Note that

$$\hat{U}_n = \frac{1}{\binom{n}{2}} \sum_{k=0}^{\infty} \lambda_k \{ \sum_{1 \leq i < j \leq n} (e_k(X_i) - a_k^*)(e_k(X_j) - a_k^*) \}. \tag{6.7.55}$$

Using equations (6.7.54) and (6.7.55), we have

$$U_n - \sum_{k=0}^{\infty} \lambda_k a_k^{*2} = \hat{U}_n + \frac{2}{n} \sum_{i=1}^{n} g(X_i) - 2 \sum_{k=0}^{\infty} \lambda_k a_k^{*2}$$

$$= \hat{U}_n + \frac{2}{n} \sum_{i=1}^{n} (g(X_i) - Eg(X_i)). \tag{6.7.56}$$

Now,

$$E(e_k(X_1) - a_k^*) = 0,$$
$$E(e_k(X_1) - a_k^*)^2 = 1 - a_k^{*2}. \tag{6.7.57}$$

Then, by (6.7.18) to (6.7.20) and an application of Theorem 6.7.3 with $e_k(x)$ replaced by $e_k(X) - a_k^*$, we get that

$$n\hat{U}_n \overset{\mathcal{L}}{\to} \sum_{k=0}^{\infty} \lambda_k (\hat{Z}_k^2 - 1 + a_k^{*2}) \text{ as } n \to \infty \tag{6.7.58}$$

where $\{\hat{Z}_k\}$ is a sequence of jointly normal random variables. In view of equation (6.7.58), we have

$$n^{\frac{1}{2}} \hat{U}_n \overset{P}{\to} 0 \text{ as } n \to \infty. \tag{6.7.59}$$

Furthermore

$$\left| E(U_n) - \sum_{k=0}^{\infty} \lambda_k a_k^{*2} \right|$$

$$= \frac{2}{n(n-1)} \left| \sum_{k=0}^{\infty} \lambda_k \sum_{1 \leq i < j \leq n} \{ E(e_k(X_i) e_k(X_j)) - a_k^{*2} \} \right|$$

$$\leq \frac{2}{n(n-1)}\Big|\sum_{k=0}^{\infty}\lambda_k \sum_{1\leq i<j\leq n}\{E(e_k(X_i)-a_k^*)(e_k(X_j))-a_k^*)\}\Big|$$

$$\leq \frac{2}{n(n-1)}\sum_{k=0}^{\infty}|\lambda_k|\sum_{1\leq i<j\leq n}|\operatorname{Cov}(e_k(X_i),e_k(X_j))|$$

$$\leq \frac{2}{n(n-1)}\sum_{k=0}^{\infty}|\lambda_k|\sup_k\sup_x(e_k'(x))^2\sum_{1\leq i<j\leq n}\operatorname{Cov}(X_i,X_j)$$

$$\leq \frac{C}{n}\sum_{k=0}^{\infty}|\lambda_k|\sum_{i=2}^{n}\operatorname{Cov}(X_1,X_i). \tag{6.7.60}$$

Therefore

$$n^{\frac{1}{2}}\Big|E(U_n)-\sum_{k=0}^{\infty}\lambda_k a_k^{*2}\Big| \to 0 \text{ as } n \to \infty \tag{6.7.61}$$

from equations (6.7.44) and (6.7.45). If $g(x)$ is monotone, then following Newman (1980) we have, as $n \to \infty$,

$$\frac{2}{\sqrt{n}}\sum_{i=1}^{n}(g(X_i)-Eg(X_i)) \overset{\mathcal{L}}{\to} N(0,4\sigma^2), \text{ as } n \to \infty. \tag{6.7.62}$$

If $g(x)$ is not monotone, then let $\tilde{g}(x) = Cx$. Since

$$\sum_{k=0}^{\infty}|a_k^*| < C < \infty, \tag{6.7.63}$$

it follows from equations (6.7.16) and (6.7.43) that $g \ll \tilde{g}$. Therefore the result given in equation (6.7.62) follows from Theorem 6.7.2. The result follows by combining equations (6.7.59), (6.7.61) and (6.7.62). $\qquad\square$

Limit Theorem for U-Statistics of Kernels of Degree 3

Suppose that $\psi(x,y,z)$ is a symmetric function from R^3 to R which is square integrable in the sense that

$$\int_{-\infty}^{\infty}\int_{-\infty}^{\infty}\int_{-\infty}^{\infty}\psi^2(x,y,z)dF(x)dF(y)dF(z) < \infty. \tag{6.7.64}$$

Assume that

$$\int_{-\infty}^{\infty}\int_{-\infty}^{\infty}\int_{-\infty}^{\infty}\psi(x,y,z)dF(x)dF(y)dF(z) = 0. \tag{6.7.65}$$

Suppose there exists an orthonormal basis $\{e_k(x),\ k \geq 0\}$ with $e_0(x) = 1$ such that

$$\psi(x,y,z) = \sum_{k,l,m\geq 0}\lambda_{klm}e_k(x)e_l(y)e_m(z), \tag{6.7.66}$$

where

$$\sum_{k,l,m \geq 0} \lambda_{klm}^2 < \infty. \tag{6.7.67}$$

Note that

$$\int_{-\infty}^{\infty} \int_{-\infty}^{\infty} \int_{-\infty}^{\infty} \psi(x,y,z)e_k(x)e_\ell(y)e_m(z)dF(x)dF(y)dF(z) = \lambda_{klm}. \tag{6.7.68}$$

Then the U-statistic corresponding to the kernel ψ of degree 3 is

$$U_n = \frac{1}{\binom{n}{3}} \sum_{1 \leq i_1 < i_2 < i_3 \leq n} \psi(X_{i_1}, X_{i_2}, X_{i_3}). \tag{6.7.69}$$

Since $e_0(x) = 1$, we get that $\lambda_{000} = 0$ from equation (6.7.68) . Then,

$$\psi(x,y,z) = \sum_{k>0} \lambda_{k00}e_k(x) + \sum_{\ell>0} \lambda_{0\ell 0}e_\ell(y)$$

$$+ \sum_{m>0} \lambda_{00m}e_m(z) + \sum_{k,\ell>0} \lambda_{k\ell 0}e_k(x)e_\ell(y)$$

$$+ \sum_{k,m>0} \lambda_{k0m}e_k(x)e_m(z) + \sum_{m,\ell>0} \lambda_{0\ell m}e_m(z)e_\ell(y)$$

$$+ \sum_{k,\ell,m>0} \lambda_{k,\ell,m}e_k(x)e_\ell(y)e_m(z). \tag{6.7.70}$$

Definition. The kernel ψ is said to be *first-order degenerate* if

$$\int_{-\infty}^{\infty} \int_{-\infty}^{\infty} \psi(x,y,z)dF(x)dF(y) = 0. \tag{6.7.71}$$

Remarks. If the kernel is first-order degenerate, then

$$\lambda_{k00} = \lambda_{0\ell 0} = \lambda_{00m} = 0 \ \text{ for all } k, \ell, m.$$

Hence

$$\psi(x,y,z) = \sum_{k,\ell>0} \lambda_{k\ell 0}e_k(x)e_\ell(y) + \sum_{k,m>0} \lambda_{k0m}e_k(x)e_m(z)$$

$$+ \sum_{m,\ell>0} \lambda_{0\ell m}e_m(z)e_\ell(y) + \sum_{k,\ell,m>0} \lambda_{k,\ell,m}e_k(x)e_\ell(y)e_m(z). \tag{6.7.72}$$

Definition. A kernel ψ is said to be *second-order degenerate* if

$$\int_{-\infty}^{\infty} \psi(x,y,z)dF(x) = 0. \tag{6.7.73}$$

Remarks. If the kernel ψ is second-order degenerate , then

$$\lambda_{k\ell 0} = \lambda_{0\ell m} = \lambda_{k0m} = 0 \ \text{ for all } k, \ell, m$$

and

$$\psi(x, y, z) = \sum_{k,\ell,m=1}^{\infty} \lambda_{k\ell m} e_k(x) e_\ell(y) e_m(z). \tag{6.7.74}$$

Theorem 6.7.5. *Let the sequence $\{X_n, \ n \geq 1\}$ be a stationary sequence of associated random variables. Let U_n be a U-statistic of degree three where the kernel $\psi(., ., .)$ is second-order degenerate and satisfies equation (6.7.64). Assume that the eigenfunctions $e_k(x)$ are differentiable and*

$$\sup_j \sup_x |e_j(x)| < \infty, \tag{6.7.75}$$

and

$$\sup_j \sup_x |e_j'(x)| < \infty. \tag{6.7.76}$$

Furthermore, assume that

$$\sum_{j=n+1}^{\infty} \mathrm{Cov}(X_1, X_j) = O(n^{-\theta}) \ \text{ for } \theta > 1 \tag{6.7.77}$$

and assume that

$$\sum_{k,\ell,m=1}^{\infty} |\lambda_{k\ell m}| < \infty.$$

Then

$$n^{\frac{3}{2}} U_n \overset{\mathcal{L}}{\to} \sum_{k,\ell,m=1}^{\infty} \lambda_{k\ell m} \tilde{Z}_k \tilde{Z}_\ell \tilde{Z}_m \ \text{ as } n \to \infty$$

where $\{\tilde{Z}_k\}$ is a sequence of mean zero correlated jointly normal random variables with covariances given by equation (6.7.34).

Proof. Since the kernel ψ satisfies equation (6.7.64) and it is second-order degenerate, we have

$$\psi(x, y, z) = \sum_{k,\ell,m=1}^{\infty} \lambda_{k\ell m} e_k(x) e_\ell(y) e_m(z). \tag{6.7.78}$$

Let A be the set $\{1 \leq k, \ell, m \leq N\}$. Let B_N denote the complement of set A . Given $\epsilon > 0$, there exists a positive integer N such that if

$$\psi_N(x, y, z) = \sum_{1 \leq k, \ell, m \leq N} \lambda_{k\ell m} e_k(x) e_\ell(y) e_m(z), \tag{6.7.79}$$

then

$$\int\int |\psi(x,y,z) - \psi_N(x,y,z)|^2 dF(x)dF(y)dF(z) = \sum_{B_N} \lambda_{k\ell m}^2 < \epsilon. \qquad (6.7.80)$$

Let $U_{n,N}$ be the U-statistic based on the kernel $\psi_N(x,y,z)$. Then

$$U_{n,N} = \frac{1}{\binom{n}{3}} \sum_{1 \le i < j < r \le n} \psi_N(X_i, X_j, X_r) \qquad (6.7.81)$$

$$= \frac{1}{\binom{n}{3}} \sum_{1 \le k,\ell,m \le N} \left(\sum_{1 \le i < j < r \le n} \lambda_{k\ell m} e_k(X_i) e_\ell(X_j) e_m(X_r) \right)$$

$$= \frac{1}{n(n-1)(n-2)} \sum_{1 \le k,\ell,m \le N} \lambda_{k\ell m} \{ \sum_{i=1}^{n} \sum_{j=1}^{n} \sum_{r=1}^{n} e_k(X_i) e_\ell(X_j) e_m(X_r)$$

$$- \sum_{i=1}^{n} \sum_{j=1}^{n} e_k(X_i) e_\ell(X_i) e_m(X_j) - \sum_{i=1}^{n} \sum_{j=1}^{n} e_k(X_i) e_\ell(X_j) e_m(X_i)$$

$$- \sum_{i=1}^{n} \sum_{j=1}^{n} e_k(X_j) e_\ell(X_i) e_m(X_i) + 2 \sum_{i=1}^{n} e_k(X_i) e_\ell(X_i) e_m(X_i) \}.$$

Therefore

$$n^{\frac{3}{2}} U_{n,N} = \frac{n^3}{n(n-1)(n-2)} \sum_{1 \le k,\ell,m \le N} \lambda_{k\ell m} \{ \sum_{i=1}^{n} \sum_{j=1}^{n} \sum_{r=1}^{n} \frac{e_k(X_i)}{\sqrt{n}} \frac{e_\ell(X_j)}{\sqrt{n}} \frac{e_m(X_r))}{\sqrt{n}}$$

$$- \sum_{i=1}^{n} \sum_{j=1}^{n} \frac{e_k(X_i) e_\ell(X_i)}{n} \frac{e_m(X_j))}{\sqrt{n}} - \sum_{i=1}^{n} \sum_{j=1}^{n} \frac{e_k(X_i) e_m(X_i)}{n} \frac{e_\ell(X_j)}{\sqrt{n}}$$

$$- \sum_{i=1}^{n} \sum_{j=1}^{n} \frac{e_k(X_j)}{\sqrt{n}} \frac{e_\ell(X_i) e_m(X_i)}{n} + \frac{2}{n^{\frac{3}{2}}} \sum_{i=1}^{n} e_k(X_i) e_\ell(X_i) e_m(X_i)) \}.$$

$$(6.7.82)$$

Following the assumptions stated in equations (6.7.75), (6.7.76) and the strong law of large numbers for associated random variables (Bagai and Prakasa Rao (1995)), we get that

$$\frac{1}{n} \sum_{i=1}^{n} e_k(X_i) e_m(X_i) \to 0 \ a.s. \ \text{as } n \to \infty, \text{ for all } k, m, \qquad (6.7.83)$$

and

$$\frac{1}{n^{\frac{3}{2}}} \sum_{i=1}^{n} e_k(X_i) e_\ell(X_i) e_m(X_i) \to 0 \ a.s. \ \text{as } n \to \infty, \text{ for all } k, \ell, m. \qquad (6.7.84)$$

From equations (6.7.28) to (6.7.32) , (6.7.82) and the discussion for deriving the convergence in (6.7.32), we get that

$$n^{\frac{3}{2}}U_{n,N} \xrightarrow{\mathcal{L}} \sum_{1 \leq k,\ell,m \leq N} \lambda_{k\ell m}\tilde{Z}_k\tilde{Z}_\ell\tilde{Z}_m \text{ as } n \to \infty. \tag{6.7.85}$$

Let $\eta_n^*(t)$, $\eta(t)$, $\eta_N(t)$ and $\eta_{n,N}^*(t)$ be the characteristic functions of

$$n^{\frac{3}{2}}U_n, \quad \sum_{k,\ell,m=1}^{\infty} \lambda_{k\ell m}\tilde{Z}_k\tilde{Z}_\ell\tilde{Z}_m, \quad \sum_{1 \leq k,\ell,m \leq N} \lambda_{k\ell m}\tilde{Z}_k\tilde{Z}_\ell\tilde{Z}_m \text{ and } n^{\frac{3}{2}}U_{n,N}$$

respectively. Then

$$|\eta_n^*(t) - \eta(t)| \leq |\eta_n^*(t) - \eta_{n,N}^*(t)| + |\eta_{n,N}^*(t) - \eta_N(t)| + |\eta_N(t) - \eta(t)|. \tag{6.7.86}$$

Relation (6.7.85) implies that , given $\epsilon > 0$, for large n depending on N, ϵ and t,

$$|\eta_{n,N}^*(t) - \eta_N(t)| \leq \frac{\epsilon}{3}. \tag{6.7.87}$$

Using Cauchy-Schwartz inequality and the fact that $\tilde{Z}_k$ has normal distribution with mean zero , we get that

$$\begin{aligned}
(E|\tilde{Z}_k\tilde{Z}_\ell\tilde{Z}_m|)^2 &\leq E(\tilde{Z}_k)^2 E|\tilde{Z}_\ell\tilde{Z}_m|^2 \\
&\leq E(\tilde{Z}_k)^2 \{E(\tilde{Z}_\ell)^4 E(\tilde{Z}_m)^4\}^{\frac{1}{2}} \\
&\leq 3E(\tilde{Z}_k)^2 \{(E(\tilde{Z}_\ell)^2)^2 (E(\tilde{Z}_m)^2)^2\}^{\frac{1}{2}} \\
&\leq C \quad \text{(by using (6.7.35))}
\end{aligned} \tag{6.7.88}$$

uniformly in k, ℓ, m. Hence

$$E|\sum_{B_N} \lambda_{k\ell m}\tilde{Z}_k\tilde{Z}_\ell\tilde{Z}_m| \leq C\sum_{B_N} |\lambda_{k\ell m}|$$

$$< \epsilon, \tag{6.7.89}$$

for large N. It follows that, for large N, depending on t,

$$\begin{aligned}
|\eta_N(t) - \eta(t)| &= |E(e^{it \sum_{1 \leq k,\ell,m \leq N} \lambda_{k\ell m}\tilde{Z}_k\tilde{Z}_\ell\tilde{Z}_m} - e^{it \sum_{k,\ell,m=1}^{\infty} \lambda_{k\ell m}\tilde{Z}_k\tilde{Z}_\ell\tilde{Z}_m})| \\
&\leq E|e^{it \sum_{B_N} \lambda_{k\ell m}\tilde{Z}_k\tilde{Z}_\ell\tilde{Z}_m} - 1| \\
&\leq |t|E|\sum_{B_N} \lambda_{k\ell m}\tilde{Z}_k\tilde{Z}_\ell\tilde{Z}_m| \\
&\leq \frac{\epsilon}{3} \quad \text{(by using (6.7.67) and (6.7.88)).}
\end{aligned} \tag{6.7.90}$$

Let us now consider

$$E|n^{\frac{3}{2}}U_n - n^{\frac{3}{2}}U_{n,N}| \tag{6.7.91}$$

$$= E|\frac{n^3}{n(n-1)(n-2)} \sum_{1 \le k,\ell,m \le N} \lambda_{k\ell m} \{\sum_{i=1}^n \sum_{j=1}^n \sum_{r=1}^n \frac{e_k(X_i)}{\sqrt{n}} \frac{e_\ell(X_j)}{\sqrt{n}} \frac{e_m(X_r))}{\sqrt{n}}$$

$$- \sum_{i=1}^n \sum_{j=1}^n \frac{e_k(X_i)e_\ell(X_i)}{n} \frac{e_m(X_j))}{\sqrt{n}} - \sum_{i=1}^n \sum_{j=1}^n \frac{e_k(X_i)e_m(X_i)}{n} \frac{e_\ell(X_j)}{\sqrt{n}}$$

$$- \sum_{i=1}^n \sum_{j=1}^n \frac{e_k(X_j)}{\sqrt{n}} \frac{e_\ell(X_i)e_m(X_i)}{n} + \frac{2}{n^{\frac{3}{2}}} \sum_{i=1}^n e_k(X_i)e_\ell(X_i)e_m(X_i))\}|.$$

In view of of (6.7.75), observe that

$$\frac{1}{n^{\frac{3}{2}}} E|\sum_{i=1}^n e_k(X_i)e_\ell(X_i)e_m(X_i)| \le C. \tag{6.7.92}$$

Using the Cauchy-Schwartz inequality, equations (6.7.40), (6.7.75)–(6.7.77) and the stationarity of the random variables, we get that

$$E|(\frac{1}{n} \sum_{i=1}^n e_k(X_i)e_\ell(X_i))(\frac{1}{\sqrt{n}} \sum_{j=1}^n e_m(X_j))|$$

$$\le \{E(\frac{1}{n} \sum_{i=1}^n e_k(X_i)e_\ell(X_i))^2\}^{\frac{1}{2}} \{E(\frac{1}{\sqrt{n}} \sum_{j=1}^n e_m(X_j))^2\}^{\frac{1}{2}}$$

$$\le CE(\frac{1}{n^2} \sum_{i=1}^n e_k^2(X_i)e_\ell^2(X_i) + \frac{1}{n^2} \sum_{1 \le i \ne j \le n} e_k(X_i)e_\ell(X_i)e_k(X_j)e_\ell(X_j))$$

$$\le C\{\frac{1}{n} + \frac{1}{n^2} \sum_{1 \le i \ne j \le n} \text{Cov}(e_k(X_i)e_\ell(X_i), e_k(X_j)e_\ell(X_j))\}$$

$$\le C\{\frac{1}{n} + \frac{1}{n^2} \sum_{1 \le i \ne j \le n} \text{Cov}(X_i, X_j)\}$$

$$\le C\{\frac{1}{n} + \frac{1}{n^2} \sum_{j=1}^n (n-j)\text{Cov}(X_i, X_j)\}$$

$$\le C. \tag{6.7.93}$$

Furthermore, using equations (6.7.75)–(6.7.77) and the Rosenthal type inequality for associated sequences (Shao and Yu (1996)), we get that

$$E|\sum_{i=1}^n \frac{e_k(X_i)}{n^{\frac{1}{2}}}|^4 \le C. \tag{6.7.94}$$

Finally, using Cauchy-Schwartz inequality, (6.7.40) and (6.7.94), we get that

$$
E\Big|\sum_{i=1}^{n}\frac{e_k(X_i)}{n^{\frac{1}{2}}}\sum_{j=1}^{n}\frac{e_\ell(X_j)}{n^{\frac{1}{2}}}\sum_{r=1}^{n}\frac{e_m(X_r)}{n^{\frac{1}{2}}}\Big|
$$

$$
\leq E\{(\sum_{i=1}^{n}\frac{e_k(X_i)}{n^{\frac{1}{2}}})^2(\sum_{j=1}^{n}\frac{e_\ell(X_j)}{n^{\frac{1}{2}}})^2\}^{\frac{1}{2}}\{E(\sum_{r=1}^{n}\frac{e_m(X_r)}{n^{\frac{1}{2}}})^2\}^{\frac{1}{2}}
$$

$$
\leq C\{E(\sum_{i=1}^{n}\frac{e_k(X_i)}{n^{\frac{1}{2}}})^4 E(\sum_{j=1}^{n}\frac{e_\ell(X_j)}{n^{\frac{1}{2}}})^4\}^{\frac{1}{4}}
$$

$$
\leq C. \tag{6.7.95}
$$

Hence, by using equations (6.7.67), (6.7.93) and (6.7.95), we get that

$$
E|n^{\frac{3}{2}}U_n - n^{\frac{3}{2}}U_{n,N}| \leq C. \tag{6.7.96}
$$

Hence, we have

$$
\begin{aligned}
|\eta_n^*(t) - \eta_{n,N}^*(t)| &= |E(e^{itn^{\frac{3}{2}}U_n} - e^{itn^{\frac{3}{2}}U_{n,N}})| \\
&\leq E|e^{itn^{\frac{3}{2}}(U_n - U_{n,N})} - 1| \\
&\leq |t|E|n^{\frac{3}{2}}U_n - n^{\frac{3}{2}}U_{n,N}| \\
&< \frac{\epsilon}{3}
\end{aligned} \tag{6.7.97}
$$

for large n, N, depending on t. The result follows by combining equations (6.7.86), (6.7.90) and (6.7.97). □

Results in this section are due to Dewan and Prakasa Rao (2001) and can be extended to kernels of arbitrary degree by similar methods.

6.8   Application to a Two-Sample Problem

Let the sequence $\{X_m, m \geq 1\}$ be a stationary sequence of associated random variables with one-dimensional marginal distribution function F and the sequence $\{Y_n, n \geq 1\}$ be another stationary sequence of associated random variables with one-dimensional marginal distribution function G. The problem of interest is to test the hypothesis $H : F = G$. We now discuss a limit theorem useful in the study of such problems.

Let $\phi(x, y)$ be a function of two variables which is square integrable in the sense that

$$
\int_{R^2} \phi^2(x, y)dF(x)dG(y) < \infty. \tag{6.8.1}
$$

Define

$$U_{m,n} = \frac{1}{mn} \sum_{i=1}^{m} \sum_{j=1}^{n} \phi(X_i, Y_j). \tag{6.8.2}$$

Under the condition stated in equation (6.8.1), there exist systems of functions $\{f_k(x)\}$ and $\{g_k(y)\}$ (with $f_0(x) = g_0(x) = 1$) which are complete and orthonormal on the spaces of square integrable functions of X and Y respectively such that

$$\phi(x, y) = \sum_{k=0}^{\infty} \lambda_k f_k(x) g_k(y), \tag{6.8.3}$$

where

$$\sum_{k=0}^{\infty} \lambda_k^2 < \infty, \tag{6.8.4}$$

and the series in (6.8.3) converges in mean square with respect to the product measure generated by the joint distribution function $F(x)G(y)$.

The functions $\{f_k(x)\}$ and $\{g_k(y)\}$ are eigenfunctions and $\{\lambda_k\}$ are the eigenvalues of ϕ in the sense that, for all $k \geq 0$,

$$\int_{-\infty}^{\infty} f_k(x)\phi(x, y)dF(x) = \lambda_k g_k(y), \tag{6.8.5}$$

and

$$\int_{-\infty}^{\infty} g_k(y)\phi(x, y)dG(y) = \lambda_k f_k(x). \tag{6.8.6}$$

Definition. The statistic U_{mn} and its kernel ϕ are called *degenerate* if

$$\int \phi(x, y)dF(x) = 0 \tag{6.8.7}$$

for all y and

$$\int \phi(x, y)dG(y) = 0 \tag{6.8.8}$$

for all x.

Theorem 6.8.1. *Let U_{mn} be a degenerate two-sample U-statistic based on two independent stationary sequences of associated random variables. Suppose that the corresponding kernel ϕ is square integrable. Assume that the eigenfunctions $f_k(x)$ and $g_k(y)$ given by equation (6.8.3) are differentiable,*

$$\sup_{k} \sup_{x} |f_k'(x)| < \infty, \tag{6.8.9}$$

and

$$\sup_{k} \sup_{y} |g_k'(y)| < \infty. \tag{6.8.10}$$

Furthermore, assume that,

$$\sum_{j=1}^{\infty} \text{Cov}(X_1, X_j) < \infty, \tag{6.8.11}$$

$$\sum_{j=1}^{\infty} \text{Cov}(Y_1, Y_j) < \infty \tag{6.8.12}$$

and

$$\sum_{k=1}^{\infty} |\lambda_k| < \infty. \tag{6.8.13}$$

Let

$$\delta_m = \frac{1}{m} \sum_{i=1}^{m-1} \text{Cov}(X_1, X_{i+1}), \tag{6.8.14}$$

and

$$\eta_n = \frac{1}{n} \sum_{j=1}^{n-1} \text{Cov}(Y_1, Y_{j+1}). \tag{6.8.15}$$

Assume that

$$mn\{\delta_m + \eta_n\} \to 0 \ \text{as} \ m, n \to \infty. \tag{6.8.16}$$

Then

$$(mn)^{\frac{1}{2}} U_{mn} \xrightarrow{\mathcal{L}} \sum_{k=1}^{\infty} \lambda_k U_k V_k \ \text{as} \ m \to \infty \ \text{and} \ n \to \infty \tag{6.8.17}$$

where $\{U_k\}$ and $\{V_k\}$ are sequences of correlated zero mean normal random variables, and are independent of each other with

$$\text{Cov}(U_k, U_{k'}) = \text{Cov}(f_k(X_1), f_{k'}(X_1)) + 2 \sum_{i=1}^{\infty} \text{Cov}(f_k(X_1), f_{k'}(X_{1+i})),$$

and

$$\text{Cov}(V_k, V_{k'}) = \text{Cov}(g_k(Y_1), g_{k'}(Y_1)) + 2 \sum_{i=1}^{\infty} \text{Cov}(g_k(Y_1), g_{k'}(Y_{1+i})).$$

Proof. Given $\epsilon > 0$, there exists N such that if

$$\phi_N(x, y) = \sum_{k=1}^{N} \lambda_k f_k(x) g_k(y), \tag{6.8.18}$$

then

$$\int_{R^2} |\phi(x, y) - \phi_N(x, y)|^2 dF(x) dG(y) = \sum_{k=N+1}^{\infty} \lambda_k^2 < \epsilon. \tag{6.8.19}$$

Let $U_{mn}^{(N)}$ be the U-statistic generated from the kernel $\phi_N(x, y)$. Then

$$U_{mn}^{(N)} = \frac{1}{mn} \sum_{i=1}^{m} \sum_{j=1}^{n} \phi_N(X_i, Y_j)$$

$$= \frac{1}{mn} \sum_{i=1}^{m} \sum_{j=1}^{n} \sum_{k=1}^{N} \lambda_k f_k(X_i) g_k(Y_j)$$

$$= \frac{1}{mn} \sum_{k=1}^{N} \lambda_k \left\{ \sum_{i=1}^{m} f_k(X_i) \right\} \left\{ \sum_{j=1}^{n} g_k(Y_j) \right\}. \tag{6.8.20}$$

Note that for $k \geq 1$,

$$E f_k(X) = E g_k(Y) = 0, \tag{6.8.21}$$

$$E f_k^2(X) = E g_k^2(Y) = 1, \tag{6.8.22}$$

and

$$E f_k(X) g_k(Y) = 0. \tag{6.8.23}$$

Consider two linear combinations

$$T_1 = \sum_{k=1}^{N} \frac{c_k}{\sqrt{m}} \sum_{i=1}^{m} f_k(X_i), \tag{6.8.24}$$

and

$$T_2 = \sum_{k=1}^{N} \frac{d_k}{\sqrt{n}} \sum_{j=1}^{n} g_k(Y_j). \tag{6.8.25}$$

In view of equations (6.8.9)–(6.8.12) and following the arguments in Theorem 6.7.3, we have

$$T_1 \overset{\mathcal{L}}{\to} N(0, \sigma_{1,N}^2) \text{ as } n \to \infty, \tag{6.8.26}$$

and

$$T_2 \overset{\mathcal{L}}{\to} N(0, \sigma_{2,N}^2) \text{ as } n \to \infty, \tag{6.8.27}$$

where

$$\sigma_{1,N}^2 = \sum_{k=1}^{N} c_k^2 + \frac{2}{m} \sum_{i=1}^{m-1} (m - i) \operatorname{Cov}(B_{1N}(X_1), B_{1N}(X_{1+i})),$$

$$\sigma_{2,N}^2 = \sum_{k=1}^{N} d_k^2 + \frac{2}{n} \sum_{j=1}^{n-1} (n - j) \operatorname{Cov}(B_{2N}(X_1), B_{2N}(X_{1+j})), \tag{6.8.28}$$

and

$$B_{1N}(X_i) = \sum_{k=1}^{N} c_k f_k(X_i),$$

$$B_{2N}(Y_j) = \sum_{k=1}^{N} d_k g_k(Y_j). \qquad (6.8.29)$$

Therefore,

$$(mn)^{\frac{1}{2}} U_{mn}^{(N)} \xrightarrow{\mathcal{L}} \sum_{k=1}^{N} \lambda_k U_k V_k \text{ as } m, n \to \infty. \qquad (6.8.30)$$

Note that

$$\text{Cov}(U_k, U_{k'}) = \text{Cov}(f_k(X_1), f_{k'}(X_1)) + 2 \sum_{i=1}^{\infty} \text{Cov}(f_k(X_1), f_{k'}(X_{1+i})), \quad (6.8.31)$$

and

$$\text{Cov}(V_k, V_{k'}) = \text{Cov}(g_k(Y_1), g_{k'}(Y_1)) + 2 \sum_{i=1}^{\infty} \text{Cov}(g_k(Y_1), g_{k'}(Y_{1+i})). \qquad (6.8.32)$$

Let $\phi_{m,n}(t)$, $\phi^*(t)$, $\phi_N^*(t)$ and $\phi_{m,n,N}(t)$ be the characteristic functions of $(mn)^{\frac{1}{2}} U_{mn}$, $\sum_{k=1}^{\infty} \lambda_k U_k V_k$, $\sum_{k=1}^{N} \lambda_k U_k V_k$ and nU_{mn}^N, respectively. In view of equation (6.8.30), for any $\epsilon > 0$, there exists $m_0(N, \epsilon, t)$ and $n_0(N, \epsilon, t)$ such that for $m \geq m_0$ and $n \geq n_0$,

$$|\phi_{m,n,N}(t) - \phi_N^*(t)| \leq \frac{\epsilon}{3}. \qquad (6.8.33)$$

As in the proof of equation (6.7.35),

$$E(U_k^2) < C, \qquad (6.8.34)$$

uniformly for $k \geq 1$, and

$$E(V_k^2) < C, \qquad (6.8.35)$$

uniformly for $k \geq 1$. In view of equations (6.8.34), (6.8.35) and the Cauchy-Schwartz inequality,

$$\begin{aligned}
E(\sum_{k=N+1}^{\infty} \lambda_k U_k V_k)^2 &\leq \sum_{k=N+1}^{\infty} \lambda_k^2 E(U_k^2) E(V_k^2) \\
&\quad + \sum_{N+1 \leq k \neq k' < \infty} |\lambda_k||\lambda_{k'}| E|U_k U_{k'}| E|V_k V_{k'}| \\
&\leq C\{\sum_{k=N+1}^{\infty} |\lambda_k|\}^2 \\
&\to 0 \text{ as } N \to \infty.
\end{aligned} \qquad (6.8.36)$$

Therefore, for large N, depending on t,

$$|\phi_N^*(t) - \phi^*(t)| \leq |t| E\left(\sum_{k=N+1}^{\infty} \lambda_k U_k V_k \right)^2$$

$$\leq \frac{\epsilon}{3}. \tag{6.8.37}$$

Let

$$\bar{\phi}(x, y) = \phi(x, y) - \phi_N(x, y). \tag{6.8.38}$$

Then, from Birkel (1986), it follows that

$$mn E |U_{m,n} - U_{m,n}^{(N)}|^2$$

$$= \frac{1}{mn} E\left(\sum_{i=1}^{m} \sum_{j=1}^{n} \bar{\phi}(X_i, Y_j) \right)^2$$

$$= \frac{1}{mn} \sum_{1 \leq i \neq i' \leq m} \sum_{1 \leq j \neq j' \leq n} E(\bar{\phi}(X_i, Y_j) \bar{\phi}(X_i', Y_j'))$$

$$\leq C \frac{1}{mn} \sum_{1 \leq i \neq i' \leq m} \sum_{1 \leq j \neq j' \leq n} \{ \mathrm{Cov}(X_i, X_i') + \mathrm{Cov}(Y_j, Y_j') \}$$

$$\leq C \frac{1}{mn} \left\{ n^2 m \sum_{i=1}^{m-1} \mathrm{Cov}(X_1, X_{i+1}) + m^2 n \sum_{j=1}^{n-1} \mathrm{Cov}(Y_1, Y_{j+1}) \right\}$$

$$= C mn \left\{ \frac{1}{m} \sum_{i=1}^{m-1} \mathrm{Cov}(X_1, X_{i+1}) + \frac{1}{n} \sum_{j=1}^{n-1} \mathrm{Cov}(Y_1, Y_{j+1}) \right\}$$

$$= mn \{ \delta_m + \eta_n \}$$

$$\to 0 \text{ as } m, n \to \infty. \tag{6.8.39}$$

Hence , for large $m \geq m_0$ and $n \geq n_0, m_0$ and n_0 depending on N, t and ϵ

$$|\phi_{m,n}(t) - \phi_{m,n,N}(t)| \leq \frac{\epsilon}{3}. \tag{6.8.40}$$

The result follows by combining equations (6.8.33), (6.8.37) and (6.8.40). $\qquad \square$

Theorem 6.8.2. *Let U_{mn} be a non-degenerate two-sample U-statistic based on stationary sequences of associated random variables. Let ϕ be the corresponding kernel. Assume that the eigenfunctions $f_k(x)$ and $g_k(y)$ given by (6.8.3) are differentiable and*

$$\sup_k \sup_x |f_k'(x)| < \infty, \tag{6.8.41}$$

$$\sup_k \sup_y |g_k'(y)| < \infty. \tag{6.8.42}$$

Furthermore, assume that

$$\sum_{j=1}^{\infty} \mathrm{Cov}(X_1, X_j) < \infty \qquad (6.8.43)$$

and

$$\sum_{k=1}^{\infty} |\lambda_k| < \infty. \qquad (6.8.44)$$

Let

$$\delta_m = \frac{1}{m} \sum_{i=1}^{m-1} \mathrm{Cov}(X_1, X_{i+1}), \qquad (6.8.45)$$

and

$$\eta_n = \frac{1}{n} \sum_{j=1}^{n-1} \mathrm{Cov}(Y_1, Y_{j+1}). \qquad (6.8.46)$$

Assume that

$$mn\{\delta_m + \eta_n\} \to 0 \ \text{as} \ m, n \to \infty. \qquad (6.8.47)$$

Let

$$g^*(x) = \int_{-\infty}^{\infty} \phi(x, y) dG(y),$$

$$f^*(y) = \int_{-\infty}^{\infty} \phi(x, y) dF(x). \qquad (6.8.48)$$

Suppose that $g^*(.)$ is monotone or g^* is Lipschitzian, and

$$\sum_{k=0}^{\infty} |Eg_k^*(X_1)| < \infty, \qquad (6.8.49)$$

and similar conditions fold for f^* with

$$\sum_{k=0}^{\infty} |Ef_k^*(Y_1)| < \infty. \qquad (6.8.50)$$

If U_{mn} has finite variance and $\frac{m}{n} \to \lambda$ as $m, n \to \infty$, then

$$(m)^{\frac{1}{2}}(U_{mn} - EU_{mn}) \xrightarrow{\mathcal{L}} N(0, \sigma_1^2 + \lambda\sigma_2^2), \qquad (6.8.51)$$

where

$$\sigma_1^2 = \mathrm{Var}(g^*(X_1)) + 2\sum_{j=1}^{\infty} \mathrm{Cov}(g^*(X_1), g^*(X_{j+1})), \qquad (6.8.52)$$

$$\sigma_2^2 = \mathrm{Var}(f^*(Y_1)) + 2\sum_{j=1}^{\infty} \mathrm{Cov}(f^*(Y_1), f^*(Y_{j+1})), \qquad (6.8.53)$$

Proof. Note that

$$EU_{m,n} = E\phi(X_1, Y_1) = Eg^*(X) = Ef^*(Y). \tag{6.8.54}$$

Since $U_{m,n}$ has finite variance, we have

$$Eg^{*2}(X_1) = E[\phi(X_1, Y_1)\phi(X_1, Y_2)] < \infty, \tag{6.8.55}$$

and

$$Ef^{*2}(Y_1) = E[\phi(X_1, Y_1)\phi(X_2, Y_1)] < \infty. \tag{6.8.56}$$

Let

$$\hat{\phi}(x, y) = \phi(x, y) - g^*(x) - f^*(y) + Eg^*(X_1). \tag{6.8.57}$$

Then $\hat{\phi}$ is square integrable and degenerate. Let $\hat{U}_{m,n}$ be the U-statistic based on $\hat{\phi}$. Then, from Theorem 6.8.1, it follows that $(mn)^{\frac{1}{2}}\hat{U}_{m,n}$ converges in distribution. Furthermore using equations (6.8.54) and (6.8.57), we get that

$$U_{m,n} - EU_{m,n} = \frac{1}{mn}\sum_{i=1}^{m}\sum_{j=1}^{n}\{\phi(X_i, Y_j) - E\phi(X_i, Y_j)\}$$

$$= \hat{U}_{m,n} + \frac{1}{m}\sum_{i=1}^{m}\{g^*(X_i) - Eg^*(X_i)\}$$

$$+ \frac{1}{n}\sum_{j=1}^{n}\{f^*(Y_j) - Ef^*(Y_j)\}. \tag{6.8.58}$$

Therefore,

$$m^{\frac{1}{2}}\{U_{m,n} - EU_{m,n}\} = m^{\frac{1}{2}}\hat{U}_{m,n} + \frac{1}{\sqrt{m}}\sum_{i=1}^{m}\{g^*(X_i) - Eg^*(X_i)\}$$

$$+ (\frac{m}{n})^{\frac{1}{2}}\frac{1}{\sqrt{n}}\sum_{j=1}^{n}\{f^*(Y_j) - Ef^*(Y_j)\}. \tag{6.8.59}$$

Therefore, from Theorem 6.8.1, we get that

$$m^{\frac{1}{2}}\hat{U}_{m,n} \xrightarrow{p} 0 \text{ as } m, n \to \infty, \tag{6.8.60}$$

$$\frac{1}{\sqrt{m}}\sum_{i=1}^{m}\{g^*(X_i) - Eg^*(X_i)\} \xrightarrow{\mathcal{L}} N(0, \sigma_1^2) \text{ as } n \to \infty, \tag{6.8.61}$$

and

$$\frac{1}{\sqrt{n}}\sum_{j=1}^{n}\{f^*(Y_j) - Ef^*(Y_j)\} \xrightarrow{\mathcal{L}} N(0, \sigma_2^2) \text{ as } n \to \infty. \tag{6.8.62}$$

The result now follows from the relation (6.8.59). □

Results in this section are from Dewan and Prakasa Rao (2001).

6.9   Limit Theorems for V-Statistics

Let $\psi(x_1, \ldots, x_k)$ be a real-valued function symmetric in its arguments. Let $E|\psi(x_{i_1}, \ldots, x_{i_k})|^r < \infty$ for some positive integer r. Then, the U-statistic of degree k based on the kernel ψ is defined as

$$U_n = \frac{1}{\binom{n}{k}} \sum_{(c)} \psi(x_{i_1}, \ldots, x_{i_k}), \qquad (6.9.1)$$

where $\sum_{(c)}$ denotes the summation over all subsets $1 \le i_1 < \ldots < i_k \le n$ of $\{1, \ldots, n\}$.

A V-statistic (Von Mises (1947)) based on the symmetric kernel ψ of degree k is defined by

$$V_n = n^{-k} \sum_{i_1=1}^{n} \cdots \sum_{i_1=1}^{n} \psi(X_{i_1}, \ldots, X_{i_k}). \qquad (6.9.2)$$

Then, one can express V_n as follows (see Lee (1990, p. 183)):

$$V_n = n^{-k} \sum_{j=1}^{k} j! S_k^{(j)} \binom{n}{j} U_n^{(j)}, \qquad (6.9.3)$$

where $U_n^{(j)}$ is a U-statistic of degree j, $S_k^{(j)}$ are Stirling numbers of the second kind (see, e.g., Abramowitz and Stegun (1965)). Also note that

$$n^k = \sum_{j=1}^{k} j! S_k^{(j)} \binom{n}{j}, \qquad (6.9.4)$$

so that

$$n^k(V_n - \theta) = \sum_{j=1}^{k} j! S_k^{(j)} \binom{n}{j} (U_n^{(j)} - \theta). \qquad (6.9.5)$$

The following theorem proves the asymptotic equivalence of the U-statistics and the V-statistics.

Theorem 6.9.1. *Let the sequence $\{X_n, n \ge 1\}$ be a stationary associated sequence. Let U_n and V_n be the U-statistic and the V-statistic respectively based on a symmetric kernel $\psi(x_1, \ldots, x_k)$ of degree k. Assume that $\psi(x_1, \ldots, x_k)$ is monotonic in x_1. Further suppose that*

$$E[|\psi(X_1, \ldots, X_k)|^{r+\delta}] < \infty \ \text{ for } r > 2, \ \delta > 0, \qquad (6.9.6)$$

and

$$u_n^* = 2 \sum_{j=n+1}^{\infty} \text{Cov}(\psi(X_1, \ldots, X_k), \psi(X_{(j-1)k+1}, \ldots, X_{jk}) = O(n^{-\frac{(r-2)(r+\delta)}{2\delta}}).$$

$$(6.9.7)$$

Then

$$E|U_n - V_n|^r = O(n^{-\frac{r}{2}}). \tag{6.9.8}$$

Proof. Let $k \geq 1$ and $p = [\frac{n}{k}]$, the greatest integer $\leq \frac{n}{k}$. Define

$$W(x_1, \ldots, x_n) = \frac{1}{p}\{\psi(x_1, \ldots, x_k) + \psi(x_{k+1}, \ldots, x_{2k}) + \ldots + \psi(x_{(p-1)k+1}, \ldots, x_{pk})\} \tag{6.9.9}$$

Then

$$\sum_{(n)} W(x_{v_1}, \ldots, x_{v_n}) = k!(n-k)! \sum_{(n,k)} \psi(x_{i_1}, \ldots, x_{i_k}), \tag{6.9.10}$$

where $\sum_{(n)}$ is summation over all the $n!$ permutations $(v_1, \ldots, v_n)$ of $\{1, \ldots, n\}$ and

$\sum_{(n,k)}$ is summation over all the $\binom{n}{k}$ subsets $(i_1, \ldots, i_k)$ of $\{1, \ldots, n\}$. Then, it is
easy to see that

$$U_n - \theta = \frac{1}{n!} \sum_{(n)} \{W(X_{v_1}, \ldots, X_{v_n}) - \theta\}. \tag{6.9.11}$$

By Minkowski's inequality and the symmetry property of W, we have

$$E[|U_n - \theta|^r] \leq E[|W(X_1, \ldots, X_n) - \theta|^r]. \tag{6.9.12}$$

Since ψ is monotone component-wise, $W(X_1, \ldots, X_n)$ is an average of p associated random variables. Then, using equations (6.9.6), (6.9.7) and the moment inequalities in Birkel (1988a), we have

$$E[|U_n - \theta|^r] = O(p^{-\frac{r}{2}}) = O(n^{-\frac{r}{2}}). \tag{6.9.13}$$

From (6.9.5), we get that

$$(V_n - \theta) = \sum_{j=1}^{k} a_j(U_n^{(j)} - \theta),$$

where

$$a_j = \frac{j! S_k^{(j)} \binom{n}{j}}{n^k}.$$

Using Minkowski's inequality, and (6.9.13), we have

$$E|V_n - \theta|^r \leq \{\sum_{j=1}^{k} a_j(E|U_n - \theta|^r)^{\frac{1}{r}}\}^r \tag{6.9.14}$$

$$\leq C\{\sum_{j=1}^{k} a_j n^{-\frac{1}{2}}\}^r\}$$

$$= Cn^{-\frac{r}{2}}\{\sum_{j=1}^{k} a_j\}^r.$$

It is easy to see that

$$a_j = S_k^{(j)} O(n^{j-k}).$$

Therefore,

$$\sum_{j=1}^{k} a_j \leq \max_{1\leq j\leq k} S_k^{(j)} O(\sum_{j=1}^{k} n^{j-k}) \tag{6.9.15}$$

$$\leq C \max_{1\leq j\leq k} S_k^{(j)}. \tag{6.9.16}$$

Hence

$$E|V_n - \theta|^r = O(n^{-\frac{r}{2}}). \tag{6.9.17}$$

The result follows using equations (6.9.13) and (6.9.15) by the c_r-inequality (Loeve (1977)).   □

Remarks. If the function ψ is partially differentiable with bounded partial derivatives, then following Birkel (1988a), the condition, given by equation (6.9.7), can be written as

$$\sum_{j=n+1}^{\infty} \sum_{i=1}^{k} \sum_{\ell=(j-1)k+1}^{jk} \text{Cov}(X_i, X_\ell) = O(n^{-\frac{(r-2)(r+\delta)}{2\delta}}). \tag{6.9.18}$$

Remarks. The condition on componentwise monotonicity of ψ can possibly be dropped by extending the results of Shao and Yu (1996) to functions of random vectors of associated random variables.

6.10   Limit Theorems for Associated Random Fields

Limit theorems for associated random fields have been discussed extensively by Bulinski and Shaskin (2007). We give a brief discussion of some recent results on the law of iterated logarithm and strong Gaussian approximation due to Shaskin (2006, 2007).

Recall that a random field $\mathbf{X} = \{X_j, \, j \in Z^d\}$ is called associated if

$$\text{Cov}(f(X_{i_1}, \ldots, X_{i_n}), g(X_{i_1}, \ldots, X_{i_n})) \geq 0$$

for any $n \geq 1$ an arbitrary collection $\{i_1, \ldots, i_n\} \subset Z^d$ of pairwise distinct indices and any bounded componentwise nondecreasing functions $f, g : R^n \to R$.

Let

$$u(r) = \sup_{j \in Z^d} \sum_{k \in Z^d : |k-j| \geq r} \text{Cov}(X_j, X_k), \quad r \geq 0 \tag{6.10.1}$$

where $|j| = \max_{i=1,\dots,d} j_i$ for $j \in Z^d$. Suppose that $u(1) < \infty$ and $u(r) \to 0$ as $r \to \infty$. The function $u(r)$ is called the *Cox-Grimmet coefficient*.

By an integer parallelpiped, we mean a set $(a, b] = ((a_1, b_1] \times \dots \times (a_d, b_d]) \cap Z^d$, where $a_i, b_i \in Z^d$ and $a_i < b_i$ for $i = 1, \dots, d$. Let $S(U) = \sum_{j \in U} X_j$ where U is an integer parallelpiped and $S_n = S((0, n])$ for $n \in N^d$. For an index $n \in N^d$ and $c \in N$, $c > 1$, we write $[n] = n_1 \dots n_d$, $c^n = (c^{n_1}, \dots, c^{n_d}) \in N^d$, $c^{n-1} = (c^{n_1-1}, \dots, c^{n_d-1}) \in N^d$. Further $\log x = \ln(\max(x, e))$, $x > 0$ and we say that $n \to \infty$, where $n \in N^d$, if $n_1 \to \infty, \dots, n_d \to \infty$.

Shaskin (2006) derived the following law of iterated logarithm for an associated random field.

Theorem 6.10.1. *Suppose that a wide sense stationary random field* $\mathbf{X}$ *is associated and* $\sup_{j \in Z^d} E|X_j|^{2+\delta} < \infty$ *for some* $\delta > 0$. *Further suppose that there is a* $\lambda > 0$ *such that* $u(r) = O(r^{-\lambda})$ *as* $r \to \infty$. *Then*

$$\limsup_{n \to \infty} = \frac{S_n}{\sqrt{2d[n] \log \log [n]}} = -\liminf_{n \to \infty} \frac{S_n}{\sqrt{2d[n] \log \log [n]}} = \sigma \quad a.s. \qquad (6.10.2)$$

where $\sigma^2 = \Sigma_{j \in Z^d} \text{Cov}(X_0, X_j)$.

For $\tau > 0$, let $G_\tau = \{n \in N^d : n_k \geq (\prod_{j \neq k} n_j)^\tau, \ k = 1, \dots, d\}$. Shaskin (2007) obtained the following result which gives a strong approximation for an associated random field.

Theorem 6.10.2. *Suppose that a wide sense stationary random field* $\mathbf{X}$ *is associated and* $\sup_{j \in Z^d} E|X_j|^{2+\delta} < \infty$ *for some* $\delta > 0$. *Further suppose that there is a* $\lambda > 0$ *such that* $u(r) = O(r^{-\lambda})$ *as* $r \to \infty$. *Then, for each* $\tau > 0$, *it is possible to construct a version of the associated random field* $\mathbf{X}$ *on a new probability space equipped with a d-parameter Wiener process* $\{W_t, t \in R_d^+\}$ *and there exists an* $\varepsilon > 0$ *such that*

$$[n]^{\varepsilon - \frac{1}{2}}(S_n - \sigma W_n) \to 0 \ as \ n \to \infty, \ n \in G_\tau \ a.s. \qquad (6.10.3)$$

We omit the proofs.

Bulinski (2011) obtained a central limit theorem for a positively associated stationary random field $X = \{X_j, \ j \in Z^d\}$ defined on an integer-valued d-dimensional lattice Z^d. The uniform integrability of the squares of normalized partial sums, taken over growing parallelpipeds in Z^d, is the main condition that is used for proving the asymptotic normality of the standardized partial sum. It extends results in Lewis (1998).

6.11 Remarks

We have not been able to discuss some other important probabilistic results for associated or negatively associated random variables due to paucity of space in the book as well as due to the fact that these results are akin to those derived for independent random variables and the proofs are similar to those in the independent case. We now point out some of these results.

Exponential inequalities for associated or positively associated random variables and their application to study rates of convergence of the strong law of large numbers were investigated in Ioannides and Roussas (1999), Oliveira (2005), Yang and Chen (2007), Sung (2007), Xing and Yang (2008), Xing et al. (2008) and Xing and Yang (2010). Exact rate in the log law for positively associated random variables was obtained in Fu (2009). Large deviation for the empirical mean for associated random variables was studied in Henriques and Oliveira (2008). Convergence rates in the strong law for associated sequences was studied in Louhichi (2000). Precise asymptotics in the law of large numbers for associated random variables were obtained by Baek et al. (2008). A non-classical law of iterated logarithm for functions of positively associated random variables was proved in Wang and Zhang (2006). Almost sure central limit theorems for associated random variables were studied by Matula (1998) and Gonchigdanzan (2002) and for positively associated random variables by Matula (2005). Weak convergence for partial sums of associated random variables was investigated in Matula (1996a). Convergence of weighted averages of associated random variables was studied in Matula (1996b). Matula and Rychlik (1990) obtained the invariance principle for non-stationary sequences of associated random variables. Limit theorems for sums of non-monotonic functions of associated random variables were discussed by Matula (2001). A bound for the distribution of the sum of discrete associated or negatively associated random variables is given in Boutsikas and Koutras (2000). Balan (2003) discussed a strong invariance principle for associated fields.

Applications of negatively associated random variables can seen in Joag-Dev and Proschan (1983). The law of iterated logarithm for negatively associated random variables was derived in Shao and Su (1999), Liu and Mei (2004)and Wang and Zhang (2007). Strong laws of large numbers for negatively associated sequences were studied by Dong and Yang (2002), Nezakati (2005) and Liu et al. (2009). Exponential inequality for negatively associated sequences was obtained by Xing et al. (2009). Kim and Ko (2003) studied almost sure convergence of averages of associated and negatively associated random variables. Probability and moment bounds for sums of negatively associated random variables were obtained by Matula (1996c). A comparison theorem on maximal inequalities between negatively associated random variables and independent random variables with the same marginal distributions was obtained by Shao (2000). Marcinkiewicz-Zygmund-Burkhoder type inequality for negatively associated sequences was proved by Zhang (2000)(cf. Lin and Bai (2010). Precise rates in the law of the logarithm for negatively associated random variables were discussed in Fu and Zhang (2007). Moment convergence rates of the law of iterated logarithm for negatively associated sequences were obtained by Fu and Hu (2010). Complete convergence for negatively associated sequences was studied in Mikusheva (2000), Liang (2000), Huang and (2002), Kuczmaszmu (2009), Zhao et al. (2010) and Liang et al. (2010).

Exponential and almost sure convergence for negatively associated random variables was studied by Han (2007), Sung (2009), Xing (2010), Jabbari et al. (2009). Complete convergence for weighted sums of row-wise negatively associated random variables was investigated in Qiu (2010). Strong laws for certain types of U-statistics based on negatively associated random variables were discussed by Budsaba et al. (2009). Convergence rates in the law of iterated logarithm for negatively associated random variables with multidimensional indices were studied in Li (2009). Strong limit theorems for weighted sums of negatively associated random variables were derived by Jing and Liang (2008). Liu (2007) and Wang et al. (2006) derived precise large deviations for negatively associated sequences. Asymptotic normality for U-statistics of negatively associated random variables was investigated in Huang and Zhang (2006). Limiting behaviour of weighted sums of negatively associated random variables was given in Baek et al. (2005). Maxima of sums and random sums of negatively associated random variables was studied in Wang et al. (2004). A non-classical law of iterated logarithm for functions of negatively associated random variables was proved in Jiang (2003) and Huang (2004). Weak convergence for functions of negatively associated random variables was discussed in Zhang (2001). An invariance principle for negatively associated sequences was obtained by Lin (1997). Huang (2003) obtained the law of iterated logarithm for geometrically weighted series of negatively associated random variables.

Chapter 7

Nonparametric Estimation for Associated Sequences

7.1 Introduction

In classical statistical inference, the observed random variables of interest are generally assumed to be independent and identically distributed. However, as was mentioned in Chapter 1, in some real life situations, the random variables need not be independent. Study of inference problems for dependent sequences of random variables is of importance due to their applications in fields such as reliability theory, finance and in time series with applications in economics. Statistical inference for stochastic processes was developed for Markov processes by Billingsley (1961) and for stochastic processes in general in Basawa and Prakasa Rao (1980) and Prakasa Rao (1983). Inference for special classes of processes, such as branching processes (cf. Guttorp (1991)), point processes (cf. Karr (1991)), diffusion type processes (cf. Prakasa Rao (1999b), Kutoyants (1984, 2004)), spatial Poisson processes (cf. Kutoyants (1998)), counting processes (cf. Jacobsen (1882)), Semimartingales (cf. Prakasa Rao (1999c)), and fractional diffusion processes (cf. Prakasa Rao (2010)) have been studied extensively. In the examples discussed in Chapter 1, the random variables of interest are not independent but are 'associated', a concept we discussed extensively in this book. We gave a review of probabilistic properties of associated sequences of random variables in Chapter 1 and in Chapter 6. One of the important problems of statistical inference is stochastic modelling. In order to understand the evolution of the observed data, it is important to estimate the probabilities of various events in the underlying mechanism which in turn leads to the problem of estimation of the distribution function or the probability density estimation whenever it exists.

We will now discuss some methods of nonparametric estimation of distribution functions or survival functions and estimation of a density function when

B.L.S. Prakasa Rao, *Associated Sequences, Demimartingales*
and Nonparametric Inference, Probability and its Applications,
DOI 10.1007/978-3-0348-0240-6_7, © Springer Basel AG 2012

the underlying process forms a strictly stationary sequence of associated random variables. Recall that the sequence of partial sums of mean zero sequence of associated random variables forms a demimartingale which is the topic of this book in Chapter 2.

It is assumed that all the expectations involved in the following discussions exist.

The following examples due to Matula (1996) discuss some methods for generating associated sequences.

(i) Let the sequence $\{Y_n, n \geq 1\}$ be a sequence of independent and identically distributed standard normal random variables . Let $I_{(-\infty,u)}(x)$ be the indicator function of the interval $(-\infty, u)$ for any fixed $u \in R$. Fix $u \in R$. Let

$$X_n = I_{(-\infty,u)}\left(\frac{Y_1 + Y_2 + \ldots + Y_n}{\sqrt{n}}\right).$$

Then the sequence $\{X_n, n \geq 1\}$ is a sequence of associated random variables with

$$\text{Cov}(X_j, X_n) = \frac{1}{\sqrt{2\pi}} \int_{-\infty}^{u} \exp[-\frac{x^2}{2}](\Phi(\frac{\sqrt{n}u - \sqrt{j}x}{\sqrt{n-j}}) - \Phi(u))dx,$$

for $j < n$, where Φ denotes the standard normal distribution function.

(ii) Let the sequence $\{Y_n, n \geq 1\}$ be a sequence of independent and identically distributed random variables with $E(Y_1) = 1$ and $E(Y_1^2) = 2$. For $n \geq 1$, let

$$X_n = \frac{1}{2^{n-1}}Y_1 + \ldots + \frac{1}{2^{n-1}}Y_{n-1} + nY_n.$$

Then the sequence $\{X_n, n \geq 1\}$ is a sequence of associated random variables with

$$\text{Cov}(X_j, X_n) = \frac{1}{2^{n-1}}(j + \frac{j-1}{2^{j-1}})$$

for $j < n$.

7.2 Nonparametric Estimation for Survival Function

Let the sequence $\{X_n, n \geq 1\}$ be a stationary sequence of associated random variables with distribution function $F(x)$, or equivalently, survival function $\bar{F}(x) = 1 - F(x)$, and density function $f(x)$. The empirical survival function $\bar{F}_n(x)$ is defined by

$$\bar{F}_n(x) = \frac{1}{n}\sum_{j=1}^{n}Y_j(x), \qquad (7.2.1)$$

where

$$Y_j(x) = \begin{cases} 1 & \text{if } X_j > x \\ 0 & \text{otherwise.} \end{cases} \tag{7.2.2}$$

It is interesting to note that, for fixed x, $Y_j's$ are increasing functions of $X_j's$ and hence are associated.

Bagai and Prakasa Rao (1991) proposed $\bar{F}_n(x)$ as an estimator for $\bar{F}(x)$ and discussed its asymptotic properties.

Theorem 7.2.1 (Bagai and Prakasa Rao (1991)). *Let the sequence $\{X_n, n \geq 1\}$ be a stationary sequence of associated random variables with bounded continuous density for X_1. Assume that, for some $r > 1$,*

$$\sum_{j=n+1}^{\infty} \{\mathrm{Cov}(X_1, X_j)\}^{1/3} = O(n^{-(r-1)}). \tag{7.2.3}$$

Then there exists a constant $C > 0$ such that, for every $\epsilon > 0$,

$$\sup_x P[|\bar{F}_n(x) - \bar{F}(x)| > \epsilon] \leq C\epsilon^{-2r} n^{-r}$$

for every $n \geq 1$. In particular, for every x,

$$\bar{F}_n(x) \to \bar{F}(x) \ a.s. \ as \ n \to \infty.$$

Proof. Observe that

$$\mathrm{Cov}(Y_i(x), Y_j(x)) = P(X_1 > x, \ X_j > x) - P(X_1 > x)P(X_j > x)$$
$$= P(X_1 \leq x, \ X_j \leq x) - P(X_1 \leq x)P(X_j \leq x)$$

which is nonnegative since the random variables $Y_1(x)$ and $Y_j(x)$ are associated. Then there exists a constant $C > 0$ such that

$$\sum_{j=n+1}^{\infty} \mathrm{Cov}(Y_1(x), Y_j(x))$$

$$\leq \sum_{j+n+1}^{\infty} [P(X_1 \leq x, \ X_j \leq x) - P(X_1 \leq x)P(X_j \leq x)]$$

$$\leq C \sum_{j=n+1}^{\infty} [\mathrm{Cov}(X_1, X_j)]^{1/3}$$

by choosing $T = [\mathrm{Cov}(X_1, X_j)]^{-1/3}$ in Theorem 6.2.14 whenever $\mathrm{Cov}(X_1, X_j) > 0$. If $\mathrm{Cov}(X_1, X_j) = 0$, then X_1 and X_j are independent as they are associated and

$$\mathrm{Cov}(Y_1(x), Y_j(x)) = P(X_1 \leq x, \ X_j \leq x) - P(X_1 \leq x)P(X_j \leq x) = 0$$

$$\leq [\text{Cov}(X_1, X_j)]^{1/3}.$$

Furthermore

$$\sup_x \sup_j |Y_j(x) - E[Y_j(x)]| \leq 2$$

and

$$u(n, x) = 2 \sum_{j=n+1}^{\infty} \text{Cov}(Y_1(x), Y_j(x)) \leq C \sum_{j=n+1}^{\infty} [\text{Cov}(X_1, X_j)]^{1/3}$$

for all x where C is independent of n and x. Hence, by Theorem 1.2.9, it follows that

$$\sup_x E| \sum_{j=1}^{n}(Y_j(x) - E[Y_j(x)])|^{2r} \leq Cn^r$$

for some $C > 0$, independent of n and x. Applying the Chebyshev inequality, we get that

$$\sup_x P[|\bar{F}_n(x) - \bar{F}(x)| > \epsilon] = \sup_x P[|\bar{F}_n(x) - \bar{F}(x)|^{2r} > \epsilon^{2r}]$$

$$\leq \sup_x \{n^{-2r}\epsilon^{-2r} E(| \sum_{j=1}^{n}(Y_j(x) - E[Y_j(x)])|^{2r})\}$$

$$\leq C\epsilon^{-2r}n^{-r}.$$

This proves the first part of the result. Observe that

$$\sum_{n=1}^{\infty} P[|\bar{F}_n(x) - \bar{F}(x)| > \epsilon] \leq Cn^{-2r}\epsilon^{-2r} \sum_{n=1}^{\infty} n^{-r} < \infty$$

since $r > 1$ by hypothesis. Hence, it follows that

$$\bar{F}_n(x) \to \bar{F}(x) \text{a.s. as } n \to \infty$$

by the Borel-Cantelli lemma. $\qquad\qquad\square$

The following theorem gives an exponential inequality for the deviation of the estimator $\bar{F}_n(x)$ from $\bar{F}(x)$.

Theorem 7.2.2. *Let the sequence $\{X_n, n \geq 1\}$ be a stationary sequence of associated random variables with one-dimensional survival function $\bar{F}(x)$. Let*

$$\gamma_n = \sum_{j=1}^{n} \text{Cov}(X_1, X_j). \tag{7.2.4}$$

Then, for any $\epsilon > 0$,

$$\sup_x P[|\bar{F}_n(x) - \bar{F}(x)| > \epsilon] \leq 2[e^{\frac{\lambda_n^2}{n} - \lambda_n \epsilon} + n^{-1/3}\lambda_n^2 e^{(2-\epsilon)\lambda_n}\gamma_n^{1/3}] \tag{7.2.5}$$

for any sequence $\{\lambda_n\}$ such that $\lambda_n \leq \frac{n}{4}$.

Proof. For all $\lambda_n > 0$,

$$E[e^{\lambda_n(\bar{F}_n(x) - \bar{F}(x))}] = E[e^{\frac{\lambda_n}{n}\sum_{j=1}^{n}(Y_j(x) - EY_j(x))}]$$

$$= E[e^{\frac{\lambda_n}{n}\sum_{j=1}^{n}(Y_j(x) - EY_j(x))}] - \prod_{j=1}^{n} E[e^{\frac{\lambda_n}{n}(Y_j(x) - EY_j(x))}]$$

$$+ \prod_{j=1}^{n} E[e^{\frac{\lambda_n}{n}(Y_j(x) - EY_j(x))}]. \qquad (7.2.6)$$

Note that for large n,

$$0 < \lambda_n \leq \frac{n}{4} \quad \text{and} \quad |\frac{\lambda_n}{n}(Y_j(x) - EY_j(x))| \leq \frac{1}{2}.$$

Furthermore, $e^u \leq 1 + u + u^2$ for $|u| \leq \frac{1}{2}$. Hence

$$\prod_{j=1}^{n} E[e^{\frac{\lambda_n}{n}(Y_j(x) - EY_j(x))}] \leq \prod_{j=1}^{n} E[1 + \frac{\lambda_n}{n}(Y_j(x) - EY_j(x)) \qquad (7.2.7)$$

$$+ \frac{\lambda_n^2}{n^2}(Y_j(x) - EY_j(x))^2]$$

$$\leq \prod_{j=1}^{n}(1 + \frac{\lambda_n^2}{n^2}) \quad \text{(since } \mathrm{Var}(Y_j(x)) \leq 1)$$

$$\leq e^{\frac{\lambda_n^2}{n}}. \qquad (7.2.8)$$

Using Theorems 6.2.9 and 6.2.14, for $T > 0$, we get that

$$|E[e^{\frac{\lambda_n}{n}\sum_{j=1}^{n}(Y_j(x) - EY_j(x))}] - \prod_{j=1}^{n} E[e^{\frac{\lambda_n}{n}(Y_j(x) - EY_j(x))}]|$$

$$\leq \frac{\lambda_n^2 e^{2\lambda_n}}{n^2} \sum_{1 \leq i < j \leq n} \mathrm{Cov}(Y_i(x), Y_j(x))$$

$$\leq \frac{\lambda_n^2 e^{2\lambda_n}}{n^2} \sum_{1 \leq i < j \leq n} \{P(X_i > x, X_j > x) - P(X_i > x)P(X_j > x)\}$$

$$\leq \frac{\lambda_n^2 e^{2\lambda_n}}{n^2} \{T^2 \sum_{1 \leq i < j \leq n} \mathrm{Cov}(X_i, X_j) + \frac{n(n-1)}{T}\}$$

$$= \frac{\lambda_n^2 e^{2\lambda_n}}{n^2} \{T^2 \sum_{j=1}^{n-1}(n-j)\,\mathrm{Cov}(X_1, X_j) + \frac{n(n-1)}{T}\}$$

$$\leq \frac{\lambda_n^2 e^{2\lambda_n}}{n}\{T^2\gamma_n^{1/3} + \frac{n}{T}\}$$

$$= n^{-1/3}\lambda_n^2 e^{2\lambda_n} \quad (\text{choosing } T = n^{1/3}\gamma_n^{-1/3}). \tag{7.2.9}$$

Using Chebyshev's inequality and combining the relations (7.2.6) to (7.2.9), we get that

$$P[\bar{F}_n(x) - \bar{F}(x) > \epsilon] \leq e^{\frac{\lambda_n^2}{n} - \lambda_n \epsilon} + n^{-1/3}\lambda_n^2 e^{(2-\epsilon)\lambda_n}\gamma_n^{1/3} \tag{7.2.10}$$

for any $\epsilon > 0$. The result stated in the theorem follows from the fact that if the members of the set of random variables $\{Y_j(x), 1 \leq j \leq n\}$ are associated, for fixed x then so are $\{-Y_j(x), 1 \leq j \leq n\}$. □

The following results are Glivenko-Cantelli type theorems for associated random variables. We do not give detailed proofs of these results.

Theorem 7.2.3 (Bagai and Prakasa Rao (1991)). *Let $\{X_n, n \geq 1\}$ be a stationary sequence of associated random variables satisfying the conditions of Theorem 7.2.1. Then for any compact subset $J \subset R$,*

$$\sup[|\bar{F}_n(x) - \bar{F}(x)| : x \in J| \to 0 \text{ a.s. as } n \to \infty.$$

Theorem 7.2.4 (Yu (1993)). *Let the sequence $\{X_n, n \geq 1\}$ be a sequence of associated random variables having the same marginal distribution function $F(x)$ for $X_n, n \geq 1$. If $F(x)$ is continuous and*

$$\sum_{n=1}^{\infty} \frac{1}{n^2} \text{Cov}(X_n, S_{n-1}) < \infty, \tag{7.2.11}$$

then, as $n \to \infty$,

$$\sup_{-\infty < x < \infty} |F_n(x) - F(x)| \to 0 \text{ a.s.} \tag{7.2.12}$$

If the sequence in the above theorem is stationary, then the condition (7.2.11) can be weakened to

$$\frac{1}{n}\sum_{i=1}^{n} \text{Cov}(X_n, X_i) \to 0. \tag{7.2.13}$$

From the central limit theorem for stationary associated random variables, the following theorem can be obtained.

Theorem 7.2.5 (Bagai and Prakasa Rao (1991)). *Let the sequence $\{X_n, n \geq 1\}$ be a stationary associated sequence of random variables with bounded continuous density for X_1 and survival function $\bar{F}(x)$. Suppose that*

$$\sum_{j=2}^{\infty} \{\text{Cov}(X_1, X_j)\}^{1/3} < \infty.$$

Define,

$$\sigma^2(x) = \bar{F}(x)[1 - \bar{F}(x)] + 2\sum_{j=2}^{\infty}\{P[X_1 > x, X_j > x] - \bar{F}^2(x)\}.$$

Then, for all x such that $0 < F(x) < 1$,

$$n^{1/2}[(\bar{F}_n(x) - \bar{F}(x))/\sigma(x)] \overset{\mathcal{L}}{\to} \Phi(x) \text{ as } n \to \infty.$$

Consider the empirical process

$$\beta_n(x) = n^{1/2}(F_n(x) - F(x)), \quad x \in R. \tag{7.2.14}$$

Weak convergence for empirical processes has been discussed by Yu (1993).

Theorem 7.2.6 (Yu (1993)). *Let the sequence $\{X_n, n \geq 1\}$ be a stationary associated sequence of random variables with bounded density for X_1 and suppose that there exists a positive constant v such that*

$$\sum_{n=1}^{\infty} n^{\frac{13}{2}+v} \text{Cov}(X_1, X_n) < \infty. \tag{7.2.15}$$

Then

$$\beta_n(.) \overset{\mathcal{L}}{\to} B(F(.)) \text{ in } D[0, 1], \tag{7.2.16}$$

where $B(.)$ is a zero-mean Gaussian process on [0,1] with covariance defined by

$$EB(s)B(t) = s \wedge t - st + 2\sum_{k=2}^{\infty}(P(X_1 \leq s, X_k \leq t) - st) \tag{7.2.17}$$

with $P\{B(.) \in C[0, 1]\} = 1$.

The above result has been improved by Oliveira and Suquet (1995) and by Shao and Yu (1996).

Theorem 7.2.7 (Oliveira and Suquet (1995)). *Let the sequence $\{X_n, n \geq 1\}$ be strictly stationary associated random variables with continuous distribution. Let $\gamma(s, t)$ be defined as*

$$\gamma(s, t) = F(s \wedge t) - F(s)F(t) \tag{7.2.18}$$

$$+ 2\sum_{k=2}^{\infty}[P(X_1 \leq s, X_k \leq t) - P(X_1 \leq s)P(X_k \leq t)].$$

Suppose the series in (7.2.18) converges uniformly on $[0, 1]^2$. Then the empirical process β_n converges weakly on $L^2(0, 1)$ to a centered Gaussian process with covariance function $\gamma(s, t)$.

In the case of uniform variables X_i, the uniform convergence of the series in (7.2.18) follows from the fact that

$$\sum_{n \geq 2} \text{Cov}^{1/3}(X_1, X_n) < \infty. \tag{7.2.19}$$

7.3 Nonparametric Density Estimation

Density estimation in the classical i.i.d. case has been extensively discussed. For comprehensive surveys, see Prakasa Rao (1983, 1999), Silverman (1986) and Efromovich (2000) among others. These results have been extended to estimation of marginal density for stationary processes which are either Markov or mixing in some sense (cf. Prakasa Rao (1983, 1999)). As has been pointed out earlier, it is of interest to study density estimation when the observations form an associated sequence, for instance, in the context of study of lifetimes of components in reliability.

Kernel Type Density Estimator

Bagai and Prakasa Rao (1995) proposed a kernel-type estimator for the unknown density function f for X_1 when the sequence $\{X_n, n \geq 1\}$ is a stationary sequence of associated random variables.

Assume that the support of f is a closed interval $I = [a, b]$ on the real line. Consider

$$f_n(x) = \frac{1}{nh_n} \sum_{j=1}^{n} K\left(\frac{x - X_j}{h_n}\right), \quad x \in I \tag{7.3.1}$$

as an estimator for $f(x)$, where $K(\cdot)$ is a suitable kernel and h_n is a bandwidth sequence.

The asymptotic behaviour of $f_n(x)$ is discussed later under the assumptions (A) listed below:

(A_1) $K(\cdot)$ is a bounded density function and of bounded variation on R satisfying

   (i) $\lim_{|u| \to \infty} |u| K(u) = 0$,

   (ii) $\int_{-\infty}^{\infty} u^2 K(u) du < \infty$.

(A_2) $K(x)$ is differentiable and $\sup_x |K'(x)| \leq c < \infty$.

Remarks. Note that standard normal density satisfies the above conditions.

In addition to the conditions (A_1) and (A_2), it is assumed that the covariance structure of $\{X_n\}$ satisfies the following condition.

(B) For all ℓ and r, $\displaystyle\sum_{j:|\ell-j|\geq r} \text{Cov}(X_j, X_\ell) \leq u(r)$, where $u(r) = e^{-\alpha r}$ for some $\alpha > 0$.

In addition to the conditions on the kernel $K(.)$ and the sequence $\{X_n, n \geq 1\}$:

(*C*) Suppose that the one-dimensional marginal density function f is thrice differentiable and the third derivative is bounded.

Let

$$\gamma_j = \int_{-\infty}^{\infty} x^j K(x)dx, \quad j = 1, 2$$

and

$$\beta_j = \int_{-\infty}^{\infty} x^j K^2(x)dx, \quad j = 0, 1, 2.$$

Under the conditions (*A*) and (*B*) given above, it can be checked that

$$B_n(x) \equiv E[f_n(x)] - f(x)]$$

$$= -h_n f'(x)\gamma_1 + \frac{h_n^2}{2} f''(x)\gamma_2 + O(h_n^3)$$

and

$$\mathrm{Var}[f_n(x)] = \frac{1}{nh_n}[f(x)\beta_0 + O(h_n)] + O(\frac{1}{nh_n^4}).$$

As a consequence of the above results, it follows that the density estimator $f_n(x)$ is an asymptotically unbiased and consistent estimator of $f(x)$ at continuity points x of f under the conditions (*A*), (*B*) and (*C*) provided $h_n \to 0$ and $nh_n^4 \to \infty$ as $n \to \infty$. It is sufficient if $u(0) < \infty$ in condition (*B*) for the weak consistency of the density estimator $f_n(x)$. The following result on the strong consistency of the estimator $f_n(x)$ can be proved using Theorem 1.2.16.

Theorem 7.3.1 (Bagai and Prakasa Rao (1995)). *Let the sequence $\{X_n, n \geq 1\}$ be a stationary sequence of associated random variables. Suppose that (A_1), (A_2), and (B) hold. Then,*

$$f_n(x) \to f(x) \ a.s. \ as \ n \to \infty$$

at continuity points x of $f(\cdot)$.

Uniform strong consistency of $f_n(x)$ follows from the following theorem.

Theorem 7.3.2 (Bagai and Prakasa Rao (1995)). *Let the sequence $\{X_n, n \geq 1\}$ be a stationary sequence of associated random variables satisfying the conditions (A) and (B). Suppose there exists $\gamma > 0$ such that*

$$h_n^{-4} = O(n^\gamma). \tag{7.3.2}$$

Further, suppose that the following condition holds:

$$(C) \qquad |f(x_1) - f(x_2)| \leq c|x_1 - x_2|, \quad x_1, x_2 \in I,$$

Then

$$\sup[|f_n(x) - f(x)|; x \varepsilon I] \to 0 \ a.s. \ as \ n \to \infty.$$

Roussas (1991) showed that for $\theta > 0$ and for any compact interval $[a, b]$,

$$\sup_{x \in [a,b]} n^{\theta} |f_n(x) - f(x)| \to 0 \text{ a.s. as } n \to \infty,$$

under some conditions. Douge (2007) obtained the rate of convergence of kernel type density estimator. We will now discuss these results from Douge (2007).

Let the sequence $\{X_n, n \geq 0\}$ be a stationary sequence of associated random variables such that X_0 and (X_0, X_j) have densities f and $f_{0,j}$ respectively. Let $K(.)$ be a probability density function and h_n be a sequence of positive numbers such that $h_n \to 0$ and $nh_n \to \infty$ as $n \to \infty$. Let

$$f_n(x) = \frac{1}{nh_n} \sum_{i=0}^{n} K(\frac{x - X_i}{h_n}) \tag{7.3.3}$$

be a kernel type density estimator of $f(x)$. Suppose the following conditions hold:

(H1) $\sup_{j \geq 1} \sup_{x,y} |f_{0,j}(x, y) - f(x)f(y)| \leq C < \infty$ and $\sup_x |f(x)| \leq C < \infty$;

(H2) the kernel $K(.)$ is bounded, differentiable and the derivative $K'(.)$ is bounded;

(H3) (i) $\sum_{k=0}^{n-1} (k+1)^{\ell} \alpha_k^{1/4} = O(n^{\gamma \ell/2})$ for all $\ell > 0$, $0 < \gamma < 1$,

(ii) $\sum_{k=0}^{\infty} \alpha_k^{1/4} < \infty$ where $\alpha_k = \sup_{|i-j| \geq k} \text{Cov}(X_i, X_j)$ and $\alpha_0 = 1$.

Theorem 7.3.3. *Under the conditions (H1)–(H3), if $h_n^{-1} = O(n^{1-\gamma})$, then there exists a positive constant $D > 0$, such that for every $\lambda > 0$, and for every x,*

$$P(|f_n(x) - E[f_n(x)]| > \lambda) \leq \sqrt{2}e^2 \exp(-D\lambda\sqrt{nh_n}). \tag{7.3.4}$$

Theorem 7.3.4. *Suppose the probability density function is Lipschitzian of order ρ and the conditions (H1)–(H3) hold for $\gamma < 1 - \frac{1}{2\rho+1}$. Furthermore, if $\lim_{n \to \infty}(\frac{nh_n}{\log\log n}) = \infty$, $h_n = (\frac{\log\log n}{n})^{1/(2\rho+1)}$ and if $\int_{-\infty}^{\infty} |u|^{\rho} K(u)du < \infty$, then, for every compact set I in R,*

$$\sup_{x \in I} |f_n(x) - f(x)| = O(\frac{\log\log n}{n})^{\rho/(2\rho+1)} \quad a.s. \tag{7.3.5}$$

We omit the proofs of these results.

The following theorem due to Dewan and Prakasa Rao (2005), extending the work in Devroye and Gyorfi (1985), gives a bound on the integrated mean deviation of the kernel type density estimator f_n from the density function f.

Theorem 7.3.5. *Let the sequence $\{X_i, i \geq 1\}$ be a stationary sequence of associated random variables with f as the probability density function of X_1. Suppose $K(.)$ is a symmetric bounded differentiable density function with compact support such that $\lim_{|u| \to \infty} |u| K(u) = 0$, $\int_{-\infty}^{\infty} u^2 K(u) du < \infty$ and $\sup_x |K'(x)| \leq C < \infty$. Further suppose that $\lim h_n = 0$, $\lim_n n h_n = \infty$ as $n \to \infty$. Let $\mathcal{F}$ be the class of density functions f with compact support such that f'' exists and is bounded. In addition, suppose that*

$$\sum_{j=1}^{n} \text{Cov}(X_1, X_j) = o(n h_n^4). \tag{7.3.6}$$

Then

$$E[\int_{-\infty}^{\infty} |f_n(x) - f(x)| dx] = J(n, h_n) + o(h_n^2 + (n h_n)^{-1/2}) + o(\frac{\sqrt{d_n}}{n h_n^2}) \tag{7.3.7}$$

where

$$J(n, h_n) = \int_{-\infty}^{\infty} \frac{\alpha \sqrt{f(x)}}{\sqrt{n h_n}} \psi(\sqrt{n h_n^5} \frac{\beta |f''(x)|}{2\alpha \sqrt{f(x)}}) \, dx. \tag{7.3.8}$$

Here

$$\psi(|a|) = |a| P(|Z| \leq |a|) + \sqrt{\frac{2}{\pi}} e^{-a^2/2}; \quad \alpha = (\int_{-\infty}^{\infty} K^2(x) dx)^{1/2} \tag{7.3.9}$$

and

$$\beta = \int_{-\infty}^{\infty} x^2 K(x) \, dx. \tag{7.3.10}$$

The proof of this theorem uses the non-uniform bound for the distribution function of partial sums of associated random variables discussed in Chapter 6. For the proof of the above theorem, see Dewan and Prakasa Rao (2005a).

Wavelet Linear Density Estimator

We now discuss a wavelet based linear density estimator for estimating the density function for a sequence of associated random variables with a common one-dimensional probability density function. Results discussed here are from Prakasa Rao (2003).

Suppose that $\{X_n, n \geq 1\}$ is a sequence of associated random variables with a common one-dimensional marginal probability density function f. The problem of interest is the estimation of probability density function f based on the observations $\{X_1, \dots, X_N\}$. We now propose an estimator based on wavelets and obtain upper bounds on the L_p-losses for the proposed estimator.

The advantages and disadvantages of the use of a wavelet based probability density estimator are discussed in Walter and Ghorai (1992) in the case of independent and identically distributed observations. The same comments continue to hold in this case. However it was shown in Prakasa Rao (1996, 1999) that one can obtain precise limits on the asymptotic mean squared error for a wavelet based linear estimator for the density function and its derivatives as well as some other functionals of the density. A short review of wavelet based estimation of density and its derivatives in the i.i.d.case is given in Prakasa Rao (1999).

Let the sequence $\{X_i,\ i \geq 1\}$ be a sequence of associated random variables with common one-dimensional marginal probability density function f. Suppose f is bounded and compactly supported. The problem is to estimate the probability density function f based on the observations $X_1, \ldots, X_N$.

Any function $f \in L_2(R)$ can be expanded in the form

$$f = \sum_{k=-\infty}^{\infty} \alpha_{j_0,k}\phi_{j_0,k} + \sum_{j=j_0}^{\infty} \sum_{k=-\infty}^{\infty} \beta_{j,k}\psi_{j,k}$$

$$= P_{j_0}f + \sum_{j=j_0}^{\infty} D_j f$$

for any integer $j_0 \geq 1$ where the functions

$$\phi_{j_0,k}(x) = 2^{j_0/2}\phi(2^{j_0}x - k)$$

and

$$\psi_{j_0,k}(x) = 2^{j_0/2}\psi(2^{j_0}x - k)$$

constitute an orthonormal basis of $L_2(R)$ (Daubechies (1988)). The functions $\phi(x)$ and $\psi(x)$ are the scale function and the orthogonal wavelet function respectively. Observe that

$$\alpha_{j_0,k} = \int_{-\infty}^{\infty} f(x)\phi_{j_0,k}(x)dx$$

and

$$\beta_{j,k} = \int_{-\infty}^{\infty} f(x)\psi_{j,k}(x)dx.$$

We suppose that the function ϕ and ψ belong to C^{r+1} for some $r \geq 1$ and have compact support contained in an interval $[-\delta, \delta]$. It follows from Corollary 5.5.2 in Daubechies (1988) that the function ψ is orthogonal to a polynomial of degree less than or equal to r. In particular

$$\int_{-\infty}^{\infty} \psi(x)x^\ell dx = 0, \quad \ell = 0, 1, \ldots, r.$$

We assume that the following conditions hold.

(A1) The sequence $\{X_n, n \geq 1\}$ is a sequence of associated random variables with

$$u(n) = \sup_{i \geq 1} \sum_{|j-i| \geq n} \mathrm{Cov}(X_i, X_j) \leq Cn^{-\alpha}$$

for some $C > 0$ and $\alpha > 0$.

(A2) Suppose the density function f belongs to the Besov class (cf. Meyer (1990))

$$F_{s,p,q} = \{f \in B_{p,q}^s, \ ||f||_{B_{p,q}^s} \leq M\}$$

for some $0 < s < r+1$, $p \geq 1$ and $q \geq 1$, where

$$||f||_{B_{p,q}^s} = ||P_0 f||_p + [\sum_{j \geq 0}(||D_j f||_p 2^{js})^q]^{1/q}.$$

(For properties of Besov spaces, see Triebel (1992) (cf. Leblanc (1996))).

Define

$$\hat{f}_N = \sum_{k \in K_{j_0}} \hat{\alpha}_{j_0,k}\phi_{j_0,k} \tag{7.3.11}$$

where

$$\hat{\alpha}_{j_0,k} = \frac{1}{N}\sum_{i=1}^N \phi_{j_0,k}(X_i) \tag{7.3.12}$$

and K_{j_0} is the set of all k such that the the intersection of the support of f and the support of $\phi_{j_0,k}$ is nonempty. Since the function ϕ has a compact support by assumption, it follows that the cardinality of the set K_{j_0} is $O(2^{j_0})$.

We now discuss the properties of the estimator $\hat{f}_N$ as an estimator of the probability density function f.

Let $p' \geq \max(2, p)$. We will now obtain bounds on $E_f||\hat{f}_N - f||_{p'}^2$. Observe that

$$E_f||\hat{f}_N - f||_{p'}^2 \leq 2(||f - P_{j_0}f||_{p'}^2 + E_f||\hat{f}_N - P_{j_0}f||_{p'}^2). \tag{7.3.13}$$

We now estimate the terms on the right-hand side of the above equation.

Lemma 7.3.6. *For any $f \in F_{s,p,q}$, $s \geq \frac{1}{p}$, there exists a constant C_1 such that*

$$||f - P_{j_0}f||_{p'}^2 \leq C_1 2^{-2s'j_0} \tag{7.3.14}$$

where

$$s' = s + \frac{1}{p'} - \frac{1}{p}. \tag{7.3.15}$$

Proof. See Leblanc (1996), p. 83. □

We will now estimate the second term in equation (7.3.13). Note that

$$E_f||\hat{f}_N - P_{j_0}f||_{p'}^2 = E_f||\sum_{k \in K_{j_0}}(\hat{\alpha}_{j_0,k} - \alpha_{j_0,k})\phi_{j_0,k}||_{p'}^2$$

$$\leq C_2 E_f\{||\hat{\alpha}_{j_0,.} - \alpha_{j_0,.}||_{\ell_{p'}(Z)}^2\}2^{2j_0(\frac{1}{2}-\frac{1}{p'})}$$

for some constant $C_2 > 0$ by Lemma 1 in Leblanc (1996), p. 82 (cf. Meyer (1990)). Here Z is the set of all integers $-\infty < k < \infty$ and the norm

$$||\lambda||_{\ell_p(Z)} = (\sum_{k \in Z} |\lambda_k|^p)^{1/p}.$$

Hence

$$E_f ||\hat{f}_N - P_{j_0} f||_{p'}^2 \le C_2 2^{2j_0(\frac{1}{2} - \frac{1}{p'})} \{ \sum_{k \in K_{j_0}} E_f |\hat{\alpha}_{j_0,k} - \alpha_{j_0,k}|^{p'} \}^{2/p'}. \qquad (7.3.16)$$

Let

$$W_i = \eta(X_i) = \phi_{j_0,k}(X_i) - E_f(\phi_{j_0,k}(X_i)), \quad 1 \le i \le N.$$

Then

$$\hat{\alpha}_{j_0,k} - \alpha_{j_0,k} = \frac{1}{N} \sum_{i=1}^{N} W_i$$

and

$$E_f |\hat{\alpha}_{j_0,k} - \alpha_{j_0,k}|^{p'} = N^{-p'} E_f |\sum_{i=1}^{N} W_i|^{p'}. \qquad (7.3.17)$$

Observe that the random variables W_i, $1 \le i \le N$ are functions of associated random variables X_i, $1 \le i \le N$. We will now estimate the term

$$E_f |\hat{\alpha}_{j_0,k} - \alpha_{j_0,k}|^{p'}$$

by applying a Rosenthal type inequality for functions of associated random variables due to Shao and Yu (1996), p. 210. Note that the sequence of random variables $\eta(X_i)$, $1 \le i \le N$ are identically distributed with mean zero. Furthermore the function $\eta(x)$ is differentiable with

$$\sup_{-\infty < x < \infty} |\eta'(x)| = \sup_{-\infty < x < \infty} |\phi'_{j_0,k}(x)| \qquad (7.3.18)$$

$$\le 2^{3j_0/2} \sup_{-\infty < x < \infty} |\phi'(2^{j_0} x - k)|$$

$$\le 2^{3j_0/2} \sup_{-\infty < x < \infty} |\phi'(x)|$$

$$\le B_0 2^{3j_0/2}$$

for some constant $B_0 > 0$ since $\phi \in C^{r+1}$ for some $r \ge 1$. In addition, for any $d \ge 0$,

$$E_f [|\eta(X_1)|^d] \le 2^d (E_f |\phi_{j_0,k}(X_1)|^d + B_1^d) \qquad (7.3.19)$$

$$\le 2^d 2^{j_0 d/2} (E_f [|\phi(2^{j_0} X_1 - k)|^d] + B_1^d)$$

$$\le 2^{d+1} 2^{j_0 d/2} B_1^d$$

$$= B_2 2^{j_0 d/2}$$

where B_1 is a bound on ϕ following the assumption that it has a compact support and that $\phi \in C^{r+1}$ and B_2 is a positive constant independent of j_0.

(A3) Suppose that $\max(2, p) \le p' < d < \infty$.

Applying Theorem 1.2.12 due to Shao and Yu (1996), it follows that for any $\varepsilon > 0$, there exists a constant D_0 depending only on ε, d, p' and α such that

$$E_f |\sum_{i=1}^{N} W_i|^{p'} \le D_0 (N^{1+\varepsilon} E|\eta(X_1)|^{p'} \tag{7.3.20}$$

$$+ (N \max_{1 \le i \le N} \sum_{\ell=1}^{N} |\operatorname{Cov}(\eta(X_i), \eta(X_\ell))|)^{p'/2}$$

$$+ N^{(d(p'-1)-p'+\alpha(p'-d))/(d-2)\vee(1+\varepsilon)}$$

$$\times \|\eta(X_1)\|_d^{d(p'-2)/(d-2)} (B_0^2 2^{3j_0} C)^{(d-p')/(d-2)})$$

where B_0 is as defined above. Note that the constants D_0 and B_0 are independent of $k \in K_{j_0}$ and j_0. Applying Newman's inequality (Newman (1984)), we obtain that

$$|\operatorname{Cov}(\eta(X_i), \eta(X_\ell))| \le \{ \sup_{-\infty < x < \infty} |\eta'(x)|\}^2 \operatorname{Cov}(X_i, X_\ell) \tag{7.3.21}$$

$$\le B_0^2 2^{3j_0} \operatorname{Cov}(X_i, X_\ell).$$

Combining the above estimates, we get that

$$E_f |\sum_{i=1}^{N} W_i|^{p'} \le D_0 (N^{1+\varepsilon} 2^{(j_0/2)p'} B_2 \tag{7.3.22}$$

$$+ (N[\max_{1 \le i \le N} \sum_{\ell=1}^{N} \operatorname{Cov}(X_i, X_\ell)] B_0^2 2^{3j_0})^{p'/2}$$

$$+ N^{(d(p'-1)-p'+\alpha(p'-d))/(d-2)\vee(1+\varepsilon)}$$

$$\times (B_0^{1/d} 2^{j_0/2})^{d(p'-2)/(d-2)} (B_0^2 2^{3j_0} C)^{(d-p')/(d-2)}).$$

Since the above estimate holds for all $k \in K_{j_0}$ and the cardinality of K is $O(2^{j_0})$, it follows that

$$E_f \|\hat{f}_N - P_{j_0} f\|_{p'}^2 \le C_2 2^{2j_0(\frac{1}{2} - \frac{1}{p'})} 2^{\frac{2j_0}{p'}} N^{-2} \{D_0 (N^{1+\varepsilon} 2^{(j_0/2)p'} B_2 \tag{7.3.23}$$

$$+ (N[\max_{1 \le i \le N} \sum_{\ell=1}^{N} \operatorname{Cov}(X_i, X_\ell)] B_0^2 2^{3j_0})^{p'/2}$$

$$+ N^{(d(p'-1)-p'+\alpha(p'-d))/(d-2)\vee(1+\varepsilon)}$$
$$\times (B_0^{1/d}2^{j_0/2})^{d(p'-2)/(d-2)}(B_0^2 2^{3j_0}C)^{(d-p')/(d-2)})\}^{2/p'}$$

from (7.3.16) and (7.3.17). Hence there exists a constant $C_3 > 0$ such that

$$E_f\|\hat{f}_N - f\|_{p'}^2 \le C_3[2^{j_0}N^{-2}\{D_0(N^{1+\varepsilon}2^{(j_0/2)p'}B_2 \tag{7.3.24}$$

$$+ (N[\max_{1\le i\le N}\sum_{\ell=1}^{N}\text{Cov}(X_i, X_\ell)]B_0^2 2^{3j_0})^{p'/2}$$

$$+ N^{(d(p'-1)-p'+\alpha(p'-d))/(d-2)\vee(1+\varepsilon)}$$

$$\times (B_0^{1/d}2^{j_0/2})^{d(p'-2)/(d-2)}(B_0^2 2^{3j_0}C)^{(d-p')/(d-2)})\}^{2/p'} + 2^{-2s'j_0}]$$

and we have the following main result.

Theorem 7.3.7. *Suppose the conditions (A1)–(A3) hold. Let* $\max(2,p) \le p' < d < \infty$. *Define the wavelet linear density estimator* $\hat{f}_N$ *by the relation (7.3.11). Then, for every* $\varepsilon > 0$, *there corresponds a constant* $C > 0$ *such that*

$$E_f\|\hat{f}_N - f\|_{p'}^2 \le C[2^{j_0}N^{-2}\{(N^{1+\varepsilon}2^{(j_0/2)p'} \tag{7.3.25}$$

$$+ (N[\max_{1\le i\le N}\sum_{\ell=1}^{N}\text{Cov}(X_i, X_\ell)]2^{3j_0})^{p'/2}$$

$$+ N^{(d(p'-1)-p'+\alpha(p'-d))/(d-2)\vee(1+\varepsilon)}$$

$$\times 2^{(j_0/2)d(p'-2)/(d-2)}2^{3j_0(d-p')/(d-2)})\}^{2/p'} + 2^{-2s'j_0}].$$

Remarks. (i) If one chooses j_0 such that $2^{j_0} = N^\gamma$, then it can be shown that the upper bound in Theorem 7.3.2 is of the order

$$O(N^{-2+\gamma a+b} + N^{-1+\gamma a}\zeta_N + N^{-2\gamma s'})$$

where a depends on d and p' and b depends on d, p', α and ε and

$$\zeta_N = \max_{1\le i\le N}\sum_{\ell=1}^{N}\text{Cov}(X_i, X_\ell).$$

(ii) Suppose $1 \le p' \le 2$. One can get upper bounds similar to those as in the Theorem 7.3.2 for the expected loss

$$E_f\|\hat{f}_N - f\|_{p'}^{p'}$$

observing that

$$E_f\|\hat{f}_N - f\|_{p'}^{p'} \le 2^{p'-1}(\|f - P_{j_0}f\|_{p'}^{p'} + E_f\|\hat{f}_N - P_{j_0}f\|_{p'}^{p'}), \tag{7.3.26}$$

$$||f - P_{j_0}f||^{p'}_{p'} \leq C_4 2^{-p's'j_0}, \tag{7.3.27}$$

and

$$E_f||\hat{f}_N - P_{j_0}f||^{p'}_{p'} \leq C' 2^{2j_0(\frac{p'}{2}-1)} \sum_{k \in K_{j_0}} E_f|\hat{\alpha}_{j_0,k} - \alpha_{j_0,k}|^{p'} \tag{7.3.28}$$

for some positive constants C_4 and C'. We will not discuss the details here.

Chaubey et al. (2006) discussed wavelet based estimation of the derivatives of a density for associated sequences with the same one-dimensional marginal distributions. Similar results were obtained for negatively associated sequences in Chaubey et al. (2008).

Density Estimator Through Delta Sequences

Suppose the sequence $\{X_n, n \geq 1\}$ is a stationary sequence of associated random variables and the marginal density f of X_1 exists. We now consider the problem of estimation of f based on $(X_1, \ldots, X_n)$. Let $\phi_n(x, y)$, $n = 1, 2, \ldots$, be a sequence of Borel-measurable functions defined on R^2. Then the empirical density function is defined as follows :

$$f^*_n(x) = \frac{1}{n} \sum_{k=1}^{n} \phi_n(x, X_k). \tag{7.3.29}$$

This function can be considered as an estimator of f. This estimator is a generalization of the histogram type density estimator, the kernel type density estimators and the density estimator obtained by the method of orthogonal series. Properties of the empirical density function or a variation of it were considered by Foldes and Revesz (1974) and Walter and Blum (1979) in the case of independent and identically distributed random variables, by Foldes (1974) for the case of stationary ϕ-mixing sequences and by Prakasa Rao (1978) for stationary Markov processes, among others (cf. Prakasa Rao (1983)).

We now study conditions leading to the exponential rate of convergence for the uniform consistency in probability of the estimator $f^*_n(x)$, that is, the conditions under which

$$\Pr(\sup_{a+\delta \leq x \leq b-\delta} |f^*_n(x) - f(x)| > \epsilon) \to 0 \text{ as } n \to \infty, \tag{7.3.30}$$

at an exponential rate. Sufficient conditions for the asymptotic property (7.3.30) to hold have been studied earlier for the case of a sequence of independent and identically distributed random variables (Foldes and Revesz (1974)), for a ϕ-mixing sequence of random variables (Foldes (1974)) and for an absolutely regular sequence of identically distributed random variables (Yoshihara (1984)).

We now derive an exponential type bound for the rate of uniform consistency for a larger class of estimators which include the kernel-type estimator as a special case.

Theorem 7.3.8. *Let the sequence $\{X_n, n \geq 1\}$ be a stationary sequence of associated random variables with the common one-dimensional marginal density function f for which*

$$|f(x_1) - f(x_2)| \leq k|x_1 - x_2| \text{ if } x_1, x_2 \in [a, b], \qquad (7.3.31)$$

$$\int_{-\infty}^{\infty} |x|^\gamma f(x)dx < \infty \text{ for some } \gamma > 0. \qquad (7.3.32)$$

Let the sequence $\{\phi_n(x,y)\}$ be a sequence of Borel measurable functions which are of bounded variation in y for a fixed x. Then,

$$\phi_n(x, y) = \phi_{1n}(x, y) - \phi_{2n}(x, y), \qquad (7.3.33)$$

where $\phi_{in}(x, y)$, $i = 1, 2$ is monotone in y for fixed x. Suppose that there exist two positive numbers α and τ and an interval $[c, d]$ containing $[a, b]$ such that for each n the interval $[c, d]$ can be divided into disjoint left-closed intervals $I_s^{(n)}$, $s = 1, 2, \ldots$, for which

$$|I_s^{(n)}| \geq \frac{1}{n^\alpha}, \quad \bigcup_{s=1}^n I_s^{(n)} = [c, d], \qquad (7.3.34)$$

$$|\phi_n(x_1, y) - \phi_n(x_2, y)| \leq n^\tau |x_1 - x_2| \qquad (7.3.35)$$

provided that x_1 and x_2 belong to the same interval $I_s^{(n)}$. Further suppose that

$$\int_a^b \phi_n(x, y)f(y)dy \to f(x) \text{ as } n \to \infty \qquad (7.3.36)$$

uniformly in $[a + \delta, b - \delta]$ for some $\delta > 0$. Suppose that for each n,

$$\text{Var}(\phi_{in}(x, X_1)) \leq h_n, \quad i = 1, 2 \qquad (7.3.37)$$

$$h_n \leq \frac{n}{w(n)\log n}, \qquad (7.3.38)$$

where $w(n) = O(n^{\beta'})$ for some $\beta' > 0$ and $w(n) \to \infty$ as $n \to \infty$, and for a positive constant C,

$$|\phi_{in}(x, y)| \leq Ch_n, \quad i = 1, 2. \qquad (7.3.39)$$

Further suppose that there exists a $v > 0$ and a sequence of positive numbers $\epsilon_n \to 0$ such that

$$|\phi_n(x_n, y_n)| \leq \epsilon_n \qquad (7.3.40)$$

whenever

$$|x_n - y_n| > n^v, \qquad (7.3.41)$$

and

$$n > n_0(\epsilon). \qquad (7.3.42)$$

Suppose that for $i = 1, 2$, $\phi_{in}(x, y)$ is differentiable with respect to y and

$$|\phi'_{in}(x, y)| \le b_n \tag{7.3.43}$$

where $\phi'_{in}(x, y)$ denotes the partial derivative of $\phi_{in}(x, y)$ with respect to y and there exists $\beta > 0$ such that

$$\frac{b_n}{h_n} = O(n^\beta). \tag{7.3.44}$$

Finally assume that

$$\frac{1}{n} \sum_{j=1}^{n} \text{Cov}(X_1, X_j) = O(e^{-n\theta}) \tag{7.3.45}$$

for some $\theta > \frac{3}{2}$. Then

$$Pr(\sup_{a+\delta \le x \le b-\delta} |f_n^*(x) - f(x)| \ge \epsilon) \le e^{-\frac{k_1 n}{h_n}} \tag{7.3.46}$$

as $n \to \infty$, where k_1 is a positive constant depending on ϵ, δ, and f.

Remarks. The list of conditions assumed above on $\phi_n(x, y)$ is long and they are similar to those of Foldes and Revesz (1974) in the i.i.d. case to include a histogram type density estimator, a kernel type density estimator and the density estimator obtained by the method of orthogonal series etc. In addition, we have assumed here that $\phi_n(x, y)$ is a function of bounded variation in y for a fixed x to deal with the dependence of association type. Covariance structure of an associated sequence plays an important role in the study of limit theorems for associated random variables. Our condition (7.3.45) on the covariance structure is of this type. The inequality (7.3.46) gives an exponential bound for uniform convergence of the density estimator f_n.

The proof of Theorem 7.3.8 is based on the following lemma which is discussed in Chapter 6.

Lemma 7.3.9. *Let $X_1, X_2, \ldots, X_n$ be associated random variables that are bounded by a constant δ'. Then, for any $\lambda > 0$,*

$$|E[e^{\lambda \sum_{i=1}^{n} X_i}] - \prod_{i=1}^{n} E[e^{\lambda X_i}]| \le \lambda^2 k^n \sum_{1 \le i < j \le n} \text{Cov}(X_i, X_j), \tag{7.3.47}$$

where

$$k = e^{\lambda \delta'}. \tag{7.3.48}$$

In view of equation (7.3.33),

$$f_n^*(x) = \frac{1}{n} \sum_{k=1}^{n} \phi_n(x, X_k)$$

$$= \frac{1}{n} \sum_{k=1}^{n} \{\phi_{1n}(x, X_k) - \phi_{2n}(x, X_k)\}$$

$$= f_{1n}(x) - f_{2n}(x) \quad \text{(say)}. \tag{7.3.49}$$

Lemma 7.3.10. *Under the conditions stated in the hypothesis of Theorem 7.3.8, there exists $\alpha' > 0$ such that for any $x \in [a, b]$,*

$$E[e^{\lambda_n(f_{1n}(x) - E(f_{1n}(x)))}] \le e^{\frac{\lambda_n^2 h_n}{n}} + cn^2 \frac{b_n^2}{h_n^2} e^{-\alpha' n}, \tag{7.3.50}$$

provided that

$$0 < \lambda_n < \frac{n}{4Ch_n}, \tag{7.3.51}$$

where C is the constant in the condition (7.3.39) and c denotes a positive constant.

Proof. Note that

$$E[e^{\lambda_n(f_{1n}(x) - E(f_{1n}(x)))}] = E[e^{\frac{\lambda_n}{n} \sum_{j=1}^n (\phi_{1n}(x, X_j) - E\phi_{1n}(x, X_j))}]$$

$$= E[e^{\frac{\lambda_n}{n} \sum_{j=1}^n Y_{nj}(x)}], \tag{7.3.52}$$

where

$$Y_{nj}(x) = \phi_{1n}(x, X_j) - E\phi_{1n}(x, X_j). \tag{7.3.53}$$

Observe that $Y_{nj}(x)$, $j = 1, 2 \ldots, n$, are increasing functions of associated random variables and hence are associated. Then

$$|E[e^{\lambda_n(f_{1n}(x) - E(f_{1n}(x)))}]|$$

$$\le |E[e^{\frac{\lambda_n}{n} \sum_{j=1}^n Y_{nj}(x)}] - \prod_{j=1}^n E[e^{\frac{\lambda_n}{n} Y_{nj}(x)}]| + \prod_{j=1}^n E[e^{\frac{\lambda_n}{n} Y_{nj}(x)}]|. \tag{7.3.54}$$

Thus, by using the inequality $e^u \le 1 + u + u^2$ for $|u| \le \frac{1}{2}$, we get

$$\prod_{j=1}^n E[e^{\frac{\lambda_n}{n} Y_{nj}(x)}] \le \prod_{j=1}^n E[1 + \frac{\lambda_n}{n} Y_{nj}(x) + \frac{\lambda_n^2}{n^2} Y_{nj}^2(x)]$$

$$\le \prod_{j=1}^n (1 + \frac{\lambda_n^2}{n^2} h_n) \quad \text{(by (7.3.37))}$$

$$\le e^{\frac{\lambda_n^2}{n} h_n}. \tag{7.3.55}$$

Further using Lemma 7.3.9 and (7.3.39) and the fact that $0 < \lambda_n < \frac{n}{4Ch_n}$, we get that

$$|E[e^{\frac{\lambda_n}{n} \sum_{j=1}^n Y_{nj}(x)}] - \prod_{j=1}^n E[e^{\frac{\lambda_n}{n} Y_{nj}(x)}]|$$

$$\le \frac{\lambda_n^2}{n^2} e^{\frac{n}{2}} \sum_{1 \le i < j \le n} \text{Cov}(Y_{ni}(x), Y_{nj}(x)) \quad \text{(by Lemma 7.3.9)}$$

$$\leq \frac{\lambda_n^2}{n^2} e^{\frac{n}{2}} b_n^2 \sum_{1 \leq i < j \leq n} \text{Cov}(X_i, X_j) \quad \text{(by Newman's (1980) inequality)}$$

$$\leq \frac{\lambda_n^2}{n^2} e^{\frac{n}{2}} b_n^2 n \sum_{j=1}^{n} \text{Cov}(X_1, X_j) \quad \text{(by stationarity of } X_j)$$

$$\leq \frac{\lambda_n^2}{n^2} e^{\frac{n}{2}} b_n^2 n^2 e^{-n\theta} \quad \text{(using (7.3.45))}$$

$$\leq n^2 \frac{b_n^2}{h_n^2} e^{-\alpha' n}, \quad \alpha' > 0. \tag{7.3.56}$$

The result follows by combining (7.3.55) and (7.3.56). □

A similar estimate holds for $f_{2n}(x)$.

Lemma 7.3.11. *Under the conditions of Theorem 7.3.8, for any $x \in [a, b]$ and for every $\epsilon > 0$,*

$$Pr(|f_n(x) - Ef_n(x)| \geq \epsilon) \leq 2[e^{-\frac{k_1(\epsilon)n}{h_n}} + n^2 \frac{b_n^2}{h_n^2} e^{-\frac{k_1(\epsilon)n}{h_n}}], \tag{7.3.57}$$

where the constant $k_1(\epsilon)$ does not depend on n and x.

Proof. Using (7.3.49), we get that

$$Pr(|f_n(x) - Ef_n(x)| \geq \epsilon) \leq Pr(|f_{1n}(x) - Ef_{1n}(x)| \geq \frac{\epsilon}{2}) \tag{7.3.58}$$

$$+ Pr(|f_{2n}(x) - Ef_{2n}(x)| \geq \frac{\epsilon}{2})$$

Note that $0 < \lambda_n < \frac{n}{4Ch_n}$ and suppose that $\frac{\lambda_n h_n}{n} \to 0$ as $n \to \infty$. Then, by the Markov inequality and Lemma 7.3.9, we get that, for $i = 1, 2$, there exists a positive constant $k_2(\epsilon)$ such that

$$Pr(f_{in}(x) - Ef_{in}(x) \geq \frac{\epsilon}{2}) \leq E[e^{\lambda_n(f_{in}(x) - E(f_{in}(x)))}]/e^{\lambda_n \epsilon/2}$$

$$= \frac{e^{\frac{\lambda_n^2 h_n}{n}} + cn^2 \frac{b_n^2}{h_n^2} e^{-\alpha' n}}{e^{\lambda_n \epsilon/2}}$$

$$\leq e^{-\frac{k_2(\epsilon)n}{h_n}} + cn^2 \frac{b_n^2}{h_n^2} e^{-\frac{k_2(\epsilon)n}{h_n}}. \tag{7.3.59}$$

One can choose $k_2(\epsilon) = \frac{\epsilon}{16C}$. The result now follows from the fact that if W_j's are associated, then so are $-W_j$'s. □

Proof of Theorem 7.3.8. The proof will be along the same lines as that given in Foldes and Revesz (1974) and Foldes (1974) in the i.i.d. case.

Let, for each positive integer n,

$$-n^T = z_0^{(n)} < z_1^{(n)} < \ldots < z_{l(n)}^{(n)} = n^T$$

(the number T will be determined later on) be a partitioning of the interval $[-n^T, n^T]$ having the following properties:

(a) $\dfrac{c_1}{n^{\alpha+2\tau}} \leq z_i^{(n)} - z_{i-1}^{(n)} \leq \dfrac{c_2}{n^{\alpha+2\tau}}, \, 0 < c_1 < c_2 < \infty, \, i = 1, \ldots, l(n),$

(b) those end points of the interval $I_s^{(n)}$ which belong to $[-n^T, n^T]$ are elements of the sequence $z_0^{(n)}, z_1^{(n)}, \ldots, z_{l(n)}^{(n)}$.

By (a), (7.3.29) and (7.3.35), we get that

$$|f_n(x) - f_n(y)| \leq c\frac{1}{n^{\alpha+\tau}} \text{ if } x, y \in [z_{i-1}^{(n)}, z_i^{(n)}], \qquad (7.3.60)$$

and

$$|Ef_n(x) - Ef_n(y)| \leq c\frac{1}{n^{\alpha+\tau}} \text{ if } x, y \in [z_{i-1}^{(n)}, z_i^{(n)}]. \qquad (7.3.61)$$

Note that

$$|f_n(x) - Ef_n(x)| \leq |f_n(x) - f_n(z_{i-1}^{(n)})| + |f_n(z_{i-1}^{(n)}) - Ef_n(z_{i-1}^{(n)})|$$
$$+ |Ef_n(z_{i-1}^{(n)}) - Ef_n(x)|. \qquad (7.3.62)$$

Therefore, using (7.3.60) and (7.3.61) we have

$$\sup_{z_{i-1}^{(n)} \leq x \leq z_i^{(n)}} |f_n(x) - Ef_n(x)| \leq \frac{c}{n^{\alpha+\tau}} + \sup_{z_{i-1}^{(n)} \leq x \leq z_i^{(n)}} |f_n(z_{i-1}^{(n)}) - Ef_n(z_{i-1}^{(n)})| + \frac{c}{n^{\alpha+\tau}}.$$

Hence,

$$\Pr(\sup_{z_{i-1}^{(n)} \leq x \leq z_i^{(n)}} |f_n(x) - Ef_n(x)| \geq \epsilon)$$

$$\leq \Pr(2\frac{c}{n^{\alpha+\tau}} \geq \frac{\epsilon}{2}) + \Pr(|f_n(z_{i-1}^{(n)}) - Ef_n(z_{i-1}^{(n)})| \geq \frac{\epsilon}{2}). \qquad (7.3.63)$$

But, for large n, $\Pr(2\frac{c}{n^{\alpha+\tau}} \geq \frac{\epsilon}{2})$ is zero. Therefore, using (7.3.62), we have

$$\Pr(\sup_{-n^T \leq x \leq n^T} |f_n(x) - Ef_n(x)| \geq \epsilon)$$

$$\leq \sum_{i=1}^{l(n)} \Pr(\sup_{z_{i-1}^{(n)} \leq x \leq z_i^{(n)}} |f_n(x) - Ef_n(x)| \geq \epsilon)\}$$

$$\leq l(n) \max_i \{\Pr(\sup_{z_{i-1}^{(n)} \leq x \leq z_i^{(n)}} |f_n(x) - Ef_n(x)| \geq \epsilon)\}$$

$$\leq l(n) \max_{i} \{\Pr(|f_n(z_{i-1}^{(n)}) - Ef_n(z_{i-1}^{(n)})| \geq \tfrac{\epsilon}{2})\}$$

$$\leq 2l(n)e^{-\frac{k_1(\epsilon)n}{h_n}} + 2l(n)n^2 \frac{b_n^2}{h_n^2} e^{-\frac{k_1(\epsilon)n}{h_n}} \quad \text{(using (7.3.57))}$$

$$\leq e^{-\frac{k_4(\epsilon)n}{h_n}} \quad \text{(using (7.3.59))} \tag{7.3.64}$$

for large n. Note that $l(n)$ is the number of partitioning intervals of an interval of length $2n^T$. Therefore $l(n) \simeq 2n^{\alpha+2\tau+T}$. Furthermore, $Ef_n(x) \to f(x)$ as $n \to \infty$ by (7.3.36). Therefore

$$\Pr(\sup_{\substack{|x| \geq n^T \\ x \in [a+\delta, b-\delta]}} |f_n(x) - Ef_n(x)| \geq \epsilon)$$

$$\leq \Pr(\sup_{\substack{|x| \geq n^T \\ x \in [a+\delta, b-\delta]}} |f_n(x)| \geq \tfrac{\epsilon}{2}) + \Pr(\sup_{\substack{|x| \geq n^T \\ x \in [a+\delta, b-\delta]}} |Ef_n(x)| \geq \tfrac{\epsilon}{2})$$

$$= \Pr(\sup_{\substack{|x| \geq n^T \\ x \in [a+\delta, b-\delta]}} |f_n(x)| \geq \tfrac{\epsilon}{2}) \tag{7.3.65}$$

for large n, since $\Pr(\sup |f(x)| \geq \tfrac{\epsilon}{2})$ is zero. Now

$$\Pr(\sup_{|x| \geq n^T} |f_n(x)| \geq \tfrac{\epsilon}{2}) \tag{7.3.66}$$

$$\leq \Pr(\sup_{|x| \geq n^T} \frac{1}{n} \sum_{k:|X_k-x|\leq\delta} |\phi_n(x, X_k)| \geq \tfrac{\epsilon}{4})$$

$$+ \Pr(\sup_{|x| \geq n^T} \frac{1}{n} \sum_{\substack{k:|X_k-x|>\delta \\ X_k \in S(x, \frac{n^T}{2})}} |\phi_n(x, X_k)| \geq \tfrac{\epsilon}{8})$$

$$+ \Pr(\sup_{|x| \geq n^T} \frac{1}{n} \sum_{\substack{k:|X_k-x|>\delta \\ X_k \notin S(x, \frac{n^T}{2})}} |\phi_n(x, X_k)| \geq \tfrac{\epsilon}{8})$$

$$\leq 2\Pr(\frac{1}{n} Ch_n \sum_{k:|X_k| \geq \frac{n^T}{2}} 1 \geq \tfrac{\epsilon}{8}) \tag{7.3.67}$$

$$+ \Pr(\sup_{|x| \geq n^T} \frac{1}{n} \sum_{\substack{k:|X_k-x|>\delta \\ X_k \notin S(x, \frac{n^T}{2})}} |\phi_n(x, X_k)| \geq \tfrac{\epsilon}{8})$$

where $S(x, \frac{n^T}{2})$ denotes the interval $[x - \frac{n^T}{2}, x + \frac{n^T}{2}]$. This inequality is a consequence of (7.3.39) and the fact that, for large n,

$$|x| \geq n^T, \ |X - x| \leq \delta \Rightarrow |X| \geq \frac{n^T}{2}$$

and

$$|x| \geq n^T, \ |X - x| > \delta, \ X \in S(x, \frac{n^T}{2}) \Rightarrow |X| \geq \frac{n^T}{2}.$$

Denote by $J_n(u)$ the following indicator function

$$J_n(u) = \begin{cases} 1 & \text{if } |u| \geq \frac{n^T}{2} \\ 0 & \text{otherwise.} \end{cases} \tag{7.3.68}$$

Note that J_n can be expressed as the sum of two monotone functions I_n and I'_n , where

$$I_n(u) = \begin{cases} 1 & \text{if } u \geq \frac{n^T}{2}, \\ 0 & \text{otherwise,} \end{cases}$$

and

$$I'_n(u) = \begin{cases} 1 & \text{if } u \leq -\frac{n^T}{2}, \\ 0 & \text{otherwise.} \end{cases}$$

Then we have to estimate the probability

$$\Pr(\sum_{k=1}^{n} J_n(X_k) \geq \frac{cn\epsilon}{h_n}) \leq \Pr(\sum_{k=1}^{n} I_n(X_k) \geq \frac{cn\epsilon}{2h_n}) + \Pr(\sum_{k=1}^{n} I'_n(X_k) \geq \frac{cn\epsilon}{2h_n}) \tag{7.3.69}$$

for some positive constant c. Since X_k's are associated , $Y_k = I_n(X_k)$, $k = 1, \ldots, n$ are associated and so are $Z_k = I'_n(X_k)$, $k = 1, \ldots, n$. Therefore, we will estimate (7.3.67) using Lemma 7.3.10 for the associated random variables $Y_1, \ldots, Y_n$ and $Z_1, \ldots, Z_n$. Now

$$\Pr(\sum_{k=1}^{n} I_n(X_k) \geq \frac{cn\epsilon}{2h_n}) \leq \Pr(\sum_{k=1}^{n} (I_n(X_k) - EI_n(X_k)) \geq \frac{cn\epsilon}{4h_n})$$

$$+ \Pr(\sum_{k=1}^{n} EI_n(X_k) \geq \frac{cn\epsilon}{4h_n}). \tag{7.3.70}$$

Note that

$$E(I_n(X_k)) = \Pr(X_k > \frac{n^T}{2})$$

$$\leq \Pr(|X_k| > \frac{n^T}{2})$$

$$= \Pr(|X_k|^\gamma > \frac{n^{T\gamma}}{2^\gamma})$$

$$\leq \frac{E(|X_k|^\gamma)2^\gamma}{n^{T\gamma}}$$

$$\leq \frac{c(\gamma)}{n^{T\gamma}} \quad \text{(using (7.3.32))} \tag{7.3.71}$$

where $c(\gamma)$ is a constant depending on γ and f. Choose T to be so large that $T\gamma - 1 > 0$. The sequence $\frac{cn\epsilon}{4h_n}$ tends to infinity by (7.3.38). Therefore

$$\Pr(\sum_{k=1}^{n} EI_n(X_k) \geq \frac{cn\epsilon}{4h_n}) = 0 \tag{7.3.72}$$

for large n. Furthermore,

$$\text{Var}(I_n(X_k)) \leq \frac{c(\gamma)}{n^{T\gamma}}. \tag{7.3.73}$$

For $0 < \lambda_n^* \leq \frac{1}{4}$,

$$\Pr(\sum_{k=1}^{n} (I_n(X_k) - EI_n(X_k)) \geq \frac{cn\epsilon}{4h_n}) \leq \frac{E(e^{\lambda_n^* \sum_{k=1}^{n} (I_n(X_k) - EI_n(X_k))})}{e^{\frac{\lambda_n^* cn\epsilon}{4h_n}}}. \tag{7.3.74}$$

Now

$$E(e^{\lambda_n^* \sum_{k=1}^{n} (I_n(X_k) - EI_n(X_k))})$$

$$= E(e^{\lambda_n^* \sum_{k=1}^{n} (I_n(X_k) - EI_n(X_k))}) - \prod_{k=1}^{n} E(e^{\lambda_n^* (I_n(X_k) - EI_n(X_k))})$$

$$+ \prod_{k=1}^{n} E(e^{\lambda_n^* (I_n(X_k) - EI_n(X_k))}). \tag{7.3.75}$$

Therefore, for $0 < \lambda_n^* \leq \frac{1}{4}$, and using the inequality $e^u \leq 1 + u + u^2$ for $|u| \leq \frac{1}{2}$, we get that

$$\prod_{k=1}^{n} E(e^{\lambda_n^* (I_n(X_k) - EI_n(X_k))})$$

$$\leq \prod_{k=1}^{n} E[1 + \lambda_n^* (I_n(X_k) - EI_n(X_k)) + (\lambda_n^*)^2 (I_n(X_k) - EI_n(X_k))^2]$$

$$= \prod_{k=1}^{n} [1 + (\lambda_n^*)^2 \text{Var}(I_n(X_k)]$$

$$\leq [1 + \frac{(\lambda_n^*)^2 c(\gamma)}{n^{T\gamma}}]^n$$

$$\leq e^{\frac{(\lambda_n^*)^2 c(\gamma)}{n^{T\gamma-1}}} \tag{7.3.76}$$

since $(1 + x_n)^n = (1 + \frac{nx_n}{n})^n \approx e^{nx_n}$. Applying Lemma 7.3.9 for the associated random variables $Y_1, \ldots, Y_n$, we get

$$|E(e^{\lambda_n^* \sum_{k=1}^{n} (I_n(X_k) - EI_n(X_k))}) - \prod_{k=1}^{n} E(e^{\lambda_n^* (I_n(X_k) - EI_n(X_k))})|$$

$$\leq (\lambda_n^*)^2 e^{2n\lambda_n^*} \sum_{1 \leq i < j \leq n} \text{Cov}(I_n(X_i), I_n(X_j)). \tag{7.3.77}$$

Then, for $0 < \lambda_n^* \leq \frac{1}{4}$, and any $t > 0$, we get that

$$|E(e^{\lambda_n^* \sum_{k=1}^n (I_n(X_k) - EI_n(X_k))}) - \prod_{k=1}^n E(e^{\lambda_n^* (I_n(X_k) - EI_n(X_k))})|$$

$$\leq c(\lambda_n^*)^2 e^{2n\lambda_n^*} \sum_{1 \leq i < j \leq n} (t^2 \operatorname{Cov}(X_i, X_j) + \frac{1}{t})$$

$$\leq c(\lambda_n^*)^2 e^{2n\lambda_n^*} (t^2 n \sum_{j=1}^n \operatorname{Cov}(X_1, X_j) + \frac{n^2}{t}) \quad \text{(by using the stationarity of } \{X_j\})$$

$$\leq c(\lambda_n^*)^2 e^{2n\lambda_n^*} n^2 (t^2 e^{-n\theta} + \frac{1}{t}) \quad \text{(by using (7.3.45))}$$

$$\leq c(\lambda_n^*)^2 n^2 e^{2n\lambda_n^*} e^{-\frac{n\theta}{3}} \quad \text{(by choosing } t = e^{\frac{n\theta}{3}}). \tag{7.3.78}$$

Using (7.3.74) and (7.3.76) in (7.3.72), we get that for $0 < \lambda_n^* \leq \frac{1}{4}$,

$$\Pr(\sum_{k=1}^n (I_n(X_k) - EI_n(X_k)) \geq \frac{cn\epsilon}{4h_n})$$

$$\leq e^{\frac{(\lambda_n^*)^2 c(\gamma)}{n^T \gamma - 1} - \frac{cn\lambda_n^* \epsilon}{2h_n}} + c(\lambda_n^*)^2 n^2 e^{2n\lambda_n^*} e^{-\frac{n\theta}{3}} e^{-\frac{cn\lambda_n^* \epsilon}{2h_n}}$$

$$\leq e^{-\frac{k_3(\epsilon)n}{h_n}} + e^{-\frac{k_4(\epsilon)n}{h_n}} \quad \text{(by using (7.3.38))}$$

$$\leq e^{-\frac{k(\epsilon)n}{h_n}}. \tag{7.3.79}$$

Substituting (7.3.72) and (7.3.77) in (7.3.68) we get an estimate for

$$\Pr(\sum_{k=1}^n I_n(X_k) \geq \frac{cn\epsilon}{2h_n}).$$

Similarly, we can get an estimate for

$$\Pr(\sum_{k=1}^n I_n'(X_k) \geq \frac{cn\epsilon}{2h_n}).$$

Combining the two estimates, we can get an estimate for the expression on the left-hand side of (7.3.67). Finally, for $T > v$ and by (7.3.40) and (7.3.41), we get that

$$\Pr(\sup_{|x| \geq n^T} \frac{1}{n} \sum_{\substack{k: |X_k - x| > \delta \\ X_k \notin S(x, \frac{n^T}{2})}} |\phi_n(x, X_k)| \geq \frac{\epsilon}{8}) \leq \Pr(\epsilon_n > \frac{\epsilon}{8}) = 0 \tag{7.3.80}$$

for large n. Using the inequalities (7.3.67) and (7.3.78) in (7.3.65), we get an estimate for (7.3.64). Choose $T > \max(v, \frac{1}{\gamma})$. Then, for large n, and from (7.3.36),

(7.3.63) and the estimate of (7.3.64) given by (7.3.78) we have the following inequality proving the theorem:

$$
\begin{aligned}
\Pr(\sup_{a+\delta \leq x \leq b-\delta} |f_n(x) - f(x)| \geq \epsilon) &\leq \Pr(\sup_{a+\delta \leq x \leq b-\delta} |f_n(x) - Ef_n(x)| \geq \frac{\epsilon}{2}) \\
&\quad + \Pr(\sup_{a+\delta \leq x \leq b-\delta} |Ef_n(x) - f(x)| \geq \frac{\epsilon}{2}) \\
&= \Pr(\sup_{a+\delta \leq x \leq b-\delta} |f_n(x) - Ef_n(x)| \geq \frac{\epsilon}{2}) \\
&\leq e^{-\frac{k_1 n}{h_n}}.
\end{aligned}
\tag{7.3.81}
$$

$\square$

Remarks. Various examples of the estimator $f_n^*(x)$ have been discussed by Foldes and Revesz (1974) in the i.i.d. case. Similar examples can be given for the associated case. For instance the standard normal density is a kernel which is a function of bounded variation and it can be checked that it satisfies all the conditions of Theorem 7.3.8 and we obtain the exponential rate for uniform convergence of the kernel type density estimator.

7.4 Nonparametric Failure Rate Estimation

The failure rate $r(x)$ is defined as

$$
r(x) = \frac{f(x)}{\bar{F}(x)}, \quad \bar{F} > 0.
\tag{7.4.1}
$$

An obvious estimate of $r(x)$ is $r_n(x)$ given by

$$
r_n(x) = \frac{f_n(x)}{\bar{F}_n(x)}
\tag{7.4.2}
$$

where $f_n(x)$ is the kernel type estimator defined in (7.3.1) and $\bar{F}_n(x)$ is as defined in (7.2.1). It is easy to see that

$$
r_n(x) - r(x) = \frac{\bar{F}(x)[f_n(x) - f(x)] - f(x)[\bar{F}_n(x) - \bar{F}(x)]}{\bar{F}(x)\bar{F}_n(x)}.
\tag{7.4.3}
$$

The following theorems give pointwise as well as uniform consistency for the estimator $r_n(x)$. For proofs, see Bagai and Prakasa Rao (1995).

Theorem 7.4.1. *Let the sequence $\{X_n, n \geq 1\}$ be a stationary sequence of associated random variables satisfying the conditions (A) and (B) in Section 7.3 and suppose that for some $r > 1$,*

$$
\sum_{j=n+1}^{\infty} \{\mathrm{Cov}(X_1, X_j)\}^{1/3} = O(n^{-(r-1)}).
$$

Then, for all $x \in S$ which are continuity points of f,

$$r_n(x) \to r(x) \text{ a.s. as } n \to \infty.$$

Theorem 7.4.2. *Let the sequence $\{X_n, n \geq 1\}$ be a stationary sequence of associated random variables satisfying the conditions of Theorem 7.2.1 and Theorem 7.3.2. Then*

$$\sup\{|r_n(x) - r(x)| : x \in J\} \to 0 \text{ a.s. as } n \to \infty.$$

Roussas (1991) has discussed strong uniform consistency of kernel type estimates of $f(.)$ and $f^{(k)}(.)$, the k-th order derivatives of f, and the hazard rate $r(.)$ for strictly stationary associated sequences and has obtained the rates of convergence. Roussas (1993) studied curve estimation for random fields of associated processes. Roussas (1995) investigated conditions for asymptotic normality of smooth estimate of the distribution function for associated random fields.

7.5 Nonparametric Mean Residual Life Function Estimation

The mean residual life function $M_F(x)$ is defined as

$$M_F(x) = E[X - x | X > x] = \frac{1}{\bar{F}(x)} \int_x^\infty \bar{F}(t) dt. \tag{7.5.1}$$

An obvious estimate of $M_F(x)$ is $M_n(x)$ given by

$$M_n(x) = \frac{1}{\bar{F}_n(x)} \int_x^\infty \bar{F}_n(t) dt, \tag{7.5.2}$$

where $\bar{F}_n(x)$ is as defined in (7.2.1). Let $T_F = \inf\{x : F(x) = 1\}$.

Theorem 7.5.1 (Shao and Yu (1996)). *Let the sequence $\{X_n, n \geq 1\}$ be a stationary sequence of associated random variables with distribution function F for X_1. If $T < T_F$ and*

$$\sum_{n=1}^\infty \frac{1}{n^2} \text{Cov}(X_n, \sum_{i=1}^n X_i) < \infty,$$

then

$$\sup_{0 \leq x \leq T} |M_n(x) - M_F(x)| \to 0 \text{ a.s. as } n \to \infty. \tag{7.5.3}$$

This result follows from the fact that the mean residual life function $M_F(x)$ can be written as a weighted function of the empirical distribution function (Shao and Yu (1996)). We omit the proof.

7.6 Remarks

Matula (1997a) derived the Glivenko-Cantelli lemma for a class of discrete associated random variables. Large deviation results for the empirical mean of associated random variables were studied in Henriques and Oliveira (2008). Kernel-type estimation of the bivariate distribution function for associated random variables was discussed in Azevedo and Oliveira (2000).

Asymptotic normality of the empirical distribution function for negatively associated random variables and its applications were discussed in Li and Yang (2006). A general method of density estimation for negatively associated random variables was investigated in Zarei et al. (2008). Asymptotic properties of nonparametric estimators for regression models based on negatively associated sequences were obtained in Liang and Zing (2007). Gu et al.(2007) studied convergence rates of fixed design regression estimators for negatively associated random variables.

7.4.3 Remark 8

Chapter 8

Nonparametric Tests for Associated Sequences

8.1 Introduction

We have studied the problem of nonparametric functional estimation, in particular density estimation, for stationary associated sequences in the last chapter. We will now discuss some nonparametric tests for comparing the one-dimensional marginal distributions for two strictly stationary associated sequences. We term such problems of comparing two populations as two-sample problems. A limit theorem for obtaining the asymptotic distribution of a test statistic, which is a U-statistic in a two-sample problem, is discussed in Chapter 6.

8.2 More on Covariance Inequalities

We have discussed generalized Hoeffding identity and some covariance inequalities in Chapter 1 and Chapter 6. We now discuss another result in this area due to Cuadras (2002). Of particular interest to us is the case when the random variables X and Y are associated. We discuss an application of this result to obtain tests for location for associated sequences.

Let (X, Y) be a bivariate random vector and suppose that $E(X^2) < \infty$ and $E(Y^2) < \infty$. Further let

$$H(x,y) = P(X \leq x,\ Y \leq y) - P(X \leq x)P(Y \leq y). \qquad (8.2.1)$$

Recall the Hoeffding identity (cf. Hoeffding (1940))

$$\mathrm{Cov}(X,Y) = \int_{R^2} H(x,y)\ dxdy, \qquad (8.2.2)$$

B.L.S. Prakasa Rao, *Associated Sequences, Demimartingales and Nonparametric Inference*, Probability and its Applications, DOI 10.1007/978-3-0348-0240-6_8, © Springer Basel AG 2012

discussed in Chapter 1. Multivariate versions of this identity were studied by
Block and Fang (1988) using the concept of a cumulant of a random vector $\mathbf{X} = (X_1, \ldots, X_k)$. Yu (1993) obtained a generalization of the covariance identity (8.2.2)
to absolutely continuous functions of the components of the random vector $\mathbf{X}$
extending the earlier work of Newman (1984). Quesada-Molina (1992) generalized
the Hoeffding identity to quasi-monotone functions $K(.,.)$ in the sense that

$$K(x, y) - K(x', y) - K(x, y') + K(x', y') \geq 0 \qquad (8.2.3)$$

whenever $x \leq x'$ and $y \leq y'$. It was proved that

$$E[K(X, Y) - K(X^*, Y^*)] = \int_{R^2} H(x, y) K(dx, dy) \qquad (8.2.4)$$

where X^* and Y^* are independent random variables independent of the random
vector (X, Y) but with X^* and Y^* having the same marginal distributions as
those of X and Y respectively. The results in Yu (1993) and Quesada-Molina
(1992) were generalized to the multidimensional case in Prakasa Rao (1998).
Cuadras (2002) proved that if $\alpha(x)$ and $\beta(y)$ are functions of bounded variation
on the support of the probability distribution of the random vector (X, Y) with
$E|\alpha(X)\beta(Y)|, E|\alpha(X)|$ and $E|\beta(Y)|$ finite, then

$$\text{Cov}(\alpha(X), \beta(Y)) = \int_{R^2} H(x, y)\, \alpha(dx)\beta(dy). \qquad (8.2.5)$$

It is clear that this result also follows as a special case of (8.2.4). For a multidimensional version of this identity, see Prakasa Rao (1998).

Suppose that $\alpha(x)$ and $\beta(y)$ are functions of bounded variation which are
mixtures of absolutely continuous component and discrete component only. Let
$\alpha^{(c)}(x)$ and $\alpha^{(d)}(x)$ denote the absolutely continuous component and the discrete
component of $\alpha(x)$ respectively. Let x_i, $i \geq 1$ be the jumps of $\alpha(x)$ with jump
sizes $\alpha(x_i + 0) - \alpha(x_i - 0) = p_i \neq 0$. Similarly let y_j, $j \geq 1$ be the jumps of $\beta(y)$
with jump sizes $\beta(y_j + 0) - \beta(y_j - 0) = q_j \neq 0$. Furthermore let $\alpha^{(c)'}(x)$ denote
the derivative of $\alpha^{(c)}(x)$ whenever it exists. Observe that the derivative of $\alpha^{(c)}(x)$
exists almost everywhere. Suppose that

$$\sup_x |\alpha^{(c)'}(x)| < \infty, \quad \sup_i |p_i| < \infty \qquad (8.2.6)$$

and

$$\sup_y |\beta^{(c)'}(y)| < \infty, \quad \sup_j |q_j| < \infty. \qquad (8.2.7)$$

Then

$$\text{Cov}(\alpha(X), \beta(Y)) = \int_{R^2} H(x, y)\alpha(dx)\beta(dy)$$

$$= \int_{R^2} H(x, y)\, \alpha^{(c)}(dx)\beta^{(c)}(dy)$$

$$+ \int_{R^2} H(x,y)\, \alpha^{(c)}(dx)\beta^{(d)}(dy)$$

$$+ \int_{R^2} H(x,y)\, \alpha^{(d)}(dx)\beta^{(c)}(dy)$$

$$+ \int_{R^2} H(x,y)\, \alpha^{(d)}(dx)\beta^{(d)}(dy) \qquad (8.2.8)$$

and hence

$$|\operatorname{Cov}(\alpha(X),\beta(Y))| \le \sup_x |\alpha^{(c)'}(x)| \sup_y |\beta^{(c)'}(y)| \int_{R^2} |H(x,y)|\, dx dy$$

$$+ \sup_x |\alpha^{(c)'}(x)| \sup_j |q_j| \sum_{j=1}^{\infty} \int_{-\infty}^{\infty} |H(x,y_j)|\, dx$$

$$+ \sup_y |\beta^{(c)'}(y)| \sup_i |p_i| \sum_{i=1}^{\infty} \int_{-\infty}^{\infty} |H(x_i,y)|\, dy$$

$$+ \sup_i |p_i| \sup_j |q_j| \sum_{i=1}^{\infty}\sum_{j=1}^{\infty} |H(x_i,y_j)|. \qquad (8.2.9)$$

Suppose the functions $\alpha(x)$ and $\beta(y)$ are purely discrete. Let x_i, $i \ge 1$ be the jumps of $\alpha(x)$ with jump sizes $\alpha(x_i + 0) - \alpha(x_i - 0) = p_i \ne 0$ and $\alpha(x)$ be a constant between different jumps. Similarly let y_j, $j \ge 1$ be the jumps of $\beta(y)$ with jump sizes $\beta(y_j + 0) - \beta(y_j - 0) = q_j \ne 0$ and $\beta(y)$ be a constant between different jumps. Then

$$\operatorname{Cov}(\alpha(X),\beta(Y)) = \int_{R^2} H(x,y)\, \alpha^{(d)}(dx)\beta^{(d)}(dy)$$

$$= \sum_{i=1}^{\infty}\sum_{j=1}^{\infty} H(x_i,y_j)p_i q_j. \qquad (8.2.10)$$

For instance, suppose that $\alpha(x) = \operatorname{sgn}(x)$ and $\beta(y) = \operatorname{sgn}(y)$ where

$$\operatorname{sgn}(x) = \begin{cases} 1 & \text{if } x > 0, \\ -1 & \text{if } x < 0 \ . \\ 0 & \text{if } x = 0 \end{cases}$$

Then, for any x_0 and y_0,

$$\operatorname{Cov}(\operatorname{sgn}(X - x_0), \operatorname{sgn}(Y - y_0)) = 4\, H(x_0, y_0) \qquad (8.2.11)$$

since the jump at x_0 is of size 2 for the function $\alpha(x)$ and for the function $\beta(y)$ (cf. Cuadras (2002)).

Suppose we choose $\alpha(x) = F(x)$ and $\beta(y) = G(y)$ where $F(.)$ and $G(.)$ are continuous marginal distribution functions of the components X and Y respectively of a bivariate random vector (X, Y). Following the result given in (8.2.5), we get that

$$\text{Cov}(F(X), G(Y)) = \int_{R^2} H(x, y) \ F(dx)G(dy). \tag{8.2.12}$$

It is easy to see that the Spearman's rank correlation coefficient ρ between X and Y is given by

$$\rho = \text{Corr}(F(X), G(Y)) = 12 \ \text{Cov}(F(X), G(Y)) = 12 \int_{R^2} H(x, y) \ F(dx)G(dy). \tag{8.2.13}$$

Note that the random variables $F(X)$ and $G(Y)$ have the standard uniform distribution.

Remarks. Suppose that X and Y are associated. Then it follows that the function $H(x, y) \geq 0$ for all x and y. Newman (1980) showed that if X and Y have finite variances, then, for any two differentiable functions h and g,

$$|\text{Cov}(h(X), g(Y))| \leq \sup_x |h'(x)| \ \sup_y |g'(y)| \ \text{Cov}(X, Y) \tag{8.2.14}$$

where h' and g' denote the derivatives of h and g, respectively. Inequality (8.2.9) extends this result to include functions of bounded variations which are mixtures of absolutely continuous component and discrete component only. We discuss an application of the inequality (8.2.14) to associated sequences in the next section.

8.3   Tests for Location

Let the sequence $\{X_n, \ n \geq 1\}$ be a stationary sequence of associated random variables. Let $F(x, \theta) = F(x - \theta)$, $F \in \Omega_s$, where $\Omega_s = \{F : F(x) = 1 - F(-x)\}$, be the distribution function of X_1 and suppose that the distribution function F is absolutely continuous with a bounded density function f.

Suppose a finite sequence of stationary associated random variables $\{X_i, 1 \leq i \leq n\}$ is observed. We wish to test the null hypothesis

$$H_0 : \theta = 0 \tag{8.3.1}$$

against the alternative hypothesis

$$H_1 : \theta > 0. \tag{8.3.2}$$

The most commonly used tests for this problem are the sign test and the Wilcoxon-signed rank test when the observations are independent and identically distributed. We now study the properties of these tests when the observations are on a stationary associated sequence of random variables.

Let C denote a generic positive constant in the sequel. Assume that

$$\sup_x f(x) < \infty. \tag{8.3.3}$$

Further assume that

$$\sum_{j=2}^{\infty} \mathrm{Cov}^{\frac{1}{3}}(X_1, X_j) < \infty. \tag{8.3.4}$$

This would imply that $\mathrm{Cov}(X_1, X_n) \to 0$ as $n \to \infty$. In particular, it follows that $\sup_n |\mathrm{Cov}(X_1, X_n)| < \infty$. Observing that $\mathrm{Cov}(X_1, X_n) > 0$; by the associative property of $X_1, \ldots, X_n$, we obtain that

$$0 \leq \mathrm{Cov}(X_1, X_j)$$
$$= [\mathrm{Cov}(X_1, X_j)]^{\frac{2}{3}} [\mathrm{Cov}(X_1, X_j)]^{\frac{1}{3}}$$
$$\leq [\sup_n \mathrm{Cov}(X_1, X_n)]^{\frac{2}{3}} [\mathrm{Cov}(X_1, X_j)]^{\frac{1}{3}}.$$

Hence

$$\sum_{j=2}^{\infty} \mathrm{Cov}(X_1, X_j) \leq [\sup_n \mathrm{Cov}(X_1, X_n)]^{\frac{2}{3}} \sum_{j=2}^{\infty} [\mathrm{Cov}(X_1, X_j)]^{\frac{1}{3}} < \infty. \tag{8.3.5}$$

Sign Test

For testing the hypothesis $H_0 : \theta = 0$ against $H_1 : \theta > 0$, the sign test is based on the statistic

$$U_n^{(1)} = \frac{1}{n} \sum_{i=1}^{n} \phi(X_i), \tag{8.3.6}$$

where

$$\phi(x) = I(x > 0), \tag{8.3.7}$$

and $I(A)$ denotes the indicator function of the set A. Observe that

$$E(\phi(X_1)) = 1 - F(0) = p \quad (\text{say}),$$
$$\mathrm{Var}(\phi(X_1)) = p - p^2,$$

and

$$\mathrm{Cov}(\phi(X_1), \phi(X_j)) = P[X_1 > 0, \ X_j > 0] - p^2. \tag{8.3.8}$$

Since the density function f of the random variable X_1 is bounded, it follows from Theorem 6.2.14 (cf. Bagai and Prakasa Rao (1991)), that

$$\sup_{x,y} |P[X_1 > x, \ Y_1 > y] - P[X_1 > x]P[Y_1 > y]| \leq C \ \mathrm{Cov}^{1/3}(X, Y). \tag{8.3.9}$$

From (8.2.10), (8.3.3) and (8.3.7) it follows that

$$\sigma^2 = \operatorname{Var} \phi(X_1) + 2 \sum_{j=2}^{\infty} \operatorname{Cov}(\phi(X_1), \phi(X_j)) < \infty. \tag{8.3.10}$$

Since $\phi(x)$ is an increasing function of x, we note that $\phi(X_1), \phi(X_2), \ldots,$ $\phi(X_n)$ are stationary associated random variables. The following theorem is an immediate consequence of the central limit theorem for associated random variables (Newman (1980)).

Theorem 8.3.1. *Let the sequence $\{X_n, n \geq 1\}$ be a sequence of stationary associated random variables with bounded density function. Then*

$$\frac{n^{-1/2} \sum_{j=1}^{n} [\phi(X_j) - E(\phi(X_j))]}{\sigma} \xrightarrow{\mathcal{L}} N(0,1) \;\; as \;\; n \to \infty. \tag{8.3.11}$$

The test procedure consists in rejecting the null hypothesis H_0 for large values of the statistic $U_n^{(1)}$ which is the proportion of positive observations.

Wilcoxon-Signed Rank Test

Let $R_1, R_2, \ldots, R_n$ be the ranks of $X_1, X_2, \ldots, X_n$. The Wilcoxon-signed rank statistic is defined by

$$T = \sum_{j=1}^{n} R_j \, \phi(X_j). \tag{8.3.12}$$

We can write T as a linear combination of two U-statistics (Hettmansperger (1984))

$$T = n U_n^{(1)} + \binom{n}{2} U_n^{(2)}, \tag{8.3.13}$$

where $n U_n^{(1)} = \sum_{i=1}^{n} \phi(X_i)$,

$$\binom{n}{2} U_n^{(2)} = \sum_{1 \leq i < j \leq n} \psi(X_1, X_j), \tag{8.3.14}$$

and

$$\psi(x, y) = I(x + y > 0). \tag{8.3.15}$$

Since the random variables $\{X_n, n \geq 1\}$ form a stationary sequence, it follows that

$$E(U_n^{(2)}) = \frac{1}{\binom{n}{2}} \sum_{1 \leq i < j \leq n} p_{ij}$$

$$= \frac{1}{\binom{n}{2}} \sum_{j=2}^{n} (n - j + 1) p_{1j} \tag{8.3.16}$$

where $p_{ij} = P[X_i + X_j > 0]$. Let

$$\theta = \int_{-\infty}^{\infty} \int_{-\infty}^{\infty} \psi(x,y) \, dF(x) dF(y),$$

$$= 1 - \int_{-\infty}^{\infty} F(-x) \, dF(x), \qquad (8.3.17)$$

$$\psi_1(x_1) = E(\psi(x_1, X_2))$$

$$= \int_{-\infty}^{\infty} \psi(x_1, x_2) \, dF(x_2)$$

$$= 1 - F(-x_1). \qquad (8.3.18)$$

$$h^{(1)}(x_1) = \psi_1(x_1) - \theta \,, \qquad (8.3.19)$$

and

$$h^{(2)}(x_1, x_2) = \psi(x_1, x_2) - \psi_1(x_1) - \psi_1(x_2) + \theta$$

$$= \psi(x_1, x_2) + F(-x_1) + F(-x_2) - 2 + \theta. \qquad (8.3.20)$$

Then the Hoeffding-decomposition (H-decomposition) for $U_n^{(2)}$ is given by (see Lee (1990))

$$U_n^{(2)} = \theta + 2H_n^{(1)} + H_n^{(2)} \qquad (8.3.21)$$

where $H_n^{(j)}$ is a U-statistic of degree j based on the kernel $h^{(j)}$, $j = 1, 2$, that is,

$$H_n^{(j)} = \frac{1}{\binom{n}{j}} \sum h^{(j)}(X_{i_1}, \ldots, X_{i_j}) \qquad (8.3.22)$$

where summation is taken over all subsets $1 \le i_1 < \ldots < i_j \le n$ of $\{1, \ldots, n\}$.

In view of the H-decomposition, we have

$$\text{Var}(U_n^{(2)}) = 4 \text{ Var}(H_n^{(1)}) + \text{Var}(H_n^{(2)}) + 4 \text{ Cov}(H_n^{(1)}, H_n^{(2)}). \qquad (8.3.23)$$

From the results discussed in Chapter 6 (cf. Dewan and Prakasa Rao (2001)), we get that

$$\text{Var}(H_n^{(1)}) = \frac{1}{n}(\sigma_1^2 + 2\sum_{j=2}^{\infty} \sigma_{1j}^2) + o(\frac{1}{n}), \qquad (8.3.24)$$

where

$$\sigma_1^2 = \text{Var}(F(-X_1)),$$

$$\sigma_{1j}^2 = \text{Cov}(F(-X_1), F(-X_{1+j})). \qquad (8.3.25)$$

Using Newman's inequality and the inequality (8.3.5), we get

$$\sum_{j=2}^{\infty} \sigma_{1j}^2 = \sum_{j=2}^{\infty} \text{Cov}(F(-X_1), F(-X_{1+j})) < \infty. \qquad (8.3.26)$$

Furthermore

$$\mathrm{Var}(H_n^{(2)}) = \binom{n}{2}^{-2} \sum_{1 \le i < j \le n} \sum_{1 \le k < \ell \le n} \mathrm{Cov}\{h^{(2)}(X_i, X_j), h^{(2)}(X_k, X_\ell)\} \quad (8.3.27)$$

where

$$\mathrm{Cov}\{h^{(2)}(X_i, X_j), h^{(2)}(X_k, X_\ell)\} \qquad\qquad (8.3.28)$$
$$= \mathrm{Cov}(\psi(X_i, X_j), \psi(X_k, X_\ell))$$
$$\quad + \mathrm{Cov}(\psi(X_i, X_j), F(-X_k)) + \mathrm{Cov}(\psi(X_i, X_j), F(-X_\ell))$$
$$\quad + \mathrm{Cov}(\psi(X_k, X_\ell), F(-X_i)) + \mathrm{Cov}(\psi(X_k, X_\ell), F(-X_j))$$
$$\quad + \mathrm{Cov}(F(-X_i), F(-X_k)) + \mathrm{Cov}(F(-X_i), F(-X_\ell))$$
$$\quad + \mathrm{Cov}(F(-X_j), F(-X_k)) + \mathrm{Cov}(F(-X_j), F(-X_\ell)). \qquad (8.3.29)$$

Using Newman's (1980) inequality, we get that

$$|\mathrm{Cov}(F(-X_i), F(-X_k))| \le \sup_x (f(x))^2 \mathrm{Cov}(X_i, X_k). \qquad (8.3.30)$$

Since the density function is bounded, it follows from Theorem 6.2.14 (cf. Bagai and Prakasa Rao (1991)) that

$$|\mathrm{Cov}(\psi(X_i, X_j), \psi(X_k, X_\ell))|$$
$$= |P[X_i + X_j > 0,\ X_\ell + X_k > 0] - P[X_i + X_j > 0]P[X_\ell + X_k > 0]|$$
$$\le C[\mathrm{Cov}(X_i + X_j, X_k + X_\ell)]^{1/3}$$
$$= C[\mathrm{Cov}(X_i, X_k) + \mathrm{Cov}(X_j, X_k) + \mathrm{Cov}(X_i, X_\ell) + \mathrm{Cov}(X_j, X_\ell)]^{1/3}. \quad (8.3.31)$$

Let $Z = X_i + X_j$. Note that the function $\psi(x_i, x_j) = I(x_i + x_j > 0) = I(z > 0)$ has a jump of size 1 at $z = 0$. Then, from equation (8.2.5), it follows that

$$|\mathrm{Cov}(\psi(X_i, X_j), F(-X_k))|$$
$$= \left| \int_{-\infty}^{\infty} P[X_i + X_j \le 0,\ X_k \le x] - P[X_i + X_j \le 0]P[X_k \le x]) \, dF(x) \right|$$
$$\le \int_{-\infty}^{\infty} |P[X_i + X_j \le 0,\ X_k \le x] - P[X_i + X_j \le 0]P[X_k \le x]| \, dF(x)$$
$$\le C \int_{-\infty}^{\infty} [\mathrm{Cov}(X_i + X_j, X_k)]^{1/3} dF(x)$$
$$= C[\mathrm{Cov}(X_i + X_j, X_k)]^{1/3}$$
$$= C[\mathrm{Cov}(X_i, X_k) + \mathrm{Cov}(X_j, X_k)]^{1/3}. \qquad (8.3.32)$$

Using equations (8.3.30), (8.3.31) and (8.3.32) in equation (8.3.29), we get that

$$|\operatorname{Cov}\{h^{(2)}(X_i,X_j),h^{(2)}(X_k,X_\ell)\}|$$

$$\leq C[\operatorname{Cov}(X_i,X_k)+\operatorname{Cov}(X_j,X_k)+\operatorname{Cov}(X_i,X_\ell)+\operatorname{Cov}(X_j,X_\ell)]^{1/3}$$

$$+[\operatorname{Cov}(X_i,X_k)+\operatorname{Cov}(X_j,X_k)]^{1/3}+[\operatorname{Cov}(X_i,X_\ell)+\operatorname{Cov}(X_j,X_\ell)]^{1/3}$$

$$+[\operatorname{Cov}(X_k,X_i)+\operatorname{Cov}(X_\ell,X_i)]^{1/3}+[\operatorname{Cov}(X_k,X_j)+\operatorname{Cov}(X_\ell,X_j)]^{1/3}$$

$$+\operatorname{Cov}(X_i,X_k)+\operatorname{Cov}(X_j,X_k)+\operatorname{Cov}(X_i,X_\ell)+\operatorname{Cov}(X_j,X_\ell)$$

$$\leq C[\operatorname{Cov}(X_i,X_k)+\operatorname{Cov}(X_j,X_k)+\operatorname{Cov}(X_i,X_\ell)+\operatorname{Cov}(X_j,X_\ell)]$$

$$+C[\operatorname{Cov}(X_i,X_k)^{1/3}+\operatorname{Cov}(X_j,X_k)^{1/3}+\operatorname{Cov}(X_i,X_\ell)^{1/3}+\operatorname{Cov}(X_j,X_\ell)^{1/3}]$$

$$=C[(\operatorname{Cov}(X_i,X_k)+\operatorname{Cov}(X_i,X_k)^{1/3})+(\operatorname{Cov}(X_j,X_k)+\operatorname{Cov}(X_j,X_k)^{1/3})$$

$$+(\operatorname{Cov}(X_i,X_\ell)+\operatorname{Cov}(X_i,X_\ell)^{1/3})+(\operatorname{Cov}(X_j,X_\ell)+\operatorname{Cov}(X_j,X_\ell)^{1/3})$$

$$=r(|i-k|)+r(|j-k|)+r(|i-\ell|)+r(|j-\ell|) \quad \text{(say)}. \tag{8.3.33}$$

From equations (8.3.4) and (8.3.5) it follows that

$$\sum_{k=1}^{\infty} r(k) < \infty. \tag{8.3.34}$$

Hence, following Serfling (1968), we have, as $n \to \infty$,

$$\operatorname{Var}(H_n^{(2)}) = o(\frac{1}{n}). \tag{8.3.35}$$

Using the Cauchy-Schwartz inequality, it follows that

$$\operatorname{Cov}(H_n^{(1)},H_n^{(2)}) = o(\frac{1}{n}). \tag{8.3.36}$$

From equations (8.3.23), (8.3.24), (8.3.35) and (8.3.36), we get that

$$\operatorname{Var}(U_n^{(2)}) = 4[\sigma_1^2 + 2\sum_{j=1}^{\infty} \sigma_{1j}^2] + o(\frac{1}{n}). \tag{8.3.37}$$

Then using the same techniques as in the results discussed in Chapter 6 (cf. Theorem 3.2, Dewan and Prakasa Rao (2002)) for obtaining the limiting distribution of U-statistics , we get the following theorem.

Theorem 8.3.2. *Let the sequence $\{X_n,\ n \geq 1\}$ be an associated sequence . Suppose equation (8.3.4) holds. Then*

$$\frac{n^{1/2}(U_n^{(2)} - \theta)}{2\sigma_U} \xrightarrow{\mathcal{L}} N(0,1) \quad \text{as } n \to \infty \tag{8.3.38}$$

where $\sigma_U^2 = \sigma_1^2 + 2\sum_{j=1}^{\infty} \sigma_{1j}^2$.

Proof. From (8.3.21), we have

$$U_n^{(2)} = \theta + 2H_n^{(1)} + H_n^{(2)}.$$

Then

$$\frac{n^{1/2}(U_n - \theta)}{2\sigma_U} = n^{-1/2} \sum_{j=1}^{n} \frac{h^{(1)}(X_j)}{\sigma_U} + n^{1/2} \frac{H_n^{(2)}}{2\sigma_U}. \tag{8.3.39}$$

It is easy to see from (8.3.35) that

$$E(n^{1/2}H_n^{(2)}) = 0 \quad \text{and} \quad n \, \text{Var} \, H_n^{(2)} \to 0 \quad \text{as } n \to \infty.$$

Hence

$$n^{1/2} \frac{H_n^{(2)}}{2\sigma_U} \overset{p}{\to} 0 \quad \text{as } n \to \infty. \tag{8.3.40}$$

Since ψ is monotonic in its arguments, $\{h^{(1)}(X_j), j \geq 1\}$ constitute a stationary associated sequence. Then, using the results from Newman (1980), it follows that

$$n^{-1/2} \sum_{j=1}^{n} \frac{h^{(1)}(X_j)}{\sigma_U} \overset{\mathcal{L}}{\to} N(0,1) \quad \text{as } n \to \infty. \tag{8.3.41}$$

Combining the relations (8.3.40) and (8.3.41), we get the result stated in Theorem 8.3.2. $\qquad \square$

Define

$$T^* = \frac{T - \gamma}{\binom{n}{2}}, \tag{8.3.42}$$

where

$$\gamma = nP[X > 0] + \binom{n}{2}\theta. \tag{8.3.43}$$

The following theorem gives the limiting distribution of the Wilcoxon signed rank statistic.

Theorem 8.3.3. *Let the sequence $\{X_n, n \geq 1\}$ be a stationary associated sequence with a bounded density function. Suppose equation (8.3.4) holds. Then*

$$\frac{n^{1/2}T^*}{2\sigma_U} \overset{\mathcal{L}}{\to} N(0,1) \quad \text{as } n \to \infty. \tag{8.3.44}$$

Proof. Note that

$$E[U_n^{(1)}] = P[X > 0], \tag{8.3.45}$$

and from equation (8.3.10)

$$\frac{1}{n} \text{Var}[U_n^{(1)}] \to 0 \quad \text{as } n \to \infty. \tag{8.3.46}$$

The result now follows using Theorem 8.3.2 and Slutsky's theorem. $\qquad \square$

Note that the Wilcoxon signed rank statistic T is the sum of ranks of positive observations. The test procedure consists in rejecting the null hypothesis H_0 for large values of the statistic T. The quantity σ_U^2 depends on the unknown distribution F even under the null hypothesis. It can be estimated using the estimators given by Peligrad and Suresh (1995) and the result of Roussas (1993) (cf. Dewan and Prakasa Rao (2005)) for estimating the variance of Wilcoxon Mann-Whitney statistic for associated sequences as discussed later in this chapter. A consistent estimator of σ_U^2 is given by

$$J_n^2 = \frac{\pi}{2}\hat{B}_n^2, \tag{8.3.47}$$

where, for $\ell = \ell_n$,

$$\hat{B}_n^2 = \frac{1}{n-\ell}\sum_{j=0}^{n-\ell}\frac{|\hat{S}_j(\ell) - \ell\hat{\bar{\psi}}_n|}{\sqrt{\ell}}, \tag{8.3.48}$$

and $\hat{S}_j(k) = \sum_{i=j+1}^{j+k}\hat{\psi}_1(X_i)$, $\hat{\bar{\psi}}_n = \frac{1}{n}\sum_{i=1}^{n}\hat{\psi}_1(X_i)$, $\hat{\psi}_1(x) = 1 - F_n(-x)$, where F_n is the empirical distribution function corresponding to F based on associated random variables $X_1, X_2, \ldots, X_n$. Note that under the null hypothesis X and $-X$ are identically distributed.

8.4 Mann-Whitney Test

Let the random variables $\{X_1, \ldots, X_m\}$ and $\{Y_1, \ldots, Y_n\}$ be two samples independent of each other, but the random variables within each sample are stationary associated with one-dimensional marginal distribution functions F and G respectively. We study the properties of the classical Wilcoxon-Mann-Whitney statistic for testing for stochastic dominance in the above setup.

Assume that the density functions f and g of F and G respectively exist . We wish to test for the equality of the two marginal distribution functions F and G. A commonly used statistic for this nonparametric testing problem is the Wilcoxon- Mann-Whitney statistic when the observations X_i, $1 \leq i \leq m$ are independent and identically distributed (i.i.d.) and Y_j, $1 \leq j \leq n$ are i.i.d. We now study the asymptotic properties of the Wilcoxon-Mann-Whitney statistic when the two samples $\{X_1, \ldots, X_m\}$ and $\{Y_1, \ldots, Y_n\}$ are from independent stationary associated stochastic processes.

We wish to test the hypothesis that

$$H_0 : F(x) = G(x) \text{ for all } x, \tag{8.4.1}$$

against the alternative

$$H_1 : F(x) \geq G(x) \text{ for all } x, \tag{8.4.2}$$

with strict inequality for some x. We can test the above hypothesis conservatively by testing

$$H_0' : \gamma = 0, \tag{8.4.3}$$

against the alternative

$$H_1' : \gamma > 0, \tag{8.4.4}$$

where

$$\gamma = 2P(Y > X) - 1 = P(Y > X) - P(Y < X). \tag{8.4.5}$$

Serfling (1980) studied the properties of the Wilcoxon statistic when the samples are from independent stationary mixing processes. Louhichi (2000) gave an example of a sequence of random variables which is associated but not mixing.

We now state some results that are used to study the properties of Wilcoxon statistic for associated random variables and discuss the asymptotic normality of the Wilcoxon statistic based on independent sequences of stationary associated variables.

Assume that

$$\sup_x f(x) < c < \infty \quad \text{and} \quad \sup_x g(x) < c < \infty. \tag{8.4.6}$$

Further assume that

$$\sum_{j=2}^{\infty} \text{Cov}^{\frac{1}{3}}(X_1, X_j) < \infty, \tag{8.4.7}$$

and

$$\sum_{j=2}^{\infty} \text{Cov}^{\frac{1}{3}}(Y_1, Y_j) < \infty. \tag{8.4.8}$$

This would imply

$$\sum_{j=2}^{\infty} \text{Cov}(X_1, X_j) < \infty, \tag{8.4.9}$$

and

$$\sum_{j=2}^{\infty} \text{Cov}(Y_1, Y_j) < \infty. \tag{8.4.10}$$

Wilcoxon Statistic

The Wilcoxon two-sample statistic is the U-statistic given by

$$U = \frac{1}{mn} \sum_{i=1}^{m} \sum_{j=1}^{n} \phi(Y_j - X_i), \tag{8.4.11}$$

where

$$\phi(u) = \begin{cases} 1 & \text{if } u > 0, \\ 0 & \text{if } u = 0, \\ -1 & \text{if } u < 0. \end{cases}$$

Note that ϕ is a kernel of degree $(1,1)$ with $E\phi(Y - X) = \gamma$. We now obtain the limiting distribution of the statistic U under some conditions. Let

$$\sigma_X^2 = 4\int_{-\infty}^{\infty} G^2(x)\, dF(x) - 4\int_{-\infty}^{\infty} G(x)\, dF(x) + 1 \qquad (8.4.12)$$

$$+ 8\sum_{j=2}^{\infty} \text{Cov}(G(X_1), G(X_j)),$$

$$\sigma_Y^2 = 4\int_{-\infty}^{\infty} F^2(x)\, dG(x) - 4\int_{-\infty}^{\infty} F(x)\, dG(x) + 1 \qquad (8.4.13)$$

$$+ 8\sum_{j=2}^{\infty} \text{Cov}(F(Y_1), F(Y_j))$$

and

$$A^2 = \sigma_X^2 + c\sigma_Y^2. \qquad (8.4.14)$$

Theorem 8.4.1. *Let the sequences $\{X_i,\ i \geq 1\}$ and $\{Y_j,\ j \geq 1\}$ be independent stationary associated sequences of random variables with one-dimensional distribution functions F and G, respectively satisfying the conditions given in equations (8.4.6) to (8.4.8). Then*

$$\sqrt{m}(U - \gamma) \overset{\mathcal{L}}{\to} N(0, A^2) \quad \text{as } n \to \infty,$$

as $m, n \to \infty$ such that $\frac{m}{n} \to c \in (0, \infty)$ where A^2 is as given by equation (8.4.16). If $F = G$, then

$$\sigma_X^2 = \sigma_Y^2$$

$$= 4\left(\frac{1}{12} + 2\sum_{j=2}^{\infty} \text{Cov}(F(X_1), F(X_j))\right), \qquad (8.4.15)$$

so that

$$A^2 = 4(1 + c)\left(\frac{1}{12} + 2\sum_{j=2}^{\infty} \text{Cov}(F(X_1), F(X_j))\right). \qquad (8.4.16)$$

Proof. Following Hoeffding's decomposition (Lee (1980)), we can write U as

$$U = \gamma + H_{m,n}^{(1,0)} + H_{m,n}^{(0,1)} + H_{m,n}^{(1,1)}, \qquad (8.4.17)$$

where

$$H_{m,n}^{(1,0)} = \frac{1}{m} \sum_{i=1}^{m} h^{(1,0)}(X_i),$$

$$h^{(1,0)}(x) = \phi_{10}(x) - \gamma, \ \phi_{10}(x) = 1 - 2G(x),$$

$$H_{m,n}^{(0,1)} = \frac{1}{n} \sum_{j=1}^{n} h^{(0,1)}(Y_j),$$

$$h^{(0,1)}(y) = \phi_{01}(y) - \gamma, \ \ \phi_{01}(y) = 2F(y) - 1,$$

and

$$H_{m,n}^{(1,1)} = \frac{1}{mn} \sum_{i=1}^{m} \sum_{j=1}^{n} h^{(1,1)}(X_i, Y_j),$$

where

$$h^{(1,1)}(x, y) = \phi(x - y) - \phi_{10}(x) - \phi_{01}(y) + \gamma.$$

It is easy to see that

$$E(\phi_{10}(X)) = \gamma,$$

$$E(\phi_{10}^2(X)) = 4 \int_{-\infty}^{\infty} G^2(x) \, dF(x) - 4 \int_{-\infty}^{\infty} G(x) \, dF(x) + 1,$$

and

$$\text{Cov}(\phi_{10}(X_i), \phi_{01}(X_j)) = 4 \ \text{Cov}(G(X_i), G(X_j)). \tag{8.4.18}$$

Since the random variables $X_1, \ldots, X_m$ are associated, so are $\phi_{10}(X_1), \ldots,$ $\phi_{10}(X_m)$ since ϕ is monotone (see, Esary, Proschan and Walkup (1967)). Furthermore conditions given by equations (8.4.6), (8.4.9) and (8.4.10) imply that

$$\sum_{j=2}^{\infty} \text{Cov}(G(X_1), G(X_j)) < \infty,$$

and

$$\sum_{j=2}^{\infty} \text{Cov}(F(Y_1), F(Y_j)) < \infty,$$

since

$$|\text{Cov}(G(X_1), G(X_j))| < (\sup_x g) \, \text{Cov}(X_1, X_j),$$

and

$$|\text{Cov}(F(Y_1), F(Y_j))| < (\sup_x f) \, \text{Cov}(Y_1, Y_j),$$

by Newman's inequality (1980). Following Newman (1980,1984), we get that

$$m^{-1/2} \sum_{i=1}^{m} (\phi_{10}(X_i) - \gamma) \overset{\mathcal{L}}{\to} N(0, \sigma_X^2) \quad \text{as } n \to \infty, \qquad (8.4.19)$$

where

$$\sigma_X^2 = 4 \int_{-\infty}^{\infty} G^2(x) \, dF(x) - 4 \int_{-\infty}^{\infty} G(x) \, dF(x) + 1 + 8 \sum_{j=2}^{\infty} \text{Cov}(G(X_1), G(X_j)).$$
$$(8.4.20)$$

Similarly, we see that

$$n^{-1/2} \sum_{j=1}^{n} (\phi_{01}(Y_j) - \gamma) \overset{\mathcal{L}}{\to} N(0, \sigma_Y^2) \quad \text{as } n \to \infty, \qquad (8.4.21)$$

where

$$\sigma_Y^2 = 4 \int_{-\infty}^{\infty} F^2(x) \, dG(x) - 4 \int_{-\infty}^{\infty} F(x) \, dG(x) + 1 + 8 \sum_{j=2}^{\infty} \text{Cov}(F(Y_i), F(Y_j)).$$
$$(8.4.22)$$

Note that $E(H_{m,n}^{(1,1)}) = 0$. Consider

$$\text{Var}(H_{m,n}^{(1,1)}) = E(H_{m,n}^{(1,1)})^2$$
$$= \frac{\Delta}{m^2 n^2}, \qquad (8.4.23)$$

where

$$\Delta = \sum_{i=1}^{m} \sum_{j=1}^{n} \sum_{i'=1}^{m} \sum_{j'=1}^{n} \Delta(i, j; i', j'), \qquad (8.4.24)$$

and

$$\Delta(i, j; i', j') = \text{Cov}(h^{(1,1)}(X_i, Y_j), h^{(1,1)}(X_{i'}, Y_{j'})). \qquad (8.4.25)$$

Following Serfling (1980),

$$\Delta(i, j; i', j') = 4 \left(E(F_{i,i'}(Y_j, Y_{j'}) - F(Y_j)F(Y_{j'})) - \text{Cov}(G(X_i, X_{i'})) \right)$$
$$= 4 \left(E(G_{j,j'}(X_i, X_{i'}) - G(X_i)G(X_{i'})) - \text{Cov}(F(Y_j, Y_{j'})), \quad (8.4.26) \right.$$

where $F_{i,i'}$ is the joint distribution function of $(X_i, X_{i'})$ and $G_{j,j'}$ is the joint distribution function of $(Y_j, Y_{j'})$.

Then, by Theorem 6.2.14 in Chapter 6, we get that there exists a constant $C > 0$ such that

$$\Delta(i, j; i', j') \leq C[\text{Cov}^{\frac{1}{3}}(X_i, X_{i'}) + \text{Cov}(X_i, X_{i'})]$$
$$= r_1(|i - i'|) \quad \text{(say)}, \qquad (8.4.27)$$

by stationarity and

$$\Delta(i,j;i',j') \le C[\mathrm{Cov}^{\frac{1}{3}}(Y_j, Y_{j'}) + \mathrm{Cov}(Y_j, Y_{j'})]$$
$$= r_2(|j - j'|) \quad (\text{say}), \tag{8.4.28}$$

by stationarity. Note that

$$\sum_{k=1}^{\infty} r_1(k) < \infty, \quad \sum_{k=1}^{\infty} r_2(k) < \infty. \tag{8.4.29}$$

by equations (8.4.7)–(8.4.10). Then, following Serfling (1980), we have

$$\Delta = o(mn^2) \tag{8.4.30}$$

as m and $n \to \infty$ such that $\frac{m}{n}$ has a limit $c \in (0, \infty)$.

Hence, from (8.4.17), we have

$$\sqrt{m}(U - \gamma) = \sqrt{m}\frac{1}{m}\sum_{i=1}^{m} h^{(1,0)}(X_i) + \sqrt{\frac{m}{n}}\frac{1}{\sqrt{n}}\sum_{j=1}^{n} h^{(0,1)}(Y_j) + \sqrt{m}H_{m,n}^{(1,1)}$$

$$\xrightarrow{\mathcal{L}} N(0, A^2), \tag{8.4.31}$$

where

$$A^2 = \sigma_X^2 + c\sigma_Y^2, \tag{8.4.32}$$

since $E(H_{m,n}^{(1,1)}) = 0$ and $\mathrm{Var}(\sqrt{m}H_{m,n}^{(1,1)}) \to 0$ as $m, n \to \infty$ such that $\frac{m}{n} \to c \in (0, \infty)$. This completes the proof of the theorem. $\qquad\square$

Estimation of the Limiting Variance

Note that the limiting variance A^2 depends on the unknown distribution F even under the null hypothesis. We need to estimate it so that the proposed test statistic can be used for testing purposes. The unknown variance A^2 can be estimated using the estimators given by Peligard and Suresh (1995).

Theorem 8.4.2 (Peligard and Suresh (1995)). *Let* $\{X_n, n \ge 1\}$ *be a stationary associated sequence of random variables with* $E(X_1) = \mu$, $E(X_1^2) < \infty$. *Let* $\{\ell_n, n \ge 1\}$ *be a sequence of positive integers with* $1 \le \ell_n \le n$. *Let* $S_j(k) = \sum_{i=j+1}^{j+k} X_i$, $\bar{X}_n = \frac{1}{n}\sum_{i=1}^{n} X_i$. *Let* $\ell_n = o(n)$ *as* $n \to \infty$. *Assume that equation* (8.4.9) *holds. Then, with* $\ell = \ell_n$,

$$B_n = \frac{1}{n-\ell}\left(\sum_{j=0}^{n-\ell} \frac{|S_j(\ell) - \ell\bar{X}_n|}{\sqrt{\ell}}\right)$$

$$\to \left(\mathrm{Var}(X_1) + 2\sum_{i=2}^{\infty} \mathrm{Cov}(X_1, X_i)\right)\sqrt{\frac{2}{\pi}} \quad \text{in } L_2\text{-mean as } n \to \infty. \tag{8.4.33}$$

In addition, if we assume that $\ell_n = O(n/(\log n)^2)$ as $n \to \infty$, then the convergence stated above holds in the almost sure sense (cf. Peligrad and Suresh (1995)).

Theorem 8.4.3 (Roussas (1993)). *Let the sequence $\{X_n, \ n \geq 1\}$ be a stationary associated sequence of random variables with bounded one-dimensional probability density function. Suppose*

$$u(n) = 2 \sum_{j=n+1}^{\infty} \mathrm{Cov}(X_1, X_j)$$

$$= O(n^{-(s-2)/2}) \quad \text{for some } s > 2. \tag{8.4.34}$$

Let ψ_n be any positive norming factor. Then, for any bounded interval $I_M = [-M, M]$, we have

$$\sup_{x \in I_M} \psi_n |F_n(x) - F(x)| \to 0, \tag{8.4.35}$$

almost surely as $n \to \infty$, provided

$$\sum_{n=1}^{\infty} n^{-s/2} \psi_n^{s+2} < \infty. \tag{8.4.36}$$

We now give a consistent estimator of the unknown variance A^2 under some conditions. Let $N = m + n$. Under the hypothesis $F = G$, the random variables $X_1, \ldots, X_m, Y_1, \ldots, Y_n$ are associated with the one-dimensional marginal distribution function F. Denote $Y_1, \ldots, Y_n$ as $X_{m+1}, \ldots, X_N$. Then $X_1, \ldots, X_N$ are associated as independent sets of associated random variables are associated (cf. Esary, Proschan and Walkup (1967)). Let $\{\ell_N, \ N \geq 1\}$ be a sequence of positive integers with $1 \leq \ell_N \leq N$. Let $S_j(k) = \sum_{i=j+1}^{j+k} \phi_{10}(X_i)$, $\bar{\phi}_N = \frac{1}{N} \sum_{i=1}^{N} \phi_{10}(X_i)$. Define $\ell = \ell_N$ and

$$B_N = \frac{1}{N-\ell} \Big[\sum_{j=0}^{N-\ell} \frac{|S_j(\ell) - \ell\bar{\phi}_N|}{\sqrt{\ell}} \Big]. \tag{8.4.37}$$

Note that B_N depends on the unknown function F. Let $\hat{\phi}_{10}(x) = 1 - 2F_N(x)$ where F_N is the empirical distribution function corresponding to F based on the associated random variables $X_1, \ldots, X_N$. Let $\hat{S}_j(k)$, $\hat{\bar{\phi}}_N$ and $\hat{B}_N$ be expressions analogous to $S_j(k)$, $\bar{\phi}_N$ and B_N with ϕ_{10} replaced by $\hat{\phi}_{10}$. Let $Z_i = \phi_{10}(X_i) - \hat{\phi}_{10}(X_i)$. Then

$$|B_N - \hat{B}_N| = \Big| \frac{1}{N-\ell} \sum_{j=1}^{\infty} \frac{|S_j(\ell) - \ell\bar{\phi}|}{\sqrt{\ell}} - \frac{1}{N-\ell} \sum_{j=1}^{\infty} \frac{|\hat{S}_j(\ell) - \ell\hat{\bar{\phi}}|}{\sqrt{\ell}} \Big|$$

$$\leq \frac{1}{(N-\ell)\sqrt{\ell}} \sum_{j=1}^{\infty} |S_j(\ell) - \hat{S}_j(\ell) - \ell(\bar{\phi} - \hat{\bar{\phi}})|$$

$$= \frac{1}{(N-\ell)\sqrt{\ell}} \sum_{j=1}^{\infty} | \sum_{i=j+1}^{j+\ell} Z_i - \ell \frac{1}{N} \sum_{i=1}^{N} Z_i |$$

$$\leq \frac{1}{(N-\ell)\sqrt{\ell}} \sum_{j=1}^{\infty} \{ \sum_{i=j+1}^{j+\ell} |Z_i| + \ell \frac{1}{N} \sum_{i=1}^{N} |Z_i| \}. \tag{8.4.38}$$

Note that

$$|Z_i| = 2|F_N(X_i) - F(X_i)|.$$

Suppose that the density function corresponding to F has a bounded support. Then, for sufficiently large $M > 0$, with probability 1,

$$\sup_{x \in R} |F_N(x) - F(x)| = \max\{ \sup_{x \in [-M,M]} |F_N(x) - F(x)|, \sup_{x \in [-M,M]^c} |F_N(x) - F(x)|\}$$

$$= \sup_{x \in [-M,M]} |F_N(x) - F(x)|. \tag{8.4.39}$$

Hence, from (8.4.39) and Theorem 8.4.3 we get

$$|B_N - \hat{B}_N| \leq \frac{2}{(N-\ell)\sqrt{\ell}} (N-\ell) \ell \sup_x |F_N(x) - F(x)|$$

$$= 2\sqrt{\ell} \ \psi_N^{-1} \sup_x \ \psi_N \ |F_N(x) - F(x)|$$

$$\to 0 \quad \text{as } N \to \infty \tag{8.4.40}$$

provided $\sqrt{\ell} \ \psi_N^{-1} = O(1)$ or $\ell_N = O(\psi_N^2)$. Therefore we get,

$$|B_N - \hat{B}_N| \to 0 \text{ a.s. as } n \to \infty. \tag{8.4.41}$$

Hence, from Theorem 8.4.2, it follows that

$$\frac{\pi}{2} \hat{B}_N^2 \to 4(\frac{1}{12} + 2 \sum_{j=2}^{\infty} \text{Cov}(F(X_1), F(X_j))) \tag{8.4.42}$$

as $n \to \infty$. Define $J_N^2 = (1+c)\frac{\pi}{2}\hat{B}_N^2$. Then $\frac{\sqrt{N}(U-\gamma)}{J_N} \xrightarrow{\mathcal{L}} N(0,1)$ as $m, n \to \infty$ such that $\frac{m}{n} \to c \in (0,\infty)$ as $n \to \infty$. Hence the statistic $\frac{\sqrt{N}(U-\gamma)}{J_N}$ can be used as a test statistic for testing $H_0' : \gamma = 0$ against $H_1' = \gamma > 0$.

On the other hand, by using Newman's inequality, one could obtain an upper bound on A^2 given by

$$4(1+c)(\frac{1}{12} + 2 \sum_{j=2}^{\infty} \text{Cov}(X_1, X_j)) \tag{8.4.43}$$

and we can have conservative tests and estimates of power based on the bound given in equation (8.4.43).

Results on tests discussed in this chapter are due to Dewan and Prakasa Rao (2005, 2006).

Chapter 9

Nonparametric Tests for Change in Marginal Density Function for Associated Sequences

9.1 Introduction

Consider a finite sequence of random variables $X_1, \ldots, X_n$. A change point is an integer k between 1 and n such that the distribution of the first k observations is different from the distribution of the last $n - k$ observations. An example is the case when the location parameter of the initial distribution has shifted for the last $n - k$ observations. Economic time series or financial time series are often affected by changes in monetary policies or forces induced by external or internal events beyond their control. Hence it is of interest to consider tests for change of the one-dimensional probability density function in a time series or for a stationary stochastic process in general. Inference problems for change point k as well as the shift in the independent case have been investigated starting with Page (1954). Chernoff and Zacks (1964) and Gardner (1969) adopted a Bayesian approach to the problem. They discussed tests and the corresponding null distributions for the null hypothesis of no shift in the mean. In the Bayesian approach, a prior distribution is assumed for the change point parameter k. Nagaraj (1990) derived locally optimal tests for the hypothesis of no change in mean assuming a prior for the change point k under the assumption of normality and independence. Nagaraj and Reddy (1993) extended these results and derived asymptotic null distribution of some test statistics for detecting a change in the mean of a process at unknown change point when the observations are strictly stationary and associated. We now

B.L.S. Prakasa Rao, *Associated Sequences, Demimartingales and Nonparametric Inference*, Probability and its Applications, DOI 10.1007/978-3-0348-0240-6_9, © Springer Basel AG 2012

discuss their results leading to locally optimal tests for change for an associated sequence with common one-dimensional probability density function under null hypothesis.

9.2 Tests for Change Point

Let the set $X_1, X_2, \ldots, X_n$ be independent Gaussian random variables with common variance σ^2 and

$$
E[X_i] = \begin{cases} \mu & i = 1, \ldots, k \\ \mu + \delta & i = k+1, \ldots, n. \end{cases}
$$

where $1 \leq k < n$. The constant k is known as the change point. We assume that k is a realization of a random variable K such that $P(K = k) = \frac{1}{n-1}$, $k = 1, \ldots, n-1$. The null hypothesis of no change can be specified as $H_0 : \delta = 0$. The alternate hypotheses of interest are $H_1 : \delta \neq 0$ and $H_2 : \delta > 0$. Let

$$
g_k = \sum_{j=k+1}^{n} X_j, \quad T_1 = \sum_{k=1}^{n-1} g_k, \quad T_4 = \sum_{k=1}^{n-1} g_k^2;
$$

$$
\hat{g}_k = \sum_{j=k+1}^{n} (X_j - \bar{X}), \quad T_2 = \sum_{k=1}^{n-1} \hat{g}_k, \quad T_5 = \sum_{k=1}^{n-1} \hat{g}_k^2;
$$

and

$$
\tilde{g}_k = [(n-1)^{-1} \sum_{j=1}^{n} (X_j - \bar{X})^2]^{-1/2} \hat{g}_k, \quad T_3 = \sum_{k=1}^{n-1} \tilde{g}_k, \quad T_6 = \sum_{k=1}^{n-1} \tilde{g}_k^2.
$$

Nagaraj (1990) proved the following:

(i) if $\mu = 0$ and $\sigma^2 = 1$, then the locally most powerful test for testing H_0 against H_2 rejects H_0 for large values of T_1;

(ii) if $\mu = 0$ and $\sigma^2 = 1$, then the locally most powerful test for testing H_0 against H_1 rejects H_0 for large values of T_4;

(iii) if μ is unknown and $\sigma^2 = 1$, then the locally most powerful invariant test for testing H_0 against H_2 rejects H_0 for large values of T_2;

(iv) if μ is unknown and $\sigma^2 = 1$, then the locally most powerful invariant test for testing H_0 against H_1 rejects H_0 for large values of T_5;

(v) if μ is unknown and σ^2 is unknown, then the locally most powerful invariant test for testing H_0 against H_2 rejects H_0 for large values of T_3; and

(vi) if μ is unknown and σ^2 is unknown, then the locally most powerful invariant test for testing H_0 against H_1 rejects H_0 for large values of T_6.

Let the sequence $\{X_k, \ k \geq 0\}$ be a stationary associated sequence. Let $\mu = E[X_k]$ and the auto-covariance function and the autocorrelation function of the process $\{X_k, \ k \geq 0\}$ be denoted by $\gamma(h) = \mathrm{Cov}(X_t, X_{t+h})$ and $\rho(h) = \mathrm{Cov}(X_t, X_{t+h}) = (\gamma(0))^{-1}\gamma(h)$ respectively. Since the process is stationary and associated, we get that $\gamma(h) \geq 0$. We assume that

$$0 < \sigma^2 = \sum_{h=-\infty}^{\infty} \gamma(h) < \infty. \tag{9.2.1}$$

This condition implies that the process $\{X_k, \ k \geq 1\}$ is ergodic from results in Lebowitz (1972) (cf. Theorem 7, Newman (1984)). In fact, the stationary associated sequence $\{X_k, \ k \geq 1\}$ is ergodic if and only if

$$\lim_{n \to \infty} n^{-1} \sum_{j=1}^{n} \mathrm{Cov}(X_1, X_j) = 0 \tag{9.2.2}$$

from Lebowitz (1972). In particular, it follows that, for $h \geq 0$,

$$\frac{1}{n-h} \sum_{i=1}^{n-h} (X_k - \mu)(X_{k+h} - \mu) \overset{a.s.}{\to} \gamma(h) \text{ as } n \to \infty. \tag{9.2.3}$$

Applying again the ergodicity and stationarity of the process $\{X_k\}$, it follows that

$$\frac{1}{n-h} \sum_{i=1}^{n-h} X_k \overset{a.s}{\to} \mu \text{ as } n \to \infty. \tag{9.2.4}$$

These observations imply that, for fixed $h \geq 0$,

$$\gamma_n(h) \equiv \frac{1}{n-h} \sum_{t=1}^{n-h} (X_t - \bar{X})(X_{t+h} - \bar{X}) \to \gamma(h) \tag{9.2.5}$$

with probability 1 where $\bar{X}$ is the sample mean. Thus the first-order sample auto-covariance is a consistent estimator for $\gamma(h)$.

Let $S_0 = 0$ and $S_n = X_1 + \cdots + X_n - n\mu$. Let

$$W_n(t) = \frac{1}{\sigma\sqrt{n}}(S_{[nt]} + (nt - [nt])X_{[nt]+1}), \quad 0 \leq t \leq 1. \tag{9.2.6}$$

It follows, from Newman and Wright (1981), that

$$W_n \overset{\mathcal{L}}{\to} W \text{ as } n \to \infty \tag{9.2.7}$$

where W is the standard Wiener process. This is the analogue of Donsker's invariance principle for stationary associated sequences.

Nagaraj and Reddy (1993) derived the asymptotic null distribution of the test statistics T_1 to T_6 under the assumption that the observations are generated by a stationary associated process. We now discuss these results.

Theorem 9.2.1. *Suppose* $\mu = 0$. *Then*

$$\sigma^{-1} n^{-3/2} T_1 \xrightarrow{\mathcal{L}} W(1) - \int_0^1 W(t)dt \ \ as \ n \to \infty.$$

Proof. Note that

$$\sigma^{-1} n^{-3/2} T_1 = \frac{1}{\sigma n^{3/2}} \sum_{k=1}^n k X_k - \frac{1}{\sigma n^{3/2}} \sum_{k=1}^n X_k$$

$$= \frac{1}{n} \sum_{k=1}^n k [W_n(\frac{k}{n}) - W_n(\frac{k-1}{n})] - \frac{1}{\sigma n^{3/2}} \sum_{k=1}^n X_k$$

$$= W_n(1) - \frac{1}{n} \sum_{k=1}^n W_n(\frac{k-1}{n}) - \frac{1}{\sigma n^{3/2}} \sum_{k=1}^n X_k$$

where $W_n(.)$ is as defined by equation (9.2.6). Let

$$R_n = W_n(1) - \int_0^1 W_n(t)dt. \tag{9.2.8}$$

Since the process $W_n \xrightarrow{\mathcal{L}} W$ and the functional $h : C[0,1] \to R$ given by $h(x(.)) = x(1) - \int_0^1 x(t)dt$ is continuous on $C[0,1]$ endowed with uniform norm, it follows that

$$R_n \xrightarrow{\mathcal{L}} W(1) - \int_0^1 W(t)dt \tag{9.2.9}$$

by the continuous mapping theorem (cf. Billingsley (1968)). Furthermore

$$E|R_n - \sigma^{-1} n^{-3/2} T_1| \le E| \int_0^1 W_n(t)dt - \frac{1}{n} \sum_{k=1}^n W_n(\frac{k-1}{n})| + \frac{1}{\sigma n^{3/2}} \sum_{k=1}^n E|X_k|$$

$$\le \sum_{k=1}^n \int_{(k-1)/n}^{k/n} E|W_n(t) - W_n(\frac{k-1}{n})|dt + \frac{1}{\sigma n^{3/2}} \sum_{k=1}^n E|X_k|$$

$$\le \sum_{k=1}^n \int_{(k-1)/n}^{k/n} \frac{2E|X_1|}{\sigma \sqrt{n}} dt + \frac{1}{\sigma n^{3/2}} \sum_{k=1}^n E|X_1|$$

$$= 3 \frac{E|X_1|}{\sigma \sqrt{n}}$$

and the last term tends to zero as $n \to \infty$. This observation, along with (9.2.9), proves the result. $\square$

Remarks. Suppose the sequence $\{X_k, k \geq 0\}$ is a stationary autoregressive process of order 1, that is,

$$X_k = \rho X_{k-1} + \epsilon_k, \quad k \geq 1$$

where $\{\epsilon_k, k \geq 1\}$ are independent standard normal random variables. Further suppose that $0 \leq \rho < 1$. Then it follows that the sequence $\{X_k, k \geq 1\}$ is stationary and associated. Applying Theorem 9.2.1, we get that

$$n^{-3/2} \sum_{k=0}^{n} k X_k \xrightarrow{\mathcal{L}} \sigma[W(1) - \int_0^1 W(t)dt]$$

as $n \to \infty$. Note that the random variable $W(1) - \int_0^1 W(t)dt$ has the Gaussian distribution with mean zero and variance $\frac{1}{3}$. Hence

$$n^{-3/2} \sum_{k=0}^{n} k X_k \xrightarrow{\mathcal{L}} N(0, \frac{\sigma^2}{3})$$

as $n \to \infty$ where $\sigma^2 = (1-\rho)^{-2}$.

Suppose μ is unknown. Then the test statistic T_2 can be used to test change in the mean. It can be checked that

$$\sigma^{-1} n^{-3/2} T_2 \xrightarrow{\mathcal{L}} \frac{1}{2} W(1) - \int_0^1 W(t)dt$$

as $n \to \infty$. From the ergodic property of the sequence $\{X_k, k \geq 1\}$, it follows that

$$\frac{1}{n} \sum_{k=1}^{n} (X_k - \mu)^2 \to E(X_1 - \mu)^2 \text{ a.s. as } n \to \infty.$$

Applying this, we can obtain the limiting distribution of the test statistic T_3 when the variance is unknown. It can be shown that

$$n^{-3/2} T_3 \xrightarrow{\mathcal{L}} \gamma[\frac{1}{2} W(1) - \int_0^1 W(t)dt]$$

as $n \to \infty$ where $\gamma^2 = \sum_{h=-\infty}^{\infty} \rho(h)$ and $\rho(h)$ is the autocorrelation function of the process $\{X_k, k \geq 0\}$.

We now study the asymptotic null distributions of the test statistics T_4, T_5 and T_6 which give locally optimal invariant tests for testing H_0 versus H_1 depending on the fact whether μ and/or σ^2 are known or unknown.

Theorem 9.2.2. *Suppose $\mu = 0$. Then*

$$n^{-2} \sigma^{-2} T_4 \xrightarrow{\mathcal{L}} \int_0^1 (W(t) - W(1))^2 dt \text{ as } n \to \infty. \tag{9.2.10}$$

Proof. Observe that

$$n^{-2}\sigma^{-2}T_4 = n^{-2}\sigma^{-2}\sum_{k=1}^{n-1}(\sum_{j=k+1}^{n}X_j)^2$$

$$= n^{-2}\sigma^{-2}\sum_{k=1}^{n}(S_n - S_k)^2$$

$$= \frac{1}{n}\sum_{k=1}^{n}[W_n(1) - W_n(\frac{k}{n})]^2.$$

Let

$$R_n = \int_0^1 [W_n(t) - W_n(1)]^2 dt. \qquad (9.2.11)$$

Then

$$E|R_n - n^{-2}\sigma^{-2}T_4| \qquad (9.2.12)$$

$$= E|\int_0^1 [W_n(t) - W_n(1)]^2 dt - \frac{1}{n}\sum_{k=1}^{n}[W_n(1) - W_n(\frac{k}{n})]^2|$$

$$= E|-2W_n(1)[\int_0^1 W_n(t)dt - \sum_{k=1}^{n}\int_{\frac{k-1}{n}}^{\frac{k}{n}} W(\frac{k}{n})dt]$$

$$+ (\int_0^1 W_n^2(t)dt - \sum_{k=1}^{n}\int_{\frac{k-1}{n}}^{\frac{k}{n}} W_n^2(\frac{k}{n})dt).$$

Applying Hölder's inequality and the triangle inequality, we get that

$$E|R_n - n^{-2}\sigma^{-2}T_4| \qquad (9.2.13)$$

$$= E|-2W_n(1)(\sum_{k=1}^{n}\int_{\frac{k-1}{n}}^{\frac{k}{n}} [W_n(t) - W_n(\frac{k}{n})]dt)$$

$$+ \sum_{k=1}^{n}\int_{\frac{k-1}{n}}^{\frac{k}{n}} [W_n^2(t) - W_n^2(\frac{k}{n})]dt|$$

$$\leq 2[E(W_n^2(1)]^{1/2}\sum_{k=1}^{n}\int_{\frac{k-1}{n}}^{\frac{k}{n}} [E|W_n(t) - W_n(\frac{k}{n})|^2]^{1/2}dt$$

$$+ \sum_{k=1}^{n}\int_{\frac{k-1}{n}}^{\frac{k}{n}} [E|W_n^2(t) - W_n^2(\frac{k}{n})|dt.$$

Since

$$\text{Var}[W_n(t)] \leq \text{Var}(S_{[nt]+1}) \leq ([nt]+1)^2\sigma^2$$

for all $n \geq 1$, we get that

$$[E|W_n(t) + W_n(\frac{k}{n})|^2]^{1/2} \leq [E(W_n^2(t))]^{1/2} + [E(W_n^2(\frac{k}{n})]^{1/2} \leq 2.$$

Hence

$$E|R_n - n^{-2}\sigma^{-2}T_4| \qquad (9.2.14)$$

$$\leq 2[E(W_n^2(1)]^{1/2} \sum_{k=1}^{n} \int_{\frac{k-1}{n}}^{\frac{k}{n}} [E|W_n(t) - W_n(\frac{k}{n})|^2]^{1/2} dt$$

$$+ \sum_{k=1}^{n} \int_{\frac{k-1}{n}}^{\frac{k}{n}} [E|W_n(t) + W_n(\frac{k}{n})|^2]^{1/2} [E|W_n(t) - W_n(\frac{k}{n})|^2]^{1/2} dt$$

$$\leq 2[E(W_n^2(1))]^{1/2} \sum_{k=1}^{n} \int_{\frac{k-1}{n}}^{\frac{k}{n}} [E|W_n(t) - W_n(\frac{k}{n})|^2]^{1/2} dt$$

$$+ 2 \sum_{k=1}^{n} \int_{\frac{k-1}{n}}^{\frac{k}{n}} [E|W_n(t) - W_n(\frac{k}{n})|^2]^{1/2} dt$$

$$\leq 4 \sum_{k=1}^{n} \int_{\frac{k-1}{n}}^{\frac{k}{n}} [E|W_n(t) - W_n(\frac{k}{n})|^2]^{1/2} dt.$$

Therefore

$$E|R_n - n^{-2}\sigma^{-2}T_4| \leq 4 \sum_{k=1}^{n} \int_{\frac{k-1}{n}}^{\frac{k}{n}} [E|W_n(t) - W_n(\frac{k}{n})|^2]^{1/2} dt \qquad (9.2.15)$$

$$= 4 \sum_{k=1}^{n} \int_{\frac{k-1}{n}}^{\frac{k}{n}} n^{-1/2} (EX_1^2)^{1/2} dt$$

$$\leq 4 \sum_{k=1}^{n} \int_{\frac{k-1}{n}}^{\frac{k}{n}} \sigma^{-1} n^{-1/2} E|X_1| dt$$

$$= \frac{E|X_1|}{\sigma\sqrt{n}}$$

and the last term tends to zero as $n \to \infty$. Consider the continuous functional

$$h(x(.)) = \int_0^1 [x(t) - x(1)]^2 dt$$

on the space $C[0,1]$. Applying arguments similar to those given earlier in the previous theorem, we prove the result stated in this theorem. $\qquad \square$

Remarks. The asymptotic null distribution of the test statistic T_5 in the unknown mean case and the test statistic T_6 in the unknown mean and variance cases can be worked out similarly. It can be proved that

$$n^{-2}\sigma^{-2}T_5 \xrightarrow{\mathcal{L}} \int_0^1 (W(t) - tW(1))^2 dt \quad \text{as } n \to \infty \qquad (9.2.16)$$

and

$$n^{-2}T_6 \xrightarrow{\mathcal{L}} \gamma^2 \int_0^1 (W(t) - tW(1))^2 dt \quad \text{as } n \to \infty \qquad (9.2.17)$$

where $\gamma^2 = \sum_{-\infty}^{\infty} \rho(h)$ and $\rho(h)$ is the auto-correlation of the process $\{X_k, k \geq 0\}$.

Remarks. Suppose the sequence $\{X_k, k \geq 0\}$ is a stationary autoregressive process of order 1, that is,

$$X_k = \rho X_{k-1} + \epsilon_k, \quad k \geq 1$$

where $\{\epsilon_k, k \geq 1\}$ are independent standard normal random variables. Further suppose that $0 \leq \rho < 1$. Then it follows that the sequence $\{X_k, k \geq 0\}$ is stationary and associated. In this case, it can be checked that

$$\gamma^2 = \frac{1+\rho}{1-\rho}$$

and

$$\hat{\gamma}^2 = \frac{1+\hat{\rho}}{1-\hat{\rho}}$$

is a consistent estimator of γ^2 where $\hat{\rho}$ is the sample first-order autocorrelation. Hence the two-sided test statistic corrected for autocorrelation is $[\hat{\gamma}]^{-1}T_6$.

9.3   Test for Change in Marginal Density Function

Following Li and Lin (2007), we now consider the problem of testing for a change of one-dimensional marginal density of a stationary associated sequence $\{X_n, n \geq 1\}$ with $E[X_1^2] < \infty$. Let H_0 denote the hypothesis that the sequence $X_1, \ldots, X_n$ has a common one-dimensional marginal density function $f(.)$ and H_1 denote the hypothesis that for some $0 < \theta < 1$,

  (i) the sequence $X_1, \ldots, X_{[n\theta]}$ has a common one-dimensional marginal density function $f_1(.)$ and

  (ii) the sequence $X_{[n\theta]+1}, \ldots, X_n$ has a common one-dimensional marginal density function $f_2(.)$ with $f_1 \neq f_2$.

We now develop a nonparametric test for testing the hypothesis H_0 against the hypothesis H_1. Here $[x]$ denotes the greatest integer less than or equal to x.

For any function $g(.)$ of bounded variation, let $V_a^b(g)$ denote the total variation of the function $g(.)$ over the closed interval $[a, b]$. Let

$$V_g(x) = \begin{cases} V_0^x(g) & \text{for } x \geq 0 \\ -V_x^0(g) & \text{for } x < 0. \end{cases}$$

(A1) Let $K(.)$ be a kernel function which is a probability density function of bounded variation such that $K(0) = 0$,

$$0 < c_K \equiv \int_{-\infty}^{\infty} K^2(x)dx < \infty$$

and

$$\int_{-\infty}^{\infty} |V_K(x)|dx < \infty.$$

(A2) Suppose the sequence $\{X_n, \; n \geq 1\}$ is a stationary associated sequence of random variables. Let $\phi(.)$ be the characteristic function and f be the probability density function of X_1 and let $\phi_{1j}(.,.)$ be the characteristic function and $f_{1j}(.,.)$ be the joint probability density function of (X_1, X_j).

(i) Suppose that the density function f is bounded and Lipshitzian of order 1, that is, there exists a constant $C > 0$ such that

$$|f(x_1) - f(x_2)| \leq C|x_1 - x_2|, \quad x_1, x_2 \in R$$

and the function $\phi(.)$ is absolutely integrable.

(ii) Suppose that the joint density function f_{ij} is uniformly Lipshitzian of order 1, that is, there exists a constant $C > 0$ such that

$$\sup_{j \geq 1; x, y} |f_{1j}(x + u, y + v)) - f_{1j}(x, y)| \leq C(|u| + |v|), \quad u, v \in R$$

and the function $\phi_{1j}(.,.)$ is absolutely integrable.

(A3) Let h_n be a sequence of positive numbers tending to zero as $n \to \infty$ such that $nh_n \to \infty$ as $n \to \infty$ and

$$\sum_{j=1}^{n} [\text{Cov}(X_1, X_n)]^{1/5} = o(h_n^{-1}).$$

Define

$$f_{[n\theta]}(x) = \frac{1}{[n\theta]h_n} \sum_{j=1}^{[n\theta]} K\left(\frac{x - X_j}{h_n}\right), \tag{9.3.1}$$

and

$$f^*_{n-[n\theta]}(x) = \frac{1}{(n - [n\theta])h_n} \sum_{j=[n\theta]+1}^{n} K(\frac{x - X_j}{h_n}). \qquad (9.3.2)$$

Note that $f_{[n\theta]}(x)$ is a kernel type density estimator of $f(x)$ based on the observations $X_1, \ldots, X_{[n\theta]}$ and $f^*_{n-[n\theta]}(x)$ is a kernel type density estimator of $f(x)$ based on the observations $X_{[n\theta]+1}, \ldots, X_n$ under the hypothesis H_0. Furthermore $f_{[n\theta]}(x) - f^*_{n-[n\theta]}(x)$ is an estimator of $f_1(x) - f_2(x)$ under the hypothesis H_1. Let

$$f_n(x) = \frac{1}{nh_n} \sum_{j=1}^{n} K(\frac{x - X_j}{h_n}). \qquad (9.3.3)$$

The function $f_n(x)$ can be considered as an estimator of the density function $f(x)$ under H_0, that is, if $\theta = 1$. Define the function

$$d_n(\theta, x) = \begin{cases} [\frac{nh_n}{c_K f_n(x)}]^{1/2} \frac{[n\theta]}{n} \frac{n-[n\theta]}{n} (f_{[n\theta]}(x) - f^*_{n-[n\theta]}(x)) & \text{if } f_n(x) \neq 0 \\ 0 & \text{if } f_n(x) = 0. \end{cases}$$
$$(9.3.4)$$

Let

$$T_n(x) = \sup_{0 \le \theta \le 1} |d_n(\theta, x)|. \qquad (9.3.5)$$

It is clear that large values of $T_n(x)$ indicate that the densities $f_1(x)$ and $f_2(x)$ are different which implies that the hypothesis H_1 is likely to hold. However, small values of $T_n(x)$ indicate that $f_1(x)$ and $f_2(x)$ might be close at the particular value of x but it will not imply that the densities f_1 and f_2 coincide. Hence we compute the value of $T_n(x)$ for different values of $x = x_i, 1 \le i \le m$. Let

$$T_{n,m} = \max_{1 \le i \le m} T_n(x_i). \qquad (9.3.6)$$

Let

$$K_+(x) = \frac{1}{2}[V_K(x) + K(x)] \qquad (9.3.7)$$

and

$$K_-(x) = \frac{1}{2}[V_k(x) - K(x)]. \qquad (9.3.8)$$

The functions $K_+(x)$ and $K_-(x)$ are nondecreasing and furthermore $K(x) = K_+(x) - K_-(x)$. Since the functions $K_+(x)$ and $K_-(x)$ are nondecreasing and the sequence $X_1, \ldots, X_n$ are associated, it follows that the random variables $\{K_+(\frac{x-X_i}{h_n}), 1 \le i \le n\}$ and $\{K_-(\frac{x-X_i}{h_n}), 1 \le i \le n\}$ form associated sequences for every $n \ge 1$.

We will now derive the asymptotic distribution of the statistic $T_{n,m}$ under some additional conditions on the kernel $K(.)$ which ensure the following.

(A4) There exist positive constants c_1 and c_2 such that

$$\text{Var}[K_+(\frac{x - X_1}{h_n})] \geq c_1 n h_n \text{ and } \text{Var}[K_-(\frac{x - X_1}{h_n})] \geq c_1 n h_n, \quad (9.3.9)$$

$$E|K_+(\frac{x - X_1}{h_n})|^3 \leq c_2(nh_n)^{3/2} \text{ and } E|K_-(\frac{x - X_1}{h_n})|^3 \leq c_2(nh_n)^{3/2}$$
$$(9.3.10)$$

and there exists a function $u(r) \to 0$ as $r \to \infty$ such that

$$\sum_{j:|k-j|\geq r} \text{Cov}(K_+(\frac{x - X_j}{h_n}), K_+(\frac{x - X_k}{h_n})) \leq n h_n u(r) \text{ for all } k, n \text{ and } r,$$
$$(9.3.11)$$

and

$$\sum_{j:|k-j|\geq r} \text{Cov}(K_-(\frac{x - X_j}{h_n}, K_-(\frac{x - X_k}{h_n})) \leq n h_n u(r) \text{ for all } k, n \text{ and } r.$$
$$(9.3.12)$$

From the definition of the estimator $f_n(x)$, we have

$$(nh_n)^{1/2}(f_n(x) - E[f_n(x)]) \qquad (9.3.13)$$

$$= (nh_n)^{-1/2} \sum_{j=1}^{n} [K(\frac{x - X_j}{h_n}) - E(K(\frac{x - X_j}{h_n}))]$$

$$= (nh_n)^{-1/2} \sum_{j=1}^{n} [K_+(\frac{x - X_j}{h_n}) - E(K_+(\frac{x - X_j}{h_n}))]$$

$$- (nh_n)^{-1/2} \sum_{j=1}^{n} [K_-(\frac{x - X_j}{h_n}) - E(K_-(\frac{x - X_j}{h_n}))]$$

$$= (nh_n)^{1/2}(f_{n+}(x) - E[f_{n+}(x)])$$
$$- (nh_n)^{1/2}(f_{n-}(x) - E[f_{n-}(x)]) \quad \text{(say)}.$$

Under the condition (A4), the central limit theorem for a double array of associated random variables due to Cox and Grimmet (1984) (see Theorem 1.2.21) implies that

$$J_{n+} \equiv (nh_n)^{1/2}(f_{n+}(x) - E[f_{n+}(x)]) \xrightarrow{\mathcal{L}} N(0, \sigma_+^2) \qquad (9.3.14)$$

and

$$J_{n-} \equiv (nh_n)^{1/2}(f_{n-}(x) - E[f_{n-}(x)]) \xrightarrow{\mathcal{L}} N(0, \sigma_-^2) \qquad (9.3.15)$$

as $n \to \infty$ where $\sigma_+^2 = f(x)c_{K_+}$ and $\sigma_-^2 = f(x)c_{K_-}$.

In addition to the conditions (A1)–(A4), we assume that the following condition holds.

(A5) The sequences $\{J_{n+}^2, n \geq 1\}$ and $\{J_{n-}^2, n \geq 1\}$ are uniformly integrable.

Define the partial sum process, for $0 < \theta \leq 1$, by the equation

$$\hat{f}_{[n\theta]}(x) = \frac{1}{(nh_n)^{1/2}} \sum_{j=1}^{[n\theta]} (K(\frac{x - X_j}{h_n}) - E(K(\frac{x - X_j}{h_n}))) \tag{9.3.16}$$

$$= \frac{[n\theta]h_n}{(nh_n)^{1/2}} [f_{[n\theta]}(x) - E(f_{[n\theta]}(x))]$$

and $\hat{f}_0(x) = 0$.

Lemma 9.3.1. *Under the conditions* (A1)–(A5), *the collection of stochastic processes* $\{\hat{f}_{[n\theta]}(x), 0 \leq \theta \leq 1\}$, $n \geq 1$ *is tight.*

Proof. In order to prove the tightness of the family $\{\hat{f}_{[n\theta]}(x), 0 \leq \theta \leq 1\}$, $n \geq 1$, it is sufficient to prove that

$$\lim_{\lambda \to \infty} \limsup_{n \to \infty} \lambda^2 P[\max_{1 \leq r \leq n} |\sum_{j=1}^{r} [K(\frac{x - X_j}{h_n}) - E(K(\frac{x - X_j}{h_n}))]| \geq \lambda(nh_n)^{1/2}] = 0 \tag{9.3.17}$$

from the remarks following Theorem 8.4 in Billingsley (1968). Applying the maximal inequality for demimartingales derived in Chapter 2 (cf. Newman and Wright (1981)), we get that

$$\lambda^2 P[\max_{1 \leq r \leq n} |\sum_{j=1}^{r} [K_+(\frac{x - X_j}{h_n}) - E(K_+(\frac{x - X_j}{h_n}))]| \geq \lambda(nh_n)^{1/2}] \tag{9.3.18}$$

$$\leq 2\lambda^2 P[\sum_{j=1}^{n} [K_+(\frac{x - X_j}{h_n}) - E(K_+(\frac{x - X_j}{h_n}))]| \geq (\lambda - \sqrt{2})(nh_n)^{1/2}]$$

$$\leq \frac{\lambda^2}{(\lambda - \sqrt{2})^2} E[J_{n+}^2 I_{[|J_{n+}| \geq (\lambda - \sqrt{2})]}].$$

Hence, condition (A5) implies that

$$\lim_{\lambda \to \infty} \limsup_{n \to \infty} \lambda^2 P[\max_{1 \leq r \leq n} |\sum_{j=1}^{r} [K_+(\frac{x - X_j}{h_n}) - E(K_+(\frac{x - X_j}{h_n}))]| \geq \lambda(nh_n)^{1/2}] = 0. \tag{9.3.19}$$

Similarly we get that

$$\lim_{\lambda \to \infty} \limsup_{n \to \infty} \lambda^2 P[\max_{1 \leq r \leq n} |\sum_{j=1}^{r} [K_-(\frac{x - X_j}{h_n}) - E(K_-(\frac{x - X_j}{h_n}))]| \geq \lambda(nh_n)^{1/2}] = 0. \tag{9.3.20}$$

Note that $K(x) = K_+(x) - K_-(x)$. In view of equations (9.3.19) and (9.3.20), we obtain the relation (9.3.17) which implies tightness stated in the lemma. $\square$

The following two results describe the asymptotic behaviour of the test statistic $T_{n,m}$ as $n \to \infty$ under the hypotheses H_0 and H_1.

Theorem 9.3.2. *Let* $x_1, \ldots, x_m$, $m \geq 1$ *be distinct continuity points of the probability density function* f. *Suppose that the conditions* (A1)–(A5) *hold. Then, under the hypothesis* H_0,

$$(d_n(\theta, x_1), \ldots, d_n(\theta, x_m)) \xrightarrow{\mathcal{L}} (B_1(\theta), \ldots, B_m(\theta)) \text{ as } n \to \infty \qquad (9.3.21)$$

where $B_1(.), \ldots, B_m(.)$ *are independent Brownian bridges. Furthermore, under the hypothesis* H_0,

$$T_{n,m} \xrightarrow{\mathcal{L}} \max_{1 \leq i \leq m} \sup_{0 \leq \theta \leq 1} |B_i(\theta)| \text{ as } n \to \infty. \qquad (9.3.22)$$

Theorem 9.3.3. *Let* $x_1, \ldots, x_m$, $m \geq 1$ *be distinct continuity points of the probability density function* f_1 *and* f_2. *Suppose that the conditions* (A1)–(A5) *hold. Then, under the hypothesis* H_1,

$$T_{n,m} \xrightarrow{p} \infty \text{ as } n \to \infty. \qquad (9.3.23)$$

We now state two lemmas which will be used to prove Theorems 9.3.2 and 9.3.3.

Lemma 9.3.4. *Under the condition* (A2), *for any* $T > 0$, $j \geq 2$,

$$\sup_{x,y} |f_{1j}(x,y) - f(x)f(y)| \leq \frac{1}{4\pi^2} \int_{-T}^{T} \int_{-T}^{T} |\phi_{1j}(s,t) - \phi(s)\phi(t)| ds dt + \frac{6\sqrt{2}C(1+A)}{T}$$
$$(9.3.24)$$

where $A = \sup_x f(x)$ *and* C *is as defined in* (A2).

For a proof of this lemma, see Lin (2003). It is related to the inequality due to Sadikova (1966) discussed in Chapter 6. The next lemma is a consequence of the Bochner's theorem (cf. Prakasa Rao (1983)) and the arguments given in Bagai and Prakasa Rao (1991).

Lemma 9.3.5. *Under the conditions* (A1) *and* (A2),

$$f_n(x) \xrightarrow{p} f(x) \text{ as } n \to \infty \qquad (9.3.25)$$

for any continuity point x *of* f.

Proof of Theorem 9.3.2. Suppose the hypothesis H_0 holds. Define the partial sum process, for $0 \leq \theta \leq 1$,

$$g_n(\theta, x) = \frac{1}{(nh_n f_n(x) c_K)^{1/2}} \sum_{j=1}^{[n\theta]} \left(K\left(\frac{x - X_j}{h_n}\right) - E\left(K\left(\frac{x - X_j}{h_n}\right)\right)\right) \qquad (9.3.26)$$

$$= \left[\frac{nh_n}{f_n(x)c_K}\right]^{1/2} \frac{[n\theta]}{n} [f_{[n\theta]}(x) - E(f_{[n\theta]}(x))]$$

if $f_n(x) > 0$ and $g_n(\theta, x) = 0$ if $f_n(x) = 0$. Observe that

$$d_n(\theta, x) = g_n(\theta, x) - \frac{[n\theta]}{n} g_n(1, x). \tag{9.3.27}$$

Hence, in order to prove (9.3.21), it is sufficient to prove that

$$(g_n(\theta, x_1), \ldots, g_n(\theta, x_m)) \overset{\mathcal{L}}{\to} (W_1(\theta), \ldots, W_m(\theta)) \text{ as } n \to \infty \tag{9.3.28}$$

where $W_1(.), \ldots, W_m(.)$ are independent standard Wiener processes. This follows from the Skorokhod construction (cf. Billingsley (1968)). By the Cramer-Wold technique, it is sufficient to prove that for any $\lambda_1, \ldots, \lambda_m \in R$,

$$\sum_{i=1}^m \lambda_i \, g_n(\theta, x_i) \overset{\mathcal{L}}{\to} \sum_{i=1}^m \lambda_i W_i(\theta) \tag{9.3.29}$$

as $n \to \infty$. Let $g_n^*(\theta, x)$ be defined in the same way as $g_n(\theta, x)$ with $f(x)$ replacing $f_n(x)$ in (9.3.26). In order to prove (9.3.29), it is sufficient to prove that

$$\sum_{i=1}^m \lambda_i \, g_n^*(\theta, x_i) \overset{\mathcal{L}}{\to} \sum_{i=1}^m W_i(\theta) \tag{9.3.30}$$

in view of Lemma 9.3.5. Let

$$\sigma_n^2 \equiv \mathrm{Var}[\sum_{i=1}^m \lambda_i \, g_n^*(\theta, x_i)] \tag{9.3.31}$$

$$= \frac{1}{nh_n c_K}([n\theta] \, \mathrm{Var}(Z_1) + 2 \sum_{j=1}^{[n\theta]}([n\theta] - j + 1) \, \mathrm{Cov}(Z_1, Z_j))$$

where

$$Z_j = \sum_{i=1}^m (\frac{\lambda_i}{f(x_i)})^{1/2} K(\frac{x_i - X_j}{h_n}). \tag{9.3.32}$$

The first term on the right-hand side of equation (9.3.31) is equal to

$$\frac{1}{h_n c_K} \frac{[n\theta]}{n} [\sum_{i=1}^m \frac{\lambda_i^2}{f(x_i)} \, \mathrm{Var}(K(\frac{x_i - X_1}{h_n})) \tag{9.3.33}$$

$$+ \sum_{1 \le i \ne k \le m} \frac{\lambda_i \lambda_k J}{(f(x_i)f(x_k))^{1/2}} \, \mathrm{Cov}(K(\frac{x_i - X_1}{h_n}), K(\frac{x_k - X_1}{h_n}))].$$

The limit of the first term on the right-hand side of equation (9.3.33) is $(\sum_{i=1}^m \lambda_i^2)\theta$. This can be shown by using Bochner's theorem (cf. Prakasa Rao (1983)). We will now show that the second term on the right-hand side of equation (9.3.33) goes to zero as $n \to \infty$. Let $\varepsilon > 0$. Choose $M > 0$ such that

$$\int_{[|x| \geq M]} K(x)dx \leq \varepsilon.$$

Since $h_n \to 0$ as $n \to \infty$, we can choose n_0 such that, for $n \geq n_0$,

$$\min_{1 \leq i \neq k \leq m} \left| \frac{x_k - x_i}{h_n} \right| > 2M$$

and $h_n \leq \varepsilon$. Then there exists a constant $C > 0$ such that, for $n \geq n_0$,

$$\left| \text{Cov}(K(\frac{x_i - X_1}{h_n}), K(\frac{x_k - X_1}{h_n})) \right| \tag{9.3.34}$$

$$\leq \int_{-\infty}^{\infty} K(\frac{x_i - x}{h_n}) K(\frac{x_k - x}{h_n}) f(x)dx$$

$$+ \int_{-\infty}^{\infty} K(\frac{x_i - x}{h_n}) f(x)dx \int_{-\infty}^{\infty} K(\frac{x_k - x}{h_n}) f(x)dx$$

$$\leq Ch_n \int_{-\infty}^{\infty} K(u) K(\frac{x_k - x_i}{h_n} + u) \, du + Ch_n^2 f(x_i) f(x_k) (\int_{-\infty}^{\infty} K(u) \, du)^2$$

$$\leq Ch_n [\int_{[|u| \geq M]} K(u)du + \int_{[|u| < M]} K(\frac{x_k - x_i}{h_n} + u)du] + Ch_n^2$$

$$\leq C\varepsilon h_n.$$

Therefore the second term on the right-hand side of equation (9.3.33) tends to zero as $n \to \infty$. We will now get a bound on $Cov(Z_1, Z_j)$ using Lemma 9.3.4 and the characteristic function inequality in Newman and Wright (1981) (see Chapter 1). Observe that

$$|\text{Cov}(Z_1, Z_j)| \tag{9.3.35}$$

$$\leq (\sum_{i=1}^{m} |\frac{\lambda_i}{(f(x_i))^{1/2}}|)^2 \max_{i,k} |\text{Cov}(K(\frac{x_i - X_1}{h_n}), K(\frac{x_k - X_1}{h_n}))|$$

$$\leq C \max_{i,k} \int_{-\infty}^{\infty} \int_{-\infty}^{\infty} |K(\frac{x_i - y_1}{h_n}) K(\frac{x_k - y_2}{h_n})| (f_{1j}(y_1, y_2) - f(y_1)f(y_2))| \, dy_1 dy_2$$

$$\leq Ch_n^2 \frac{1}{4\pi^2} [\int_{-T}^{T} \int_{-T}^{T} |\phi_{1j}(s, t) - \phi(s)\phi(t)| \, dsdt + \frac{6\sqrt{2}C(1 + A)}{T}]$$

$$\leq Ch_n^2 [\frac{T^4}{4\pi^2} \text{Cov}(X_1, X_j) + \frac{6\sqrt{2}C(1 + A)}{T}]$$

$$\leq Ch_n^2 [\text{Cov}(X_1, X_j)]^{1/5}$$

by choosing $T = [\text{Cov}(X_1, X_j)]^{1/5}$. Note that $\text{Cov}(X_1, X_j) \geq 0$ since X_1 and X_j are associated random variables. Therefore the second term on the right-hand side of equation (9.3.31) given by

$$\frac{2}{nh_nc_K} \sum_{j=1}^{[n\theta]} ([n\theta] - j + 1) \operatorname{Cov}(Z_1, Z_j)) \leq 2Ch_n \sum_{j=2}^{n} [\operatorname{Cov}(X_1, X_j)]^{1/5}. \quad (9.3.36)$$

Note that the term on the right-hand side of equation (9.3.36) tends to zero as $n \to \infty$ by the condition (A3). Combining the above results, we get that

$$\sigma_n^2 \to (\sum_{i=1}^{n} \lambda_i^2)\theta \quad \text{as} \quad n \to \infty. \quad (9.3.37)$$

Note that

$$\sum_{i=1}^{m} \lambda_i \, g_n^*(\theta, x_i) \quad (9.3.38)$$

$$= \frac{1}{(nh_nc_K)^{1/2}} \sum_{i=1}^{m} \lambda_i \sum_{j=1}^{[n\theta]} \frac{1}{\sqrt{f(x_i)}} (K(\frac{x_i - X_j}{h_n}) - E[K(\frac{x_i - X_j}{h_n})])$$

$$= \sum_{j=1}^{[n\theta]} \frac{1}{(nh_nc_K)^{1/2}} \sum_{i=1}^{m} \frac{\lambda_i}{\sqrt{f(x_i)}} (K(\frac{x_i - X_j}{h_n}) - E[K(\frac{x_i - X_j}{h_n})])$$

$$= \sum_{j=1}^{[n\theta]} \eta_{nj} \quad \text{(say)}$$

where

$$\eta_{nj} = \frac{1}{(nh_nc_K)^{1/2}} \sum_{i=1}^{m} \frac{\lambda_i}{\sqrt{f(x_i)}} (K(\frac{x_i - X_j}{h_n}) - E[K(\frac{x_i - X_j}{h_n})]). \quad (9.3.39)$$

We will now prove that

$$\sum_{j=1}^{[n\theta]} \eta_{nj} \xrightarrow{\mathcal{L}} N(0, (\sum_{i=1}^{n} \lambda_i^2)\theta) \quad \text{as} \quad n \to \infty. \quad (9.3.40)$$

Let

$$H(x) = \sum_{i=1}^{m} \frac{\lambda_i}{\sqrt{f(x_i)}} (K(\frac{x_i - X_j}{h_n}) - E[K(\frac{x_i - X_j}{h_n})]). \quad (9.3.41)$$

It is easy to see that the function $H(x)$ is a function of bounded variation. Furthermore

$$|E[\exp\{it \sum_{j=1}^{k} \eta_{nj}\}] - E[\exp\{it \sum_{j=1}^{k-1} \eta_{nj}\}] E[it \exp\{\eta_{nk}\}]| \quad (9.3.42)$$

$$\le \frac{4t^2}{nh_nc_K} \sum_{j=1}^{k-1} |\operatorname{Cov}(T_H(X_1), T_H(X_j))|.$$

By repeated application of this inequality, it follows that

$$|E[\exp\{it\sum_{j=1}^{[n\theta]} \eta_{nj}\}] - \Pi_{j=1}^{[n\theta]} E[it\exp\{\eta_{nj}\}]| \tag{9.3.43}$$

$$\le \frac{4t^2}{nh_nc_K} \sum_{1\le i\ne j\le[n\theta]} |\operatorname{Cov}(T_H(X_i), T_H(X_j))|$$

$$\le \frac{4t^2}{h_nc_K} \sum_{j=2}^{[n\theta]} |\operatorname{Cov}(T_H(X_1), T_H(X_j))|$$

$$\le \frac{4t^2 h_n}{c_K} [\sum_{i=1}^{m} \frac{|\lambda_i|}{\sqrt{f(x_i)}}]^2 (\int_{-\infty}^{\infty} |T_K(x)|dx) \sum_{j=2}^{[n\theta]} [\operatorname{Cov}(X_1), X_j)]^{1/5}$$

$$\le Ch_n \sum_{j=2}^{[n\theta]} [\operatorname{Cov}(X_1, X_j)]^{1/5}$$

by noting that

$$T_H(x) \le \sum_{i=1}^{m} |\frac{\lambda_i}{\sqrt{f(x_i}} T_K(\frac{x_i - x}{h_n})| \tag{9.3.44}$$

and

$$\int_{-\infty}^{\infty} |T_K(x)|dx < \infty.$$

The last term in equation (9.3.43) tends to zero as $n \to \infty$ by condition (A3). Hence

$$|E[\exp\{it\sum_{j=1}^{[n\theta]} \eta_{nj}\}] - \Pi_{j=1}^{[n\theta]} E[it\exp\{\eta_{nj}\}]| \to 0 \text{ as } n \to \infty. \tag{9.3.45}$$

As a consequence of the central limit theorem for a double array of sums of independent random variables (cf. Loeve (1963)), it follows that

$$\sum_{j=1}^{[n\theta]} \eta_{nj} \overset{\mathcal{L}}{\to} N(0, (\sum_{i=1}^{m} \lambda_i^2)\theta) \text{ as } n \to \infty. \tag{9.3.46}$$

This in turn proves (9.3.29). This result can also be seen as a consequence of the central limit theorem for double array of associated random variables due to Cox and Grimmet (1984) stated in Chapter 1. The tightness of the sequence of processes $\{(g_n^*(\theta, x_1), \ldots, g_n^*(\theta, x_m)), 0 \le \theta \le 1\}$, $n \ge 1$ follows from Lemma 9.3.1. From the comments made earlier, it follows that the sequence of processes $\{(g_n(\theta, x_1), \ldots, g_n(\theta, x_m)), 0 \le \theta \le 1\}$, $n \ge 1$ converge in distribution to the m-dimensional process $\{W_1(\theta), \ldots, W_m(\theta)), 0 \le \theta \le 1\}$ from the results in Billingsley (1968). The results stated in Theorem 9.3.2 follow by an application of the continuous mapping theorem (cf. Billingsley (1968)).   □

Proof of Theorem 9.3.3. Suppose the hypothesis H_1 holds. Without loss of generality, suppose that $f_1(x_1) \ne f_2(x_1)$ for some $x_1 \in R$. Let $0 < \theta < 1$. Since

$$f_{[n\theta]}(x_1) \xrightarrow{p} f_1(x_1) \quad \text{as } n \to \infty \tag{9.3.47}$$

and

$$f^*_{n-[n\theta]}(x_1) \xrightarrow{p} f_2(x_1) \quad \text{as } n \to \infty, \tag{9.3.48}$$

under the conditions (A1) and (A2), and since $nh_n \to \infty$ by hypothesis, it follows that

$$d_n(\theta, x_1) \xrightarrow{p} \infty \quad \text{as } n \to \infty. \tag{9.3.49}$$

Furthermore observe that

$$T_{nm} \ge d_n(\theta, x_1). \tag{9.3.50}$$

Hence

$$T_{nm} \xrightarrow{p} \infty \quad \text{as } n \to \infty \tag{9.3.51}$$

under the hypothesis H_1.   □

Bibliography

Abramowitz, M. and Stegun, J. (1965) *Handbook of Mathematical Functions*, Dover, New York.

Agbeko, N. (1986) Concave function inequalities for sub-(super-)martingales, *Ann. Univ. Sci. Budapest, Sect. Math.*, **29**, 9-17.

Alsmeyer, G. and Rosler, U. (2006) Maximal ϕ-inequalities for nonnegative submartingales, *Theor. Probab. Appl.*, **50**, 118-128.

Amirdjanova, A. and Linn, M. (2007) Stochastic evolution equations for nonlinear filtering of random fields in the presence of fractional Brownian sheet observation noise, arXiv:0707.3856v1 [math.PR] 26 July 2007.

Arjas, E. and Norros, I. (1984) Life lengths and association : a dynamic approach, *Math. Operat. Res.*, **9**, 151-158.

Azevedo, C. and Oliveira, P.E. (2000) Kernel-type estimation of bivariate distribution function for associated random variables, *New Trends in Probability and Statistics*, Vol.5 (Tartu, 1999), pp.17-25, VSP, Utrecht.

Azuma, K. (1967) Weighted sums of certain dependent random variables, *Tohoku Math. J.*, **19**, 357-367.

Baek, Jong Li., Park, Sung Tae., Chung, Sung Mo. and Seo, Hye Young. (2005) On the almost sure convergence of weighted sums of negatively associated random variables, *Commun. Korean Math. Soc.*, **20**, 539-546.

Baek, Jong Li., Seo, Hye-Young., and Lee, Gil Hwan. (2008) On precise asymptotics in the law of large numbers of associated random variables, *Honam Math. J.*, **30**, 9-20.

Bagai, I. and Prakasa Rao, B.L.S. (1991) Estimation of the survival function for stationary associated processes, *Statist. Probab. Lett.*, **12**, 385-391.

Bagai, I. and Prakasa Rao, B.L.S. (1992) Analysis of survival data with two dependent competing risks, *Biom. J.*, **7**, 801-814.

Bagai, I. and Prakasa Rao, B.L.S. (1995) Kernel-type density and failure rate estimation for associated sequences, *Ann. Inst. Statist. Math.*, **47**, 253-266.

Balan, R. (2005) A strong invariance principle for associated random fields, *Ann. Probab.*, **33**, 823-840.

Barlow, R.E. and Proschan, F. (1975) *Statistical Theory of Reliability and Life Testing : Probability Models*, Holt, Rinehart and Winston.

Basawa, I.V. and Prakasa Rao, B.L.S. (1980) *Statistical Inference for Stochastic Processes*, Academic Press, London.

Batty, C.J.K. (1976) An extension of an inequality of R. Holley, *Quart. J. of Math.*, **27**, 457-461.

Becker, N.G. and Utev, S. (2001) Threshold results for U-statistics of dependent binary variables, *J. Theoret. Probab.*, **14**, 97-114.

Billingsley, P. (1961) *Statistical Inference for Markov Processes*, University of Chicago Press, Chicago.

Billingsley, P. (1968) *Convergence of Probability Measures*, Wiley, New York.

Birkel, T. (1986) *Momentenabschatzungen und Grenzwertsatze fur Partialsummen Assoziierter Zufallsvariablen.* Doctoral thesis, Univ. of Koln.

Birkel, T. (1988a) Moment bounds for associated sequences, *Ann. Probab.*, **16**, 1184-1193.

Birkel, T. (1988b) On the convergence rate in the central limit theorem for associated processes, *Ann. Probab.*, 16, 1689-1698.

Birkel, T. (1988c) The invariance principle for associated processes, *Stoch. Process. Appl.*, **27**, 57-71.

Birkel, T. (1989) A note on the strong law of large numbers for positively dependent random variables, *Statist. Probab. Lett.*, **7**, 17-20.

Block, H.W. and Fang, Z. (1988) A multivariate extension of Hoeffding's Lemma, *Ann. Probab.*, **16**, 1803-1820.

Boutsikas, Michael V. and Koutras, Markos V. (2000) A bound for the distribution of the sum of discrete associated or negatively associated random variables, *Ann. Appl. Probab.*, **10**, 1137-1150.

Budsaba, K., Chen, P., Panishkan, K. and Volodin, A. (2009) Strong laws for certain types of U-statistics based negatively associated random variables, *Siberian Adv. Math.*, **19**, 225-232.

Bulinski, A.V. (1993) Inequalities for the moments of sums of associated multi-index variables, *Theor. Probab. Appl.* , **38**, 342-349.

Bulinski, A.V. (1995) Rates of convergence in the central limit theorem for fields of associated random variables, *Theor. Probab. Appl.* , **40**, 136-144.

Bulinski, A.V. (1996) On the convergence rates in the CLT for positively and negatively dependent random fields, In *Probability Theory and Mathematical Statistics*, Ed. I.A. Ibragimov and A. Yu. Zaitsev, Gordon and Breach, UK.

Bulinski, A.V. (2011) On the Newman conjecture, arXiv: 1104.4180v1 [math.PR] 21 2011.

Bulinski, A.V. and Shabanovich, E. (1998) Asymptotical behaviour for some functionals of positively and negatively dependent random fields, *Fundam. Prikl. Mat.* , **4**, 479-492 (in Russian).

Bulinski, A.V. and Shaskin, A. (2007) *Limit Theorems for Associated Random Fields and Related Systems*, World Scientific, Singapore.

Burton, R.M., Dabrowski, A.R. and Dehling, H. (1986) An invariance principle for weakly associated random variables, *Stoch. Proc. Appl.*, **23**, 301-306.

Burton, R.M. and Waymire, E. (1985) Scaling limits for associated random measures, *Ann. Probab.*, **13**, 1267-1278.

Burton, R.M. and Waymire, E. (1986) The central limit problem for infinitely divisible random measures. In *Dependence in Probability and Statistics*, Ed. Taqqu, M. Eberlein, E., Birkhauser, Boston.

Cairoli, R. and Walsh, J. (1975) Stochastic integrals in the plane, *Acta Math.*, **134**, 111-183.

Chaubey, Y.P., Doosti, H. and Prakasa Rao, B.L.S. (2006) Wavelet based estimation of the derivatives of a density with associated variables, *Inter. J. Pure and App. Math.*, **27**, 97-106.

Chaubey, Y.P., Doosti, H. and Prakasa Rao, B.L.S. (2008) Wavelet based estimation of the derivatives of a density for a negatively associated process, *J. Statist. Theory and Practice*, **2**, 453-463.

Chernoff, H. and Zacks, S. (1964) Estimating the current mean of a normal distribution which is subjected to changes in time, *Ann. Math. Statist.*, **35**, 999-1018.

Chow, Y.S. (1960) A martingale inequality and the law of large numbers, *Proc. Amer. Math. Soc.*, **11**, 107-111.

Chow, Y.S. and Teicher, H. (1988) *Probability Theory: Independence, Interchangeability, Martingales*, Second Edition, Springer Verlag, new York.

Christofides, T.C. (2000) Maximal inequalities for demimartingales and a strong law of large numbers, *Statist. Probab. Lett.*, **50**, 357-363.

Christofides, T.C. (2003) Maximal inequalities for N-demimartingales, *Arch. Inequal. Appl.*, **1**, 387-397.

Christofides, T.C. (2004) U-statistics on associated random variables, *J. Statist. Plann. Infer.*, **119**, 1-15.

Christofides, T.C. and Hadjikyriakou, M. (2009) Exponential inequalities for N-demimartingales and negatively associated random variables, *Statist. Probab. Lett.*, **79**, 2060-2065.

Christofides, T.C. and Hadjikyriakou, M. (2010) Maximal inequalities for multi-dimensionally indexed demimartingales and the Hajek-Renyi inequality for associated random variables, Preprint, University of Cyprus, Nicosia.

Christofides, T.C. and Vaggelatou, E. (2004) A connection between supermodular ordering and positive/negative association, *J. Multivariate. Anal.*, **88**, 138-151.

Chung, K.L. (1974) *A Course in Probability Theory*, Academic Press, New York.

Chung, M., Rajput, B.S. and Tortrat, A. (1982) Semistable laws on topological vector spaces. *Z. Wahrsch. Theor. verw Gebiete* , **60**, 209-218.

Cox, J.T. and Grimmett, G. (1981) Central limit theorems for percolation models, *J. Statist. Phys.*, **25**, 237-251.

Cox, J.T. and Grimmett, G. (1984) Central limit theorems for associated random variables and the percolation model, *Ann. Probab.*, **12**, 514-528.

Cuadras, C.M. (2002) On the covariance between functions, *J. Multivariate Anal.*, **81**, 19-27.

Dabrowski, A.R. (1985) A functional law of the iterated logarithm for associated sequences, *Statist. Probab. Lett.*, **3**, 209-212.

Dabrowski, A.R. and Dehling, H. (1988) A Berry-Esseen theorem and a functional law of the iterated logarithm for weakly associated random variables, *Stochastic Process. Appl.*, **30**, 247-289.

Daley, D.J. (1968) Stochastically monotone Markov chains, *Z. Wahrsch. Theor. verw Gebiete*, **10**, 305-317.

Daubechies, I. (1988) Orthogonal bases of compactly supported wavelets, *Comm. Pure Appl. Math.*, **41**, 909-996.

Denker, M. and Keller, G. (1983) On U-statistics and von-Mises statistics for weakly dependent processes, *Z. Wahrsch. Theor. verw. Gebiete* , **64**, 505-522.

Devroye, L. and Gyorfi, L. (1985) *Nonparametric Density Estimation – The L_1 View*, Wiley, New York.

Dewan, I. and Prakasa Rao, B.L.S. (1997a) Remarks on the strong law of large numbers for a triangular array of associated random variables, *Metrika* , **45**, 225-234..

Dewan, I. and Prakasa Rao, B.L.S. (1997b) Remarks on Berry-Esseen type bound for stationary associated random variables, *Gujarat Statist. Rev.* , **24**, 19-20.

Dewan, I. and Prakasa Rao, B.L.S. (1999) A general method of density estimation for associated random variables, *J. Nonparametric Statist.*, **10**, 405-420.

Dewan, I. and Prakasa Rao, B.L.S. (2001) Asymptotic normality of U-statistics of associated random variables, *J. Statist. Plann. Infer.*, **97**, 201-225.

Dewan, I. and Prakasa Rao, B.L.S. (2002) Central limit theorem for U-statistics of associated random variables, *Statist. Probab. Lett.*, **57**, 9-15.

Dewan, I. and Prakasa Rao, B.L.S. (2005) Wilcoxon-signed rank test for associated sequences, *Statist. Probab. Lett.*, **71**, 131-142.

Dewan, I. and Prakasa Rao, B.L.S. (2005a) Non-uniform and uniform Berry-Esseen type bounds for stationary associated sequences, *J. Nonparametric Stat.*, **17**, 217-235.

Dewan, I. and Prakasa Rao, B.L.S. (2006) Hajek-Renyi type inequality for some non-monotonic functions of associated random variables, *Journal of Inequalities and Applications*, Volume 2006, Article ID 58317, 8 pp.

Dewan, I. and Prakasa Rao, B.L.S. (2006) Mann-Whitney test for associated sequences, *Ann. Inst. Statist. Math.*, **55**, 111-119.

Dong, Zhi Shan., and Yang, Xiao Yun. (2002) A strong law of large numbers for negatively associated random variables without stationary distribution *Chinese J. Appl. Probab. Statist.*, **18**, 357-362 (in Chinese).

Doob, J.L. (1953) *Stochastic Processes*, Wiley, New York.

Douge, L. (2007) Vitesses de convergence dans la loi forte des grands nombres et dans l'estimation de la densite pour des variables aleatoires associees, *C. R. Acad. Sci. Paris. Ser. I*, **344**, 515-518.

Dufresnoy, J. (1967) Autour de l'inegalite de Kolmogorov, *C. R. Acad. Sci. Paris.*, **264 A**, 603.

Eagleson, G.K. (1979) Orthogonal expansions and U-statistics, *Austral. J. Stat.*, **21**, 221-237.

Efromovich, S. (2000) *Nonparametric Curve Estimation: Methods, Theory and Applications*, Springer, New York.

Esary, J., and Proschan, F. (1970) A reliability bound for systems of maintained, interdependent components, *J. Amer. Statist. Assoc.*, **65**, 329-338.

Esary, J., Proschan, F. and Walkup, D. (1967) Association of random variables with applications, *Ann. Math. Statist.*, **38**, 1466-1474.

Esseen, C.G. (1956) A moment inequality with an application to the central limit theorem, *Skand. Aktuarie.*, **39**, 160-170.

Evans, S.N. (1990) Association and random measures, *Probab. Th. Re. Fields*, **86**, 1-19.

Fazekas, I. and Klesov, O. (2001) A general approach to the strong law of large numbers, *Theor. Probab. Appl.* , **45**, 436-449.

Feller, W. (1977) *An Introduction to Probability Theory and its Applications*, Vol.II, Wiley Eastern, New Delhi.

Foldes, A. (1974) Density estimation for dependent samples, *Studia Sci. Math. Hungar.*, **9**, 443-452.

Foldes, A. and Revesz, P. (1974) A general method of density estimation , *Studia Sci. Math. Hungar.*, **9**, 82-92.

Foley, R.D. and Kiessler, P.C. (1989) Positive correlations in a three node Jackson queuing network, *Adv. Appl. Probab.*, **21**, 241-242.

Fortuin, C., Kastelyn, P. and Ginibre, J. (1971) Correlation inequalities on some partially ordered sets, *Comm. Math. Phys.*, **22**, 89-103.

Fu, Ke-Ang (2009) Exact rates in log law for positively associated random variables, *J. Math. Anal. Appl.*, **356**, 280-287.

Fu, Ke-Ang and Hu, Li-Hua (2010) Moment convergence rates of LIL for negatively associated sequences, *J. Korean Math. Soc.*, **47**, 263-275.

Fu, Ke-Ang and Zhang, Li-Xin (2007) Precise rates in the law of the logarithm for negatively associated random variables, *Comput. Math. Appl.*, **54**, 687-698.

Gardner Jr., L.A. (1969) On detecting changes in the mean of a normal variates, *Ann. Math. Statist.*, **40**, 116-126.

Glasserman, P. (1992) Processes with associated increments, *J. Appl. Probab.*, **29**, 313-333.

Gonchigdanzan, Khurelbaatar. (2002) Almost sure central limit theorems for strongly mixing and associated random variables, *Int. J. Math. Math. Sci.*, **29**, 125-131.

Gregory, G.G. (1977) Large sample theory for U-statistics and tests of fit, *Ann. Statist.*, **5**, 110-123.

Gu, Wentao., Roussas, G.G., and Tan, L.T. (2007) On the convergence rate of fixed design regression estimators for negatively associated random variables, *Statist. Probab. Lett.*, **77**, 1214-1224.

Guttorp, Peter (1991) *Statistical Inference for Branching Processes*, Wiley, New York.

Gyires, B. (1981) Linear forms in random variables defined on a homogeneous Markov chain, In *The First Pannonian Symposium on Mathematical Statistics*, Ed. P. Revesz et al., Lecture Notes in Statistics No.8, Springer Verlag, New York, pp. 110-121.

Hadjikyriakou, M. (2010) *Probability and Moment Inequalities for Demimartingales and Associated Random Variables*, Ph.D Dissertation, University of Cyprus, Nicosia.

Hdajikyriakou, M. (2011) Marcinkiewicz-Zygmund inequality for nonnegative N-demimartingales and related results, *Statist. Probab. Lett.*, **81**, 678-684.

Hajek, J. and Renyi, A. (1955) A generalization of an inequality of Kolmogorov, *Acta. Math. Acad. Sci. Hung.*, **6**, 281-284.

Hall, P. (1979) An invariance theorem for U-statistics, *Stoch. Proc. Appl.*, **9**, 163-174.

Hall, P. (1982) *Rates of Convergence in the Central Limit Theorem*, Pitman, Boston.

Han, Kwang-Hee (2007) Exponential inequality and almost sure convergence for the negatively associated sequence, *Honam Math. J.*, **29**, 367-375.

Harremoës, P. (2008) Some new maximal inequalities, *Statist. Probab. Lett.*, **78**, 2776-2780.

Harris, T.E. (1960), A lower bound for the critical probability in a certain percolation process, *Proc. Camb. Phil. Soc.*, **59**, 13-20.

Henriques, C. and Oliveira, P.E. (2008) Large deviations for the empirical mean of associated random variables, *Statist. Probab. Lett.*, **78**, 594-598.

Herrndorf, N. (1984) An example on the central limit theorem for associated sequences, *Ann. Probab.*, **12**, 912-917.

Hettsmansperger, T.P. (1984) *Statistical Inference Based on Ranks*, Wiley, New York.

Hjort, N.L., Natvig, B. and Funnemark, E. (1985) The association in time of a Markov process with application to multistate reliability theory, *J. Applied Probab.*, **22**,473-479.

Hoeffding, W. (1940) Masstabinvariante Korrelations-theorie, *Schr. Math. Inst., University Berlin*, **5**, 181-233.

Hoeffding, W. (1948) A class of statistics with asymptotically normal distribution, *Ann. Math. Statist.*, **19**, 293-325.

Holley, R. (1974) Remarks on the FKG inequalities, *Comm. Math. Phys.*, **36**, 227-231.

Hu, S.H. and Hu, M. (2006) A general approach to strong law of large numbers, *Statist. Probab. Lett.*, **76**, 843-851.

Hu, S.H., Chen, G.J., and Wang, X.J. (2008) On extending the Brunk-Prohorov strong law of large numbers for martingale differences, *Statist. Probab. Lett.*, **78**, 3187-3194.

Hu, Shuhe., Shen, Yan., Wang, Xuejun., and Yang, Wenzhi (2010) A note on the inequalities for N-demimartingales and demimartingales, Preprint, Anhui University, Hefei.

Hu, T. and Hu , J. (1998), Comparison of order statistics between dependent and independent random variables, *Statist. Probab. Letters*, **37**, 1-6.

Huang, Wei. (2003) A law of the iterated logarithm for geometrically weighted series of negatively associated random variables, *Statist. Probab. Lett.*, **63**, 133-143.

Huang, Wei. (2004) A nonclassical law of the iterated logarithm for functions of negatively associated random variables, *Stochastic Anal. Appl.*, **22**, 657-678.

Huang, Wei. and Zhang, Lin-Xi. (2006) Asymptotic normality for U-statistics of negatively associated random variables, *Statist. Probab. Lett.*, **76**, 1125-1131.

Huang Wen-Tao., and Xu, Bing. (2002) Some maximal inequalities and complete convergences of negatively associated random sequences, *Statist. Probab. Lett.*, **57**, 183-191.

Ioannides, D.A. and Roussas, G.G. (1999) Exponential inequality for associated random variables, *Statist. Probab. Lett.*, **42**, 423-431.

Jabbari, H., Jabbari, M. and Azarnoosh, H.A. (2009) An exponential inequality for negatively associated random variables, *Electron J. Stat.*, **3**, 165-175.

Jacobsen, M. (1982) *Statistical Analysis of Counting Process*, Lecture Notes in Statistics, No. 12, Springer-Verlag, New York.

Jiang, Ye. (2003) A nonclassical law of iterated logarithm for negatively associated random variables, it Appl. Math. J. Chinese Univ. Ser. B, **18**, 200-208.

Jing, Bing-Yi. and Liang, Han-Ying. (2008) Strong limit theorems for weighted sums of negatively associated random variables, *J. Theoret. Probab.*, **21**, 890-909.

Joag-Dev, K. (1983) Independence via uncorrelatedness under certain dependence structures, *Ann. Probab.*, **11**, 1037-1041.

Joag-Dev, K. and Proschan, F. (1983) Negative association of random variables with applications, *Ann. Statist.*, **11**, 286-295.

Karlin, S. and Rinott, Y. (1980) Classes of orderings of measures and related correlation inequalities .1. Multivariate totally positive distributions, *J. Multivariate. Anal.*, **10**, 467-498.

Karr, A.F. (1991) *Point Processes and their Statistical Inference*, Marcel Dekker, New York.

Keilson, J. and Kester, A. (1977) Monotone matrices and monotone Markov processes, *Stochastic Processes. Appl.*, **5**, 231-241.

Kemperman, J.H.B. (1977) On the FKG inequalities for measures on a partially ordered space, *Indag. Math.*, **39**, 313-331.

Khoshnevisan, D. and Lewis, T. (1998) A law of the iterated logarithm for stable processes in random scenery, *Stochastic Processes. Appl.*, **74**, 89-121.

Kim, Tae-Sung. and Ko, Mi-Hwa. (2003) On the almost sure convergence of weighted averages of associated and negatively associated random variables, *Stochastic analysis and applications*, Vol.3, 69-76, Nova Sci. Publ., Hauppauge, NY.

Kirstein, B.M. (1976) Monotonicity and comparability of time-homogeneous Markov processes, *Math. Operationsforsch. Statist.*, **7**, 151-168.

Kuber, S. and Dharamadikari, A. (1996) Association in time of a finite semi-Markov process, *Statist. Probab. Lett.*, **26**, 125-133.

Kuczmaszewska, Anna. (2009) On complete convergence for arrays of row wise negatively associated random variables, *Statist. Probab. Lett.*, **79**, 116-124.

Kutoyants, Yu. A. (1984) *Parameter Estimation for Stochastic Processes* (trans. and ed. B.L.S. Prakasa Rao), Heldermann, Berlin.

Kutoyants, Yu. A. (1998) *Statistical Inference for Spatial Poisson Processes*, Lecture Notes in Statistics No. 134, Springer-Verlag, New York.

Kutoyants, Yu. A. (2004) *Statistical Inference for Ergodic Diffusion Processes*, Springer-Verlag, London.

Leblanc, F. (1996) Wavelet linear density estimator for a discrete-time stochastic process: L_p-losses, *Statist. Probab. Lett.*, **27**, 71-84.

Lebowitz, J. (1972) Bounds on the correlations and analyticity properties of ferromagnetic Ising spin systems, *Comm. Math. Phys.*, **28**, 313-321.

Lee, A.J. (1990) *U-statistics*, Marcel Dekker, New York.

Lee, M.T., Rachev, S.T. and Samorodnitsky, G. (1990) Association of stable random variables, *Ann. Probab.*, **18**, 387-397.

Lefevre, C. and Milhaud, X. (1990) On the association of the life lengths of components subjected to a stochastic environment, *Adv. App. Probab.*, **22**, 961-964.

Lehmann, E.L. (1966) Some concepts of dependence, *Ann. Math. Statist.*, **37**, 1137-1153.

Lewis, T. (1998) Limit theorems for partial sums of quasi-associated random variables, In *Asymptotic Methods in Probability and Statistics*, Ed. B. Szysskowicz, Elsevier, pp.31-48.

Li, Degui., and Lin, Zhengyan. (2007) A nonparametric test for the change of the density function under association, *J. Nonparametric Stat.*, **19**, 1-12.

Li, Yong-ming., and Yang, Shan-chao. (2006) Asymptotic normality the empirical distribution under negatively associated sequences and its applications, *J. Math. Res. Exposition*, **26**, 457-464.

Liang, Han-Ying. (2000) Complete convergence for weighted sums of negatively associated random variables, *Statist. Probab. Lett.*, **48**, 317-325.

Liang, Han-Ying., and Jing, Bing-Yi. (2005) Asymptotic properties of estimates of nonparametric regression models based on negatively associated sequences, *J. Multivariate Anal.*, **95**, 227-245.

Liang, Han-Ying., Li, De Li., and Rosalsky, Andrew (2010) Complete moment and integral convergence for sums of negatively associated random variables, *Acta Math. Sin. (Eng. Ser.)*, **26**, 419-432.

Lin, Zhengyan. (1997) An invariance principle for negatively associated random variables, *Chinese Sci. Bull.*, **42**, 359-364.

Lin, Zhengyan. (2003) Asymptotic normality of the kernel estimate of a probability density function under association dependence, *Acta Mathematica Scientia*, **23 B**, 345-350.

Lin, Zhengyan., and Bai, Zhidong,. (2010) *Probability Inequalities*, Science Press, Beijing and Springer-Verlag, Berlin.

Lindqvist, B.H. (1988) Association of probability measures on partially ordered spaces, *J. Multivariate. Anal.*, **26**, 111-132.

Liu, Li-Xin. and Cheng, Shi Hong. (2008) Strong law of large numbers for negatively associated variables and their applications, *Acta Math. Sinica (Chin. Ser.)*, **51**, 275-280.

Liu, Li-Xin. and Mei, C.-L. (2004) A general law of the iterated logarithm for negatively associated random variables, *J. Math. Sci. (N.Y.)*, **123**, 3767-3775.

Liu, Yan. (2007) Precise large deviations for negatively associated random variables with consistently varying tails, *Statist. Probab. Lett.*, **77**, 181-189.

Loeve, M. (1977) *Probability Theory*, Springer Verlag, New York.

Louhichi, S. (2000) Convergence rates in the strong laws for associated random variables, *Probab. Math. Statist.*, **20**, 203-214.

Majerak, D., Nowak, W. and Zieba, W. (2005) Conditional strong law of large numbers, *Inter. J. Pure and Appl. Math.*, **20**, 143-157.

Marshall, A.W. and Olkin, I. (1967) A multivariate exponential distribution, *J. Amer. Statist. Assoc.*, **62**, 30-44.

Marshall, A.W. and Olkin, I. (1979) *Inequalities: Theory of Majorization and its Applications*, Academic Press, New York.

Matula, P. (1996a)A remark on the weak convergence of sums of associated random variables, *Ann. Univ. Mariae Curie-Sklodowska Sect.A*, **50**, 115-123.

Matula, P. (1996b) Convergence of weighted averages of associated random variables, *Probab. Math. Statist.*, **16**, 337-343.

Matula, P. (1997) Probability and moment bounds for sums of negatively associated random variables, *Theory Probab. Math. Statist.*, **55**, 135-141.

Matula, P. (1997a) The Glivenko-Cantelli lemma for a class of discrete associated random variables, *Ann. Univ. MarCurie-Sklodowska Sect. A*, **51**, 129-132.

Matula, P. (1998) The almost sure central limit theorem for associated random variables, *Probab. Math. Statist.*, **18**, 411-416.

Matula, P. (2001) Limit theorems for sums of non-monotonic functions of associated random variables, *J. Math. Sci. (New York)*, **105**, 2590-2593.

Matula, P. (2005) On almost sure limit theorems for positively dependent random variables, *Statist. Probab. Lett.*, **74**, 59-66.

Matula, P. and Rychlik, Z. (1990) The invariance principle for non-stationary sequences of associated random variables, *Ann. Inst. Henri Poincaré Probab. Statist.*, **26**, 387-397.

Meyer, Y. (1990) *Ondolettes et Operateurs*, Hermann, Paris.

Mikusheva, A.E. (2000) On the complete convergence of sums of negatively associated random variables, *Math. Notes*, **68**, 355-362.

Nagaraj, N.K. (1990) Two sided tests for change in level of correlation data, *Commun. Statist.* **B**, **19**, 896-878.

Nagaraj, N.K. and Reddy, C.S. (1993) Asymptotic null distributions of tests for change in level in correlated data, *Sankhya A*, **55**, 37-48.

Nandi, H.K. and Sen, P.K. (1963) On the properties of U-statistics when the observations are not independent II, *Calcutta Statist. Assoc. Bull.*, **12**, 125-143.

Newman, C.M. (1980) Normal fluctuations and the FKG inequalities, *Comm. Math. Phys.*, **74**, 119-128.

Newman, C.M. (1983) A general central limit theorem for FKG systems, *Comm. Math. Phys.*, **91**, 75-80.

Newman, C.M. (1984) Asymptotic independence and limit theorems for positively and negatively dependent random variables. In : *Inequalities in Statistics and Probability*, Tong, Y.L. (ed.), 127-140, IMS, Hayward.

Newman, C.M. and Wright, A.L. (1981) An invariance principle for certain dependent sequences, *Ann. Probab.*, **9**, 671-675.

Newman, C.M. and Wright, A.L. (1982) Associated random variables and martingale inequalities, *Z. Wahrsch. Theorie und Verw. Gebiete*, **59**, 361-371.

Nezakati, A. (2005) A note on the strong law of large numbers for associated sequences, *Int. J. Math. Math. Sci.*, **19**, 3195-3198.

Oliveira, P.E. (2005) An exponential inequality for associated variables, *Statist. Probab. Lett.*, **78**, 189-197.

Oliveira, P.E. and Suquet, C. (1995) $L^2(0,1)$ weak convergence of the empirical process for dependent variables, In : *Wavelets and Statistics.*, Ed. P. Bickel et al., 331-344, Lecture Notes in Statistics 103, Springer Verlag, New York.

Osekowski, A. (2007) Inequalities for dominated martingales, *Bernoulli*, **13**, 54-79.

Page, E.S. (1954) Continuous inspection schemes, *Biometrika*, **41**, 100-115.

Peligard, M. and Suresh, R. (1995) Estimation of variance of partial sums of an associated sequence of random variables, *Stochastic Processes. Appl.*, **56**, 307-319.

Petrov, V.V. (1975) *Sums of Independent Random Variables*, Springer, New York.

Pitt, L. (1982) Positively correlated normal variables are associated, *Ann. Probab.*, **10**, 496-499.

Prakasa Rao, B.L.S. (1978) Density estimation for Markov processes using delta sequences, *Ann. Inst. Statist. Math.*, **30**, 321-328.

Prakasa Rao, B.L.S. (1983) *Nonparametric Functional Estimation*, Academic Press, Orlando.

Prakasa Rao, B.L.S. (1987) Characterization of probability measures by linear functions defined on a homogeneous Markov chain, *Sankhya Ser. A*, **49**, 199-206.

Prakasa Rao, B.L.S. (1993) Bernstein-type inequality for associated sequences, In : *Statistics and Probability : A Raghu Raj Bahadur Festschrift*, Ed. J. K. Ghosh, S.K. Mitra, , K.R. Parthasarathy, and B.L.S. Prakasa Rao, 499-509, Wiley Eastern, New Delhi.

Prakasa Rao, B.L.S. (1996) Nonparametric estimation of the derivatives of a density by the method of wavelets, *Bull. Inform. Cyb.*, **28**, 91-100.

Prakasa Rao, B.L.S. (1998) Hoeffding identity, multivariance and multicorrelation, *Statistics*, **32**, 13-29.

Prakasa Rao, B.L.S. (1999) Estimation of the integrated squared density derivative by wavelets, *Bull. Inform. Cyb.*, **31**, 47-65.

Prakasa Rao, B.L.S. (1999a) Nonparametric functional estimation : An overview, In *Asymptotics, Nonparametrics and Time Series*, Ed. Subir Ghosh, Marcel Dekker, New York, pp. 461-509.

Prakasa Rao, B.L.S. (1999b) *Statistical Inference for Diffusion Type Processes*, Arnold, London and Oxford University Press, New York.

Prakasa Rao, B.L.S. (1999c) *Semimartingales and their Statistical Inference*, Chapman and Hall, London /CRC Press, Boca Raton.

Prakasa Rao, B.L.S. (2002) Hajek-Renyi-type inequality for associated sequences, *Statist. Probab. Lett.*, **57**, 139-143.

Prakasa Rao, B.L.S. (2002) Whittle type inequality for demisubmartingales, *Proc. Amer. Math. Soc.*, **130**, 3719-3724.

Prakasa Rao, B.L.S. (2002a) Negatively associated random variables and inequalities for negative demisubmartingales, Preprint, Indian Statistical Institute, New Delhi.

Prakasa Rao, B.L.S. (2003) Wavelet linear density estimation for associated sequences, *J. Indian Statist. Assoc.*, **41**, 369-379.

Prakasa Rao, B.L.S. (2004) On some inequalities for N-demimartingales, *J. Indian Soc. Agricultural Statist.*, **57**, 208-216.

Prakasa Rao, B.L.S. (2007) On some maximal inequalities for demisubmartingales and N-demisupermartingales, *Journal of Inequalities in Pure and Applied Mathematics*, **8**, Article 112, pp. 1-17.

Prakasa Rao, B.L.S. (2009) Conditional independence, conditional mixing and conditional association, *Ann. Inst. Statist. Math.*, **61**, 441-460.

Prakasa Rao, B.L.S. (2010) *Statistical Inference for Fractional Diffusion Processes*, Wiley, Chichester.

Prakasa Rao, B.L.S. and Dewan, I. (2001) Associated sequences and related inference problems, In *Handbook of Statistics: Stochastic Processes, Theory and Methods*, ed. D.N. Shanbhag and C.R. Rao, **19**, North-Holland, Amsterdam, pp. 693-728.

Preston, C.J. (1974) A generalization of the FKG inequalities, *Comm. Math. Phys.*, **36**, 233-241.

Qiu, De Hua. (2010) Complete convergence for weighted sums of arrays of row wise negatively associated random variables, *J. Math. Res. Exposition*, **30**, 149-158.

Qiying, W. (1995) The strong law of U statistics with ϕ^*-mixing samples, *Statist. Probab. Lett.*, **23**, 151-155.

Queseda- Molina, J.J. (1992) A generalization of an identity of Hoeffding and some applications, *J. Ital. Statist. Soc.*, **3**, 405-411.

Resnick, S.I. (1988) Association and multivariate extreme value distributions. In *Gani Festschrift: Studies in Statistical Modelling and Statistical Science*, Ed. C.C. Heyde, 261-271. Statist. Soc. of Australia.

Rio, E. (2009) Moment inequalities for sums of dependent random variables under projective conditions, *J. Theoret. Probab.*, **22**, 146-163.

Roberts, A.W. and Verbeg, D.E. (1973) *Convex Functions*, Academic Press, New York.

Roussas, G.G. (1991) Kernel estimates under association : strong uniform consistency, *Statist. Probab. Letters*, **12**, 393-403.

Roussas, G.G. (1993) Curve estimation in random fields of associated processes, *J. Nonparametr. Statist.*, **2**, 215-224.

Roussas, G.G. (1994) Asymptotic normality of random fields of positively or negatively associated processes, *J. Multivariate Anal.*, **50**, 152-173.

Roussas, G.G. (1995) Asymptotic normality of smooth estimate of a random field distribution function under association, *Statist. Probab. Lett.*, **24**, 77-90.

Roussas, G.G. (1999) Positive and negative dependence with some statistical applications, In *Asymptotics, Nonparametrics and Time Series*, Ed. Subir Ghosh, Marcel Dekker, New York, pp. 757-788.

Roussas, G.G. (2008) On conditional independence, mixing and association, *Stoch. Anal. Appl.*, **26**, 1274-1309.

Sadikova S.M. (1966) Two dimensional analogues of an inequality of Esseen with application to the central limit theorem, *Theor. Probab. Appl.*, **11**, 325-335.

Samorodnitsky, G. (1995) Association of infinitely divisible random vectors, *Stochastic Processes. Appl.*, **55**, 45-55.

Sen, P.K. (1963) On the properties of U-statistics when the observations are not independent, *Calcutta Statist. Assoc. Bull.*, **12**, 69-92.

Sen, P.K. (1972) Limiting behaviour of regular functionals of empirical distributions for stationary-mixing processes, *Z. Wahrsch. Theorie und Verw. Gebiete*, **25**, 71-82.

Serfling, R.J. (1968) The Wilcoxon two-sample statistic on strongly mixing processes, *Ann. Math. Statist.*, **39**, 1202-1209.

Serfling, R.J. (1980) *Approximation Theorems of Mathematical Statistics*, Wiley, New York.

Sethuraman, S. (2000) Central limit theorems for additive functionals of the simple exclusion process, *Ann. Probab.*, **28**, 277-302; Correction: *Ann. Probab.*, **34** , 427-428.

Shaked, M. and Shantikumar, J.G. (1990) Parametric stochastic convexity and concavity for stochastic processes, *Ann. Inst. Statist. Math.*, **42**, 509-531.

Shaked, M., and Tong, Y.L. (1985) Some partial orderings of exchangeable random variables by positive dependence, *J. Multivariate Anal.*, **17**, 333-349.

Shao, Qi-Man. (2000) A comparison theorem on maximal inequalities between negatively associated and independent random variables, *J. Theoret. Probab.*, **13**, 343-356.

Shao, Qi-Man., and Su, Chun. (1999) The law of the iterated logarithm for negatively associated random variables, *Stochastic Processes. Appl.*, **83**, 139-148.

Shao, Qi-Man., and Yu, H. (1996) Weak convergence for weighted empirical processes of dependent sequences, *Ann. Probab.*, **24**, 2098-2127.

Shaskin, A.P. (2005) On Newman's central limit theorem, *Theor. Probab. Appl.*, **50**, 330-337.

Shaskin, A.P. (2006) The law of the iterated logarithm for an associated random field, *Russian Math. Surveys*, **61**, 359-361.

Shaskin, A.P. (2007) Strong Gaussian approximation of an associated random field, *Russian Math. Surveys*, **62**, 1012-1014.

Shepp, L.A. (1964) A local limit theorem, *Ann. Math. Statist.*, **35**, 419-423.

Silverman, B.W. (1986) *Density Estimation for Statistics and Data Analysis*, Chapman and Hall, London.

Sung, Soo Hak. (2007) A note on the exponential inequality for associated random variables, *Statist. Probab. Lett.*, **77**, 1730-1736.

Sung, Soo Hak. (2008) A note on the Hajek-Renyi inequality for associated random variables, *Statist. Probab. Lett.*, **78**, 885-889.

Sung, Soo Hak. (2009) An exponential inequality for negatively associated random variables, *J. Inequal. Appl.*, Art. ID 649427, 7 pp.

Szekli, R. (1995) *Stochastic Ordering and Dependence in Applied Probability*, Lecture notes in Statistics, No. 97, Springer-Verlag, NewYork.

Triebel, H. (1992) *Theory of function Spaces II*, Birkhauser, Berlin.

Van Beek, P. (1972) An application of Fourier methods to the problem of sharpening the Berry-Esseen inequality, *Z. Wahrsch. Theorie und Verw. Gebiete*, **23**, 187-196.

Von Mises, R. (1947) On the asymptotic distribution of differentiable statistical functions, *Ann. Math. Statist.*, **18**, 301-348.

Walter, G and Ghorai, J. (1992) Advantages and disadvantages of density estimation with wavelets, In *Proceedings of the 24th Symp. on the Interface*, Ed. H. Joseph Newton, Interface FNA,VA 24: 234-243.

Wang, Dingcheng. and Tang, Qihe. (2004) Maxima of sums and random sums for negatively associated random variables with heavy tails, *Statist. Probab. Lett.*, **68**, 287-295.

Wang, Jian Feng.(2004) Maximal inequalities for associated random variables and demimartingales, *Statist. Probab. Lett.*, **66**, 347-354.

Wang, Jian Feng., and Zhang, Li-Xin. (2006) A nonclassical law of the iterated logarithm for functions of positively associated random variables, *Metrika*, **64**, 361-378.

Wang, Jian Feng., and Zhang, Li-Xin. (2007) A Berry-Esseen theorem and a law of the iterated logarithm for asymptotically associated sequences, *Acta Math. Sin. (Engl. Ser.)*, **23**, 127-136.

Wang, XueJun., and Hu, ShuHe (2009) Maximal inequalities for demimartingales and their applications, *Science in China Series A: Mathematics*, **52**, 2207-2217.

Wang, XueJun., Hu, ShuHe., and Prakasa Rao, B.L.S. (2011) Maximal inequalities for N-demimartingales and strong law of large numbers, *Statist. Probab. Lett.*, **81**, 1348-1353.

Wang, XueJun., Hu, ShuHe., Zhao, Ting., and Yang, WenZhi. (2010) Doob's type inequality and strong law of large numbers for demimartingales, *Journal of Inequalities and Applications*, Volume 2010, Article ID 838301, pp. 1-11.

Wang, XueJun., Prakasa Rao, B.L.S., Hu, ShuHe., and Yang, WenZhi. (2011a) On some maximal inequalities for demimartingales and N-demimartingales based on concave Young functions, Preprint, Anhui University, Hefei, P.R. China and University of Hyderabad, Hyderabad, India.

Wang, Yue Bao., Wang, Kai Yong. and Cheng, Dong Ya. (2006) Precise large deviations for sums of negatively associated random variables with common dominatedly varying tails, *Acta Math. Sin. (Engl. Ser.)*, **22**, 1725-1734.

Weron, A. (1984) Stable processes and measures: A survey,. In *Probability Theory on Vector Spaces*, III. Lecture Notes in Math, 1080, 306-364, Springer, Berlin.

Whittle, P. (1969) Refinements of Kolmogorov's inequality, *Theor. Probab. appl.*, **14**, 310-311.

Wood, T.E. (1983) A Berry Esseen theorem for associated random variables, *Ann. Probab.*, **11**, 1042-1047.

Wood, T.E. (1984) Sample paths of demimartingales, In *Probability Theory on Vector Spaces*, III. Lecture notes in Math, No. 1080, 365-373, Springer, Berlin.

Wood, T.E. (1985) A local limit theorem for associated sequences, *Ann. Probab.*, **13**, 625-629.

Xing, Guodong. (2009) On the exponential inequalities for strictly and negatively associated random variables, *J. Statist. Plann. Inference*, **139**, 3453-3460.

Xing, Guodong., Yang, Shanchao. (2008) Notes on the exponential inequalities for strictly stationary and positively associated random variables, *J. Statist. Plann. Inference*, **138**, 4132-4140.

Xing, Guodong., Yang, Shanchao. (2010) An exponential inequality for strictly stationary and negatively associated random variables, *Comm. Statist. Theory Methods*, **39**, 340-349.

Xing, Guodong., Yang, Shanchao. (2010) Some exponential inequalities for positively associated random variables and rates of convergence of the strong law of large numbers, *J. Theoret. Probab.*, **23**, 169-192.

Xing, Guodong., Yang, Shanchao., and Liu, Ailin. (2008) Exponential inequalities for positively associated random variables and applications, *J. Inequal. Appl.*, Art. ID 385362, 11 pp.

Xing, Guodong., Yang, Shanchao., Liu, Ailin., and Wang, Xiangping. (2009) A remark on the exponential inequality for negatively associated random variables, *J. Korean Statist. Soc.*, **38**, 53-57.

Yang, Shanchao., and Chen, Min. (2007) Exponential inequalities for associated random variables and strong law of large numbers, *Sci. China Ser. A*, **50**, 705-714.

Yoshihara, K. (1976) Limiting behaviour of *U*-statistics for stationary absolutely regular processes, *Z. Wahrsch. Verw. Gebiete*, **35**, 237-252.

Yoshihara, K. (1984) The Berry-Esseen theorems for *U*-statistics generated by absolutely regular processes, *Yokohama Math. J.*, **32**, 89-111.

Yu, H. (1993) A Glivenko-Cantelli lemma and weak convergence for empirical processes of associated sequences, *Probab. Th. Rel. Fields*, **95**, 357-370.

Yu, H. (1996) A strong invariance principle for associated sequences, *Ann. Probab.*, **24**, 2079-2097.

Yuan, De-Mei., An, Jun., and Wu, Xiu-Shan. (2010) Conditional limit theorems for conditionally negatively associated random variables, *Monatsh. Math.*, **161**, 449-473.

Yuan, De-Mei., and Yang, Yu-Kun. (2011) Conditional versions of limit theorems for conditionally associated random variables, *J. Math. Anal. Appl.*, **376**, 282-293.

Zarei, H., Jabbari, H., Dewan, I., and Azarnoosh, H.A. (2008) A general method of density estimation for negatively associated random variables, *JPSS J. Probab. Stat. Sci.*, **6**, 39-51.

Zhang, Li-Xin.(2001) The weak convergence for functions of negatively associated random variables, *J. Multivariate. Anal.*, **78**, 272-298.

Zhao, Yuexu., Qiu, Zheyong., Zhang, Chunguo., and Zhao, Yehua. (2010) The rates in complete convergence for negatively associated sequences, *Comput. Appl. Math.*, **29**, 31-45.

Index